# CONNECTED, INTELLIGENT, AUTOMATED

# CONNECTED, INTELLIGENT, AUTOMATED

## The Definitive Guide to
## Digital Transformation and Quality 4.0

## N. M. RADZIWILL

**First Edition**

Quality Press
Milwaukee, Wisconsin

Printed in the United States of America
23  22  21  20    5  4  3  2  1
Publisher's Cataloging-in-Publication Data

Names: Radziwill, Nicole M., author.
Title: Connected, intelligent, automated : the definitive guide to digital
transformation and quality 4.0, first edition / Nicole Radziwill.
Description: Includes bibliographical references and index. | Milwaukee,
WI: Quality Press, 2020.
Identifiers: LCCN: 2019953974 | ISBN: 978-1-951058-005 (pbk.) | 978-1-951058-01-2
(ebook) | 978-1-951058-02-9 (pdf)
Subjects: LCSH Total quality management. | Organizational effectiveness. |
Production management. | Artificial intelligence—Industrial applications. |
Computer integrated manufacturing systems. | System analysis. | Decision making. |
Information technology. | BISAC TECHNOLOGY & ENGINEERING / Automation |
COMPUTERS / Artificial Intelligence / General | TECHNOLOGY & ENGINEERING /
Quality Control | BUSINESS & ECONOMICS / Total Quality Management |
TECHNOLOGY & ENGINEERING / Technical & Manufacturing Industries & Trades
Classification: LCC HD62.15 .R335 2020 | DDC 658.4/013—dc23

Publisher: Seiche Sanders
Managing Editor: Sharon Woodhouse
Sr. Creative Services Specialist: Randall L. Benson

ASQ advances individual, organizational, and community excellence worldwide
through learning, quality improvement, and knowledge exchange.

Attention bookstores, wholesalers, schools, and corporations: Quality Press books, are
available at quantity discounts with bulk purchases for business, trade, or educational
uses. For information, please contact Quality Press at 800-248-1946 or books@asq.org.

To place orders or browse the selection of ASQExcellence and Quality Press titles, visit
our website at http://www.asq.org/quality-press.

♾ Printed on acid-free paper

Quality Press
600 N. Plankinton Ave.
Milwaukee, WI 53203-2914
Email: books@asq.org
ASQ    Excellence Through Quality™

# TABLE OF CONTENTS

## CHAPTER TWO

# ABOUT THIS BOOK

Quality 4.0 is the secret ingredient for a successful digital transformation (Chapter 1). Quality 4.0 initiatives support digital transformation, enhancing connectedness (Chapter 2), intelligence (Chapter 3), and automation (Chapter 4) with emerging technologies. These three aspects come together to solve practical problems across many industries (Chapter 5).

Data is the foundation of a digital transformation that successfully enhances quality and performance. It fuels the advanced analytics that are produced by machine learning and other algorithms (Chapter 6). Data science is the practice that ties it all together, linking the mechanics of acquiring and assimilating the data with analysis and communication of outcomes to the business (Chapter 7). For data science efforts to yield useful results, an organization must have a solid foundation in data quality and data management (Chapter 8). With this foundation in place, software systems can be implemented to digitize manual processes and add a layer of checks and balances (Chapter 9). Blockchain can augment these systems, especially when data quality and immutability are imperative, but relational databases should always be considered first (Chapter 10).

To turn digital strategy into action, start by understanding your business and the competitive environment it is embedded within. Clear roles, responsibilities, data-driven decision-making processes, and standard work should be defined. The Baldrige Excellence Framework (BEF), total quality management (TQM), Toyota Production System (TPS), and kaizen can all be used (jointly or individually) to make this happen (Chapter 11). Then, articulate your impacts on the environment, protect the health and safety of your workforce, and protect your business by investing in cybersecurity (Chapter 12). Once the foundations are in place, seek out the voice of the customer (VoC)—a process that is greatly supported by using digital systems, social media, and other Industry 4.0 technologies (Chapter 13).

When your organizational backbone to coordinate people, processes, and technologies is in place, define the smart products and services you will offer (if any). Make sure that the workforce (and members of the customer and supplier ecosystem, if appropriate) has access to accurate, timely, and complete knowledge assets (Chapter 14). Finally, you will be ready to create and execute your own strategy for digital transformation. Anchoring it with a quality-driven foundation and focus greatly increases your likelihood of success (Chapter 15).

Peter Merrill, author of *Innovation Generation* and *Innovation Never Stops*, and chair of the Canadian National Committee on Innovation for the International Organization for Standardization (ISO), describes innovation as "quality for tomorrow." This book will show you and your team, regardless of your industry or level of quality maturity, how to use digital transformation and emerging technologies to innovate with quality—for tomorrow and the years ahead.

# THE UNKNOWN CITIZEN

He was found by the Bureau of Statistics to be
One against whom there was no official complaint,
And all the reports on his conduct agree
That, in the modern sense of an old-fashioned word, he was a saint,
For in everything he did he served the Greater Community.
Except for the War till the day he retired
He worked in a factory and never got fired,
But satisfied his employers, Fudge Motors Inc.
Yet he wasn't a scab or odd in his views,
For his Union reports that he paid his dues,
(Our report on his Union shows it was sound)
And our Social Psychology workers found
That he was popular with his mates and liked a drink.
The Press are convinced that he bought a paper every day
And that his reactions to advertisements were normal in every way.
Policies taken out in his name prove that he was fully insured,
And his Health-card shows he was once in hospital but left it cured.
Both Producers Research and High-Grade Living declare
He was fully sensible to the advantages of the Instalment Plan
And had everything necessary to the Modern Man,
A phonograph, a radio, a car and a frigidaire.
Our researchers into Public Opinion are content
That he held the proper opinions for the time of year;
When there was peace, he was for peace: when there was war, he went.
He was married and added five children to the population,
Which our Eugenist says was the right number for a parent of his generation.
And our teachers report that he never interfered with their education.
Was he free? Was he happy? The question is absurd:
Had anything been wrong, we should certainly have heard.

# PREFACE

## Connected, Intelligent, Automated
### *The Definitive Guide to Digital Transformation and Quality 4.0*

The moral is that it is necessary to innovate, to predict needs of the
customer, give him more. He that innovates and is lucky
will take the market.
—W. EDWARDS DEMING, *THE NEW ECONOMICS* (1993)

I f you were to ask any thought leader what they consider to be today's most
important technology trend, they'd likely say digital transformation. The
popularity of this term probably exceeds most people's understanding of
it, but that hasn't prevented it from becoming sine qua non for almost every
organization in the world today. Essentially, *digital transformation* is the application of digital technology to enhance an organization's abilities to meet its
strategic objectives, build capabilities, and enhance agility.

Since 2016, many organizations have started a digital transformation initiative but are struggling to achieve desired levels of operational excellence and
customer experience. According to a 2019 survey by Celonis (https://tinyurl
.com/yyhbp85m), 73% of C-suite execs are jumping straight into the tactics of
launching artificial intelligence (AI), machine learning, and automation initiatives and experiencing disappointing results. Even though more than a third of
businesses surveyed report spending more than $500,000 on digital transfor-
mation over the past year, 45% of C-suite executives "don't know where to start
when developing their transformation strategy."

A 2018 McKinsey survey (https://www.mckinsey.com/business-functions
/organization/our-insights/unlocking-success-in-digital-transformations) revealed that success rates for digital transformation efforts are dismal—below
26% for technology, media, and telecom firms, and between 4% and 11% for the oil
and gas, pharmaceutical, and automotive industries. Company size also matters.
Small organizations (fewer than 100 employees) are nearly three times as likely
as larger organizations to report a successful initiative.

Building and executing a quality-driven digital strategy can help solve these issues and mitigate the overall risks of digital transformation initiatives. While lack of clarity can lead to disappointing improvement initiatives, delayed product introductions, and wasted investments, Celonis notes that "the rush to transform is threatening to derail the potential for success" (Celonis 2019).

A quality-driven approach lays the foundations for success, and this book will help you get started. It provides a conceptual framework for understanding how emerging technologies can be leveraged to improve quality and performance—and a practical, actionable guide for moving forward. Through compelling, pragmatic examples, it demonstrates that Industry 4.0 and Quality 4.0 are not new but have a deep history with templates for success. Had W. H. Auden known (when he wrote *The Unknown Citizen* in 1940) that data collection on people and systems would become even *more* prevalent in our time, he may have used even starker language to describe data-driven innovation.

The Quality 4.0 banner reflects the consolidation of research and practice around connected, intelligent, automated technologies for more effectively managing and improving our organizations. This book presents a Quality 4.0-based approach for ensuring digital transformation success.

## WHO THIS BOOK IS FOR

- CEOs, CIOs, CTOs, and business leaders who want to drive excellence, value, and growth by embracing technology to enhance performance
- Chief strategy officers, chief digital officers, and SVP/VPs of digital transformation who are leading digital transformation efforts
- Anyone participating in digital transformation efforts, including data science, AI, and machine learning leaders
- Senior leaders and executives who want to create a Quality 4.0 strategy
- Senior leaders and executives outside the quality profession who have started digital transformation efforts but have not improved performance or customer experience
- Anyone in any industry who wants to learn how Industry 4.0 and Quality 4.0 can help improve quality and performance in their team or organization

## QUALITY 4.0 IS FOR ALL INDUSTRIES

Although many publications and research articles have focused on the smart-factory aspects of Industry 4.0, Quality 4.0 is for everyone. Any business system that has hierarchies or networks, where data can be collected at any level, can benefit from the new technologies that support connectedness. AI and machine learning are more accessible now than ever, and even small businesses can explore the potential of Quality 4.0 to improve processes. Thanks to cloud services, you can be up and running and create your own machine learning models in minutes—at a fraction of what you'd pay if you had to set up and maintain the infrastructure yourself (as we had to do 10 and 20 years ago).

Similar to how the practice and profession of quality expanded beyond manufacturing into all other industries, the new Quality 4.0 techniques and philosophies will expand far beyond manufacturing. This book will provide you with a foundation to understand them and to begin to explore ways to leverage them as you plan and execute digital transformation initiatives.

## WHY I WROTE THIS BOOK

This book is the culmination of two and a half decades of research and practice: exploring, implementing, and critically examining the quality and performance improvement aspects of (what we now call) Industry 4.0 technologies. The changes ahead are powerful, exciting, and overwhelming—and we can draw on the lessons from past work to mitigate the risks of digital transformation today.

My goal is to share what I've learned to help you sift through the noise, find the signals for *your* organization, and drive successful digital transformation initiatives grounded in quality.

## MY INDUSTRY 4.0/QUALITY 4.0 STORY

When I graduated from college in the mid-1990s with a meteorology degree, I was more interested in the information technology part than the weather forecasting part. After working at an e-commerce consultancy for a few years post-graduation, building storefronts and websites when it was still a technically complex task, I moved to Boulder, Colorado, to work at the Forecast Systems Lab (FSL) of the National Oceanic and Atmospheric Administration. It was my first introduction to quality and Industry 4.0 technologies.

It's fascinating to look back at your life and career to find that some of the smallest revelations led to the biggest shifts in your professional identity. The FSL was one of those experiences, thanks mostly to my office mate, a software engineer and member of the American Society for Quality (ASQ). He was a Certified Software Quality Engineer (CSQE), and *much* older and more mature than I, so I really looked up to him. (He was in his mid-30s at the time; I was 22.)

The job we were working on was also my first introduction to what we now call Industry 4.0 technologies. We were building a sensor network to take "junk data" from Global Positioning System (GPS) receivers, scattered across the United States and sited in about half the states, and find water vapor content in the atmosphere. We built connections to each of the sites, developed a data acquisition and processing framework, and wrote programs to assess the quality of the data and make our data collection and troubleshooting processes more intelligent.

My office mate loved to code, and he loved to tinker with instrumentation. But most important, *nothing ever stressed him out*. He trusted the process! And he realized that in order to trust the process, there had to *be* a process that everyone was aware of, that everyone was brought into, and that was continually improving. I observed the zen-like way in which he moved through his day, always learning, and always improving. I admired that inner professional peace and wanted some for myself.

I wanted to be like him, so I joined ASQ!

Over the next decade, my love of quality *and* my love of data deepened. I helped enterprises successfully implement their Customer Relationship Management (CRM) applications and integrate them with other business systems, using lean principles. I led a software development team at a national lab, working on monitor, control, and data analysis soon after. I helped them implement agile practices and ISO 9001–based practices to set expectations, improve product and service quality, and reduce stress—not only for ourselves but for our customers too.

Next, I worked with an executive team for a few years on data management, answering questions like when and how to archive data, how to support pipelines for processing data in motion, how to use machine learning to understand the data you've collected, and how to analyze data when it's too big, can't fit in one file (or even on one machine), or can't even be downloaded due to its size. I learned how to balance compliance and effectiveness, using standards and frameworks to support business objectives rather than threaten them. Then, I went to a university, and for the next decade worked to distill all this knowledge into an undergraduate program focused on data-driven production systems.

All these things I spent time with—working with big data, using AI and machine learning, and building pipelines to draw insights from data in motion—are currently hot topics across industries. Depending on your specific interests or industry, these initiatives might be referred to as digital transformation, Industry 4.0, smart factory, smart manufacturing, quality intelligence, smart quality, or Quality 4.0. Although I didn't know it at the time, I was (along with everyone else in the research community working in these areas) one of the pioneers. We were learning how to bring these new technologies and capabilities into practice.

I had a feeling that the digital transformation of quality would eventually happen, that automation and data-driven intelligence would become a normal and natural part of most work environments. And indeed, that time is now. We are on the cusp of a broad transformation that will likely last decades.

Much as the manufacturing workforce needs to be retooled to develop more skills in visualizing and interpreting data, the quality workforce will benefit from a deeper understanding of technologies to drive quality and performance. How will technology impact quality management, product life cycle management, or process management? How do all these emerging technologies work together? Should you invest in AI or machine learning, and if so, when? How can you use connectedness, intelligence, and automation pragmatically to avoid the deep financial pits that can come from chasing hype?

## HOW THIS BOOK IS STRUCTURED

- **Part 1: The Quality Revolution**—The first five chapters describe the tools, techniques, and emerging technologies that are available to help you be more competitive and achieve long-term success. Chapter 2 explains how technologies connect people, machines, materials, and data, and why these connections matter. Chapter 3 shows the profound link between quality management and intelligent systems (including machine learning) and can help you determine whether (and how) to apply those methods to your problems. Chapter 4 examines the spectrum of possibilities for automating processes, tasks, and data analysis. Chapter 5 brings all these concepts together to describe performance improvement in cross-industry use cases.

- **Part 2: Turn Data into Intelligence**—Data is the life-giving blood of the connected, intelligent, automated enterprise. Without available high-quality data when people or machines need it, processes will be less effective and decisions will be less robust. Chapters 6 and 7 provide background and context about algorithms, analytics, and data science, including guidance on how to manage those efforts. Chapter 8 covers data management foundations that should be in place to protect those investments, while Chapter 9 examines the relationships between software systems like quality management systems (QMS), enterprise resource planning (ERP), manufacturing execution systems (MES), and environment, health, and safety (EHS). Chapter 10 explains blockchain and when you should (and should not) consider using it.
- **Part 3: Turn Digital Strategy into Action**—Armed with the background and conceptual framework from Parts 1 and 2, the remaining chapters provide actionable guidance for developing a visionary, customer-focused, fact-driven, agile learning organization that can excel in its digital transformation efforts. Excellence frameworks are covered in Chapter 11, because for intelligent agents to function, basic questions about how the business runs and how decisions are made must be answered. Chapter 12 explains the convergence of environment, health, safety, and quality (EHSQ), and how to get deeper and richer information about quality and performance by looking across traditional system boundaries. Chapter 13 examines how VoC is changing in the digital age. Chapter 14 ties these concepts together and outlines the requirements for a quality-driven strategy for digital transformation, while Chapter 15 provides ways to realize these principles in your organization.

Rather than simply describing each of the technologies and concepts, examples are provided to tie them to quality and performance improvement. Each chapter addresses social and political issues that can impact the success of your Quality 4.0 strategy or digital transformation initiatives. Finally, the main points from each chapter are summarized at the end in a section called "The Bottom Line."

With the information you learn in this book, you will understand how emerging technologies can contribute to quality and performance improvement, and be able to apply these lessons to create a successful strategy for digital transformation that meets quality and performance goals.

## MOVING FORWARD

If you're older than 25, you probably remember what the web was like when it was fresh: lots of static pages to read, and maybe a form here or there where you could send information to someone over the internet, but not much more than that. By the mid-2000s, most web pages were dynamic and driven by databases. People using the web could interact with the content—leaving comments, adding bookmarks, and creating user-generated content (for example, in blogs and wikis)—and with each other. The enabling technologies of Web 2.0 expanded connectedness, leveraged the intelligence of crowds, and automated content updating and delivery.

Quality management and organizational excellence are about to experience a similar shift. Here's an example: today, QMS display information about risks, controls, and events on dashboards. In some cases, forecasts and projections are available to aid in planning and decision making. But in general, there is little interaction with the data, little exploration, and nearly no collaboration with other people or organizations that is enabled by software and systems. The QMS of the future will guide us toward better ways to manage our organizations, and may even do it for us autonomously, having learned the best decisions to make under different circumstances.

The Industry 4.0 and Quality 4.0 labels will fade from the vernacular as we become acclimated to hyperconnectivity in our work environments, ubiquitous and embedded intelligent systems, and automated processes. What will *not* fade is the new attention to sourcing, analyzing, and interpreting information in real time, and using it to make better decisions, implement smoother processes, and accelerate innovation. Data is about to fuel the revolution.

## REFERENCES

Celonis. (2019, March 21). "Half of C-suite executives admit to launching transformation initiatives without a clear strategy." https://www.celonis.com /press/celonis-study-almost-half-of-c-suite-executives-admit-to-launching -transformation-initiatives-without-a-clear-strategy

## CHAPTER ONE

# QUALITY 4.0 AND THE FOURTH INDUSTRIAL REVOLUTION

Any sufficiently advanced technology is indistinguishable from magic.
—ARTHUR C. CLARKE

C overing 140,000 miles, the privately owned and maintained U.S. rail network transported nearly 1.1 million carloads of freight each month in 2017, and again in 2018. Rail transportation carries coal to power plants, food to grocery stores, cars to dealerships, and many other things to many other places. This massive industry reports approximately $74 billion in revenue each year. The efficiency of rail transport is impressive too: a single train can carry the load of a hundred trucks. Because the rails are nearly frictionless, one train can, on average, move a ton of freight 470 miles, with 75% fewer greenhouse gas emissions than if the same cargo were transported by truck (Association of American Railroads [AAR], 2019).

But rail disasters can be expensive and deadly and, additionally, can disrupt supply chains. On April 4, 2019, a relatively small derailment on the Palmerston North line in New Zealand interrupted deliveries to 70 companies and cost the forestry and logging sector more than $115,000 each day (Lawrence & Mitchell, 2019). In 2005, a derailment on one track in western Wyoming stalled coal deliveries and was associated with price increases for electric utilities (Gedik et al., 2014). Natural disasters can also lead to railway incidents. In 1993, for example, flooding on the Mississippi and Missouri rivers was associated with $182 million of supply chain disruptions (Haefner et al., 1996).

Like many industries, rail transportation is experiencing unprecedented digital transformation, focused on risk assessment, asset management, and improving safety. The most exciting performance improvements are being driven by what we now refer to as Quality 4.0: improving connectedness, intelligence, and automation to enhance performance and promote organizational excellence.

An ounce of prevention is worth a pound of cure. For example, people and cargo are now protected by real-time diagnostic monitoring that supports more effective maintenance programs. This is made possible by the emergence of condition-based maintenance (CBM) from corrective and preventive maintenance approaches. Corrective (or reactive) maintenance is initiated once a defect or failure occurs. Preventive maintenance anticipates future issues and addresses potential problems through scheduled inspections, replacements, renewals, and preplanned equipment overhauls. CBM, a special type of predictive maintenance, focuses on using data for "discovering those components where maintenance is required so that the maintenance cost is greatly reduced" (Ghofrani et al., 2018). The foundations of CBM are big data and machine learning models.

Today, many of the activities related to railway inspection and traffic monitoring are performed using new sensor technologies, machine vision, and some machine learning. In January 2019, at the annual conference for the AAR, I was able to see some of this CBM technology in action, including a system for structural health monitoring on the underbody of the railcars. As the train cars are in motion (referred to as "rolling stock"), images are recorded at multiple wavelengths, including in the visible and infrared bands, by stationary cameras mounted in a shallow pit between the rails. The images are segmented, compared with prior expectations about how frequently problems are typically detected, and used to train machine learning classifiers that flag when problems are emerging (Schlake et al., 2009).

The multispectral approach makes it possible for the automated system to detect issues that human inspectors, performing regular visual inspections, would not be able to identify. This enables more thorough and efficient inspections, opens up the possibility to inspect continuously rather than occasionally, and improves safety by eliminating the physical risks of safety inspections. By reducing the costs of inspections and audits, real-time quality management enhances the efficiency of the rail network while substantially improving safety.

There are many other examples in the research, all of which use new technologies to increase connections, enhance intelligence, and automate decision making or operational processes. Each of these examples improves quality in some way. For example, Quality 4.0 approaches can be used to:

- Detect wheel defects using neural networks and support vector machines (SVMs) to classify flat spot, shelling, and nonroundness faults before they cause accidents or loss to assets (Krummenacher et al., 2017)
- Reduce maintenance time by a factor of 10 and improve rail noise emission and train integrity, protect against derailment, and reduce maintenance costs and downtime by implementing standards for CBM (Pfaff et al., 2017)
- Alert operators to rail fractures, scoring, and wear by generating 25,000 image profiles per second using 3D laser cameras, with data analyzed using Principal Component Analysis (PCA) and random forests (a machine learning technique) to classify tracks as healthy or faulty (Santur et al., 2016)
- Detect track anomalies by applying deep learning, which addresses the complexities associated with large numbers of failure modes and the high likelihood of false alarms (Gibert et al., 2016)
- Detect moving obstacles at railway crossings using machine vision and image processing, which improves safety at crossings while protecting assets from costly damage (Pu et al., 2014)
- Predict the degree of remedial action required to prepare a segment of track for transport of dangerous goods (like pollutants), using multilayer perceptrons, classification and regression trees (CARTs), and SVM (Matías et al., 2007)

The most interesting thing I learned at this railroad conference, though, was that despite all these exciting advancements, the rail professionals I spoke with didn't feel as though they're on the cutting edge of Industry 4.0 or Quality 4.0. "Most of us still keep track of our inspections and corrective actions in spreadsheets, so I think we're pretty far behind," one told me.

The moral of this story is: sometimes, even if you're on the cutting edge of Quality 4.0 innovation like many in the rail industry, it may be difficult to see and appreciate if you're close to it—because these new capabilities have been slowly becoming part of our lives for several years.

## THE FOURTH INDUSTRIAL REVOLUTION

Every April since 1947, engineers and innovators have gathered at the Hannover Messe trade fair in Germany to share new technologies and breakthrough techniques in manufacturing. The event is widely recognized as the premier forum for industrial technology. By 2019, there were over 250,000 attendees and 6,500 exhibitors at the weeklong show.

In 2011, Hannover Messe chose the "Industrie 4.0" theme to advertise the event, emphasizing the increasing intelligence and interconnectedness of "smart" manufacturing systems (Kagermann et al., 2011). In addition to the new technologies, the conference organizers wanted to make sure that attendees also recognized the social and workforce benefits of the emerging technology-enabled ecosystem:

> Industrie 4.0 will address and solve some of the challenges facing the world today such as resource and energy efficiency, urban production and demographic change. Industrie 4.0 enables continuous resource productivity and efficiency gains to be delivered across the entire value network. It allows work to be organized in a way that takes demographic change and social factors into account. Smart assistance systems release workers from having to perform routine tasks, enabling them to focus on creative, value-added activities. In view of the impending shortage of skilled workers, this will allow older workers to extend their working lives and remain productive for longer. Flexible work organization will . . . [promote] a better work-life balance. (Kagermann et al., 2013)

The Americanized "Industry 4.0" label was chosen to reflect a turning point informed by industrial revolutions of the past. How did we get to where we are now, and what does it reveal? We can understand the evolution better by tracing the development of key technologies from the 1700s to the present time. During the first industrial revolution (late 1700s and early 1800s), innovations in steam and water power made it possible for production facilities to scale up—and expanded the potential locations in which production facilities could be constructed. Before that time, manufacturing facilities had to be constructed along rivers so that water wheels could be used to generate power and water transport could be used to move whatever was being produced.

By the late 1800s, the second industrial revolution was under way. The discovery of electricity, establishment of the infrastructure for delivering it to plants and facilities, and development of rail infrastructure enabled engineers to develop machinery for mass production. Iron ore production increased, so the machines themselves could be mass produced. In the United States, the expansion of railways also made it easier to obtain supplies and deliver finished goods. In Europe, railways and an expanding canal infrastructure were both used to facilitate transport. In addition, electric light made it possible to extend operating hours, which directly contributed to more productivity—especially on assembly lines.

The widespread availability of reliable power also led to a renaissance in computing: from predominantly analog to digital methods. By the end of World War II, this process had begun in earnest. For example, digital calculators gradually replaced slide rules for common computations, and computer programs replaced tools like nomograms for determining the relationships between inputs and outputs.

The third industrial revolution occurred between 1969 and 1972, with the invention and rapid adoption of the Programmable Logic Controller (PLC). With PLCs, common processes like filling and reloading tanks, turning engines on and off, and controlling sequences of events could be automated, and no longer required human monitoring or intervention. The rapid adoption was due in part to PLCs specifically not being marketed as computers—at the time, many people had a mental image of "computer" that included an air-conditioned room packed with cabinets and wires, and operations fraught with reliability problems. PLCs, which were designed for reliability and relatively easy to program using ladder logic, evolved throughout the 1980s and 1990s and remain ubiquitous in industrial environments (Segovia & Theorin, 2012).

Throughout the 1980s, the cost of computing continued to decrease and personal computers entered most workplaces in the early to mid-1990s. The advent of the internet led to another revolution in connecting people to information, but it wasn't enough to fundamentally transform the way people live and work until interactive capabilities ("Web 2.0") became more prevalent. The expansion of mobile devices, the introduction of mobile apps, and the increasing reliability of cloud computing led to a convergence of services. Multiple customer touch points (phone, fax, web, tablets) gradually blended into the "single view of the customer" that most organizations now have. Less than twenty years ago, companies had a hard time keeping track of customer service phone calls, e-mails, and web queries—if you wrote an e-mail to check on the status of a prior call, they might not be able to figure out you were the same person. Now, this single view is taken for granted.

And now, we are on the cusp of the fourth industrial revolution, one that introduces intelligent cyber-physical systems to the mix (Figure 1.1). Cyber-physical systems link objects in the physical world to people, data sources, and other objects, and communicate via local and global networks. In 2015, the

World Economic Forum (WEF) launched its Digital Transformation Initiative to coordinate research to anticipate the impact on business and society. Recognizing that digital transformation has been ongoing since the emergence of digital computing in the 1950s—first with mainframes, then client-server computing and personal computers, followed by the advent of the web and the early e-commerce sites—the WEF aimed to explore the next phase.

By examining previous patterns of convergence, the WEF anticipated an even broader convergence among physical, digital, and biological worlds. Klaus Schwab (2016), executive chairman of the WEF, explains:

> We have yet to grasp fully the speed and breadth of this new revolution. Consider the unlimited possibilities of having billions of people connected by mobile devices, giving rise to unprecedented processing power, storage capabilities and knowledge access. Or think about the staggering confluence of emerging technology breakthroughs, covering wide-ranging fields such

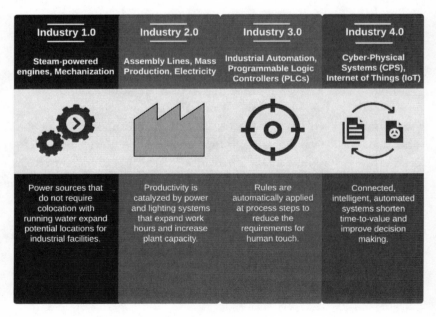

| Industry 1.0 | Industry 2.0 | Industry 3.0 | Industry 4.0 |
|---|---|---|---|
| Steam-powered engines, Mechanization | Assembly Lines, Mass Production, Electricity | Industrial Automation, Programmable Logic Controllers (PLCs) | Cyber-Physical Systems (CPS), Internet of Things (IoT) |
| Power sources that do not require colocation with running water expand potential locations for industrial facilities. | Productivity is catalyzed by power and lighting systems that expand work hours and increase plant capacity. | Rules are automatically applied at process steps to reduce the requirements for human touch. | Connected, intelligent, automated systems shorten time-to-value and improve decision making. |

FIGURE 1.1. The four industrial revolutions.

as artificial intelligence (AI), robotics, the internet of things (IoT), autonomous vehicles, 3D printing, nanotechnology, biotechnology, materials science, energy storage, and quantum computing. . . . The changes are historic in terms of their size, speed, and scope. . . . The changes are so profound that, from the perspective of human history, there has never been a time of greater promise or potential peril.

The first industrial revolution was characterized by steam-powered machines, and the second by electricity and assembly lines. Innovations in computing and industrial automation defined the third. The fourth industrial revolution brings us machine intelligence, pervasive computing, affordable storage, and robust connectivity. How can we leverage them to improve quality and performance? That's the domain of Quality 4.0, the pursuit of performance excellence during this disruptive era of physical, digital, and social transformation.

## WHY NOW?

Although the growth and expansion of the internet accelerated innovation, there are many reasons why the revolution is just beginning now:

- **Cloud Computing:** Until recently, each organization had to create its own information technology (IT) infrastructure, buying and setting up servers and firewalls and hiring staff to maintain connections to the internet. With reliable cloud computing, companies can outsource these tasks, significantly shortening setup time and allowing organizations to focus on their own core competencies. Software as a Service (SaaS), Platform as a Service (PaaS), and Infrastructure as a Service (IaaS) can optimize resource utilization, provide better responsiveness, increase resilience, and result in cost savings.
- **More Data Is More Readily Available:** IoT devices and people are producing data at greater rates than ever before. Enabling technologies like sensors and actuators, and devices like Arduino and Raspberry Pi are now affordable, accessible, and powerful. Together, they are catalyzing innovation.
- **IP Inventory and Global Connectivity:** Since 1988, computer scientists have known that, eventually, the bank of internet protocol (IP) addresses would be depleted (Huston, 2008). Billions of new IoT devices exacerbated this problem. Fortunately, IP version 6 (IPv6), which does have enough address space, is now in use worldwide. In addition, improved network infrastructure is expanding availability

and robustness of connections, and innovations in this field are making it easier to work fully or partially off-line and automatically synchronize.

- **Intelligent Processing:** Affordable data storage, computing capabilities, and processing power are available to generate insights in near real time for decision making. Technologies for augmenting and enhancing human performance (e.g., exoskeletons and brain-computer interfaces) will reveal new mechanisms for innovation. Software reuse, a difficult proposition until recently, means that more options for intelligent processing are available. High-performance software libraries for advanced processing and visualization of data are now often free, easy to find, and, in many cases, easy to use.

- **New Modes of Interacting with People and Data:** Touchscreens, voice-activated interfaces, and personal assistants are now common. New interfaces like virtual reality (VR), augmented reality (AR), and mixed reality (MR)—collectively referred to as XR in many publications—expand possibilities for training and navigating a hybrid physical-digital environment with greater ease, and also for delivering training (especially in dangerous or always-on industrial environments).

- **New Modes of Production:** Technologies like 3D printing, nanotechnology, and gene editing (CRISPR) are poised to change the nature of business models and means of production across industries. Utilities like blockchain may challenge ingrained centralized perceptions of trust, control, consensus, and value creation, opening up new ways to reliably produce goods and deliver services.

Many of these technologies have been in existence for years but have not been reliable, accessible, or affordable enough. Now that these barriers to entry have been lessened, the technologies can be more readily applied across industries to improve quality and performance.

## THE IMPACT OF INDUSTRY 4.0 ON QUALITY

With so many new developments beginning to change the way we live and work, it is not surprising that quality management and quality engineering may also need to adapt. The impact of Industry 4.0 on the quality profession was first described, albeit indirectly, in the 2015 American Society for Quality (ASQ) *Future of Quality* report. This study, which is conducted every few years, aims to uncover key issues related to quality that might emerge in the upcoming five- to ten-year period.

Although the term *Quality 4.0* would not be used in print for another year, the findings in this report were clearly intended to prepare the quality community for the challenges of the Industry 4.0 future. The new reality of quality, according to the experts whose perspectives were captured in the study, would focus not so much on the interests of individual companies but on the health and viability of the entire industrial ecosystem. They projected that between 2015 and 2025, there would be:

- A *change in the nature of boundaries*, both within and between organizations, and how information is shared between them. These shifting boundaries will result from increased availability and transparency of data, making it possible to work efficiently at any distance. Technology will help us break down communication silos.
- A greater focus on *customer experience, participative markets* (e.g., where customers both consume and produce energy), and *prosumerism* (where customers engage in co-creation to design and develop the products they want). As a result of shifting organizational boundaries, customers, employees, suppliers, and partners may assume multiple roles. Voice of the customer (VoC) will expand to include the voice of things (VoT), which will provide information about our customers from the connected objects around them (Goasduff, 2017).
- Enhanced *visibility into business processes*, with the ability to monitor and respond to any element in near real time. Supply chain "omniscience" will provide information in the gaps that were previously obscured, opening up new opportunities for optimization.
- A shift to *prioritizing continuous learning and adaptation* over traditional factors like efficiency, effectiveness, usability, and satisfaction. The overall goals of quality will not change, but there will be more and different ways to use technology to achieve them.

Although these are not all the issues raised in the report, the key themes are clear. The nature of "organization" as a concept is changing, and the nature of "customer" is changing as well. Organizations will no longer be defined solely by their employees and business partners but also by the customers who participate—without even explicitly being aware of their integral involvement—in ongoing dialogues that shape the evolution of product lines and new services. New business models will not have to rely on ownership, consumption, or centralized production. The value-based approach will accentuate the importance of trust, transparency, and security; and new technologies like blockchain will help us implement and deploy systems to support those changes.

## QUALITY 4.0: THE NEXT GENERATION OF EXCELLENCE

The themes described in the 2015 ASQ *Future of Quality* report suggest that quality and performance improvement will be driven by connectedness, intelligence, and automation. In many cases, this will be enabled by Industry 4.0 technologies like IoT and machine learning. In the organizational ecosystems built on these technologies, humans and machines will cooperate to achieve shared goals, and use data to generate insights and deliver value in near real time.

Examples of these three elements will be presented in the following sections. Although the technologies and external environment will change over the upcoming decades, the role of the quality professional will remain the same as always: to harmonize people, processes, and technologies to help organizations achieve their goals and deliver business results sustainably.

### Connectedness

Every hour of every day, drivers consult Google Traffic to find traffic jams and plan better routes. Some people use the raw data to decide, on their own, whether to dynamically reroute (Figure 1.2). Others trust Google to choose a route for them and guide them to their destination using voice prompts. In its most basic form, Google Traffic provides nothing more than data aggregation: it looks at the total number of cars, how fast they're traveling, and how the speeds compare with the speed limits on those roads to make its determinations. Although other services like Waze and Mapquest also provide traffic information, Google data is more comprehensive and accurate, thanks to its user base. The more users providing information about their location and speed, the more accurate the traffic reports will be (Stenovec, 2015).

This example demonstrates the Quality 4.0 principle of connectedness: by gathering and aggregating data from thousands of people, the real-time state of the entire system can be inferred. This enables drivers to make more informed (and often better) decisions about their travel. Even without intelligent agents to select candidates for best routes, or implementing automation to choose the best one, there is value in connecting people to data like this.

FIGURE 1.2. Google Traffic aggregates data from thousands of drivers to visualize traffic jams.

Large-scale data aggregation can also reveal unexpected insights. A screen capture taken from Google Traffic at 11:30 p.m. local time on Monday, May 27, 2019, shows traffic jams north of Trotwood, Ohio, in the immediate vicinity of Northridge, and just west of Byron (Figure 1.3).

Why were there three jams this late at night in suburban areas just north of a major city? The traffic jams were not due to accidents or congestion. Approximately 45 minutes earlier, a large, violent tornado had started ripping a path through Trotwood (National Weather Service, 2019). Nearly two-thirds of a mile wide, and with winds approaching 200 mph, it moved east across the metro area for the next half hour. The winds leveled homes and businesses, and wrapped cars around utility poles with extreme force. Because vehicles were unable to navigate through roadways covered with debris, the path of the tornado was clearly marked by Google Traffic.

The moral of the story is: it's not just network connectivity that drives insights, but *connectedness*. By connecting people to data (or other people), they can get the information they need to perform.

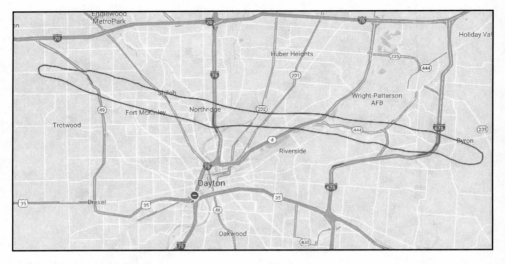

FIGURE 1.3. Path of the May 27, 2019, Dayton, OH, EF-4 tornado on Google Traffic (Peter Forister, Twitter: @forecaster25).

## Intelligence

In 1956, when the first group of researchers and practitioners who had been exploring artificial intelligence (AI) gathered on the campus of Dartmouth for the world's first workshop on this subject, they didn't really know what it was. Their goal was to define it, by looking for common threads through the problems each one had been working on, and then figuring out how to form solid collaborations.

The problems these people were exploring were fundamental and compelling, especially during the 1950s. Allen Newell, a cognitive psychologist, wanted to explore the problem of teaching a machine to play chess, and also to understand the concept of search. Herbert Simon, who would eventually win the 1978 Nobel Prize in Economics, was interested in decision making in organizations—in particular, the cognitive processes associated with rational decision making. John McCarty, a cognitive scientist, developed the Lisp family of programming languages that supported much of the early AI research. Ross Ashby, a psychiatrist, wanted to understand mechanical and biological control systems. Over 40 other people, with interests and pursuits just as diverse as these, also attended.

All the participants in the Dartmouth workshop shared a unifying goal: to create machines that could match or exceed the cognitive performance of humans. Intelligence could be defined in many ways, including deliberating on the next move in a game, making a well-reasoned recommendation for a new purchase, or understanding the meaning of a paragraph of text. Although deep learning has started to approach human performance on tasks like image recognition, speech recognition, and handwriting analysis, more complex tasks like detecting emotions, emulating beliefs, and interpreting meanings are still not in reach. Despite some practical success, the research landscape for AI today is still rather broad, deep, and aspirational.

## Automation

Automating a process means reducing or eliminating the need for human intervention. Credit card fraud detection, for example, is no longer a manual task: sophisticated machine learning algorithms use datasets describing patterns of behavior for each cardholder and compare new transactions with the baseline to determine whether there may be a problem. Automation makes it possible for the credit card company to provide millions of customers with accurate, timely handling of potential criminal incidents, minimizing losses for both parties.

Automation isn't an all-or-nothing commitment. On the path to autonomy, there are many degrees of automation that can be implemented. For example, an operator can define a process that a computer or intelligent agent executes, the computer can make and execute all decisions, or a combination (or context-dependent) choice can be made between the strategies. Machine intelligence fits seamlessly into this spectrum: an algorithm can provide advice, take action with approvals or adjustments, or take action entirely on its own. As part of process design, we have to decide what value to deliver through various degrees of intelligence and automation.

This was explained in more detail by Sheridan and Verplank (1978), who outlined ten degrees of automation. Their list can be used to select appropriate automation approaches for a given problem context:

1. Human specifies process, and computer directly executes the instructions
2. Computer assists human by determining options, and human selects the desired option
3. Computer assists human by determining options and suggesting a choice; human selects an option that may or may not be what was recommended
4. Computer assists human by determining options *and* selecting a choice; human has the option to follow the computer's recommendation or not

5. Computer selects and implements option, but requires human approval prior to executing it
6. Computer selects the best option and automatically implements it, but gives the human the chance to stop the process
7. Computer selects and implements options automatically, then reports results to the human
8. Computer selects and implements options automatically, telling the human about the results only if asked, and reports comprehensive results
9. Computer selects and implements options automatically, telling the human about the results only if asked, and reports only some information
10. Computer selects options, implements options, and automatically performs the whole job; it may or may not tell the human anything that has transpired, although logs may be collected to keep track of what occurred

Keeping a human in the loop can improve the quality of the automated solution. For example, on June 18, 2019, Jean-Francois Bonnefon (a behavioral scientist at the MIT Media Lab) shared a story on Twitter about his recent experience submitting a scientific paper for review. The journal had recently implemented an automated system for the first step of the process—checking the submission to make sure it fits the scope of the journal, is the right length, has the right kind of references, and is in general reviewable. This is usually a task performed manually by the editor or associate editors, and if the paper doesn't pass that initial quality check, it's a *desk reject* that does not proceed to peer review.

When Bonnefon submitted his paper, the automated system quickly did its thing and sent him an automated desk reject less than two minutes later. The paper could not be accepted, the e-mail explained, because of "a high level of textual overlap with previous literature." The author was confused because neither he nor his coauthors had engaged in any plagiarism. Fortunately, the editor-bot sent a copy of his paper with all the so-called plagiarized areas highlighted. About half the references, written in the standard citation format that the journal requires, had been flagged for bad behavior. "It would have taken 2 minutes," he said, "for a human to realize the bot was acting up. But there is obviously no human in the loop here. We're letting bots make autonomous decisions to reject scientific papers. I'm so excited to be at the forefront of this new era! (a little pissed, too)."

But like connectedness and intelligence, automation can be valuable on its own. For example, when I signed up for automatic bill pay for my power bill, I authorized the utility company to automate the payments on the same day each month. The service doesn't require me to be connected to the internet or to connect with the utility company in any way—the payment just automatically happens, triggered by a business rule that specifies what day to issue the payment. Executing business rules does not typically rise to the level of an intelligent system.

## DISCOVERY: THE NEW ROLE OF QUALITY

Connectedness, intelligence, and automation improve performance by helping us discover patterns and insights without having to explicitly define them, and accelerate the process of acting on data through automation. My credit card company, for example, implements intelligent agents to monitor my purchases, and detects when potential fraud has occurred. I get an automated text message listing the last three transactions, and asking me to review them for legitimacy. If I respond YES, no action is taken, and the fraud detection algorithms are updated (having learned more about my purchasing behaviors). If I text NO, my card is immediately shut down to avoid further losses, and a new card is issued. Delivery of value and prevention of loss are nearly instantaneous depending only on how quickly I can respond to the text message.

Today's quality profession began in the early 1900s, during the middle of the second industrial revolution. Scientific management, introduced contemporaneously by Henri Fayol in France and Frederick Winslow Taylor in the United States, modeled production systems as machines and aimed to use data and observations to make them perform better. Factories needed methods to make sure assembly lines ran smoothly, to produce artifacts to specifications, to train workers to perform accurately and consistently, and to control costs by standardizing work. The methods for statistical process control introduced by statistician Walter A. Shewhart (1891–1967) helped operators determine whether variation was due to random or special causes.

As industrial production matured, methods expanded to encompass design. Processes were consciously constructed to be able to produce to specifications. Joseph M. Juran (1904–2008) introduced the concept of Quality by Design in the 1960s and formalized it in 1986. Since then, it has been formalized in pharmaceutical manufacturing through the International Council for Harmonisation of Technical Requirements for Pharmaceuticals for Human Use (ICH) Guidelines Q8 through Q11 (DeFeo, 2019). In the 1980s and 1990s, the adoption of personal computing once again changed the landscape. Business productivity software established a foothold, in particular, spreadsheets and word processing. Organizations regrouped quality efforts around the value of culture and active engagement in quality—and total quality management (TQM), lean, and Six Sigma gained in popularity.

The progression can be summarized through four themes:

- **Quality by Inspection:** In the earliest days of quality as a practice and profession, quality control relied on inspecting bad quality out of the total items that had been produced. By the 1950s, the broader management focus of quality assurance had emerged.
- **Quality by Design:** Inspired by statistician W. Edwards Deming's recommendation to "cease dependence on inspection" and Juran's Quality by Design, holistic methods emerged for designing quality into processes, to prevent quality problems before they could occur.
- **Quality by Empowerment:** TQM and the philosophy of Six Sigma advocated a holistic approach to quality, making it everyone's responsibility and empowering individuals to contribute to continuous improvement.
- **Quality by Discovery:** In a smart, hyperconnected environment, quality will depend on how quickly data can be assimilated, aggregated, and applied. It will result in discovering and addressing root causes quickly, and rapidly capturing opportunities for growth and improvement.

As adaptive systems that are connected, intelligent, and automated are more widely implemented, opportunities for breakthrough performance improvement will be revealed—and there will once again be a renaissance in quality tools and methods. The pace will depend on how well we can discover (and act on) new insights about ourselves, our products, and our organizations.

## VALUE PROPOSITIONS FOR QUALITY 4.0

A value proposition explains the benefits a product or activity will deliver. Quality 4.0 initiatives that enhance connectedness, add intelligence, or advance automation tend to offer one or more of these value propositions (Radziwill, 2018):

1. Augment (or improve on) human intelligence
2. Increase the speed and quality of decision making
3. Improve transparency, traceability, and auditability

4. Anticipate changes, reveal biases, and adapt to new circumstances and knowledge
5. Reveal opportunities for continuous improvement and new business models
6. Learn how to learn; cultivate self-awareness and other-awareness as a skill

For example, predictive maintenance can help operators anticipate equipment failures and proactively reduce downtime. Intelligent algorithms can assess supply chain risk on an ongoing basis and provide recommendations about whether to take corrective action. Traditional quality practices can also be used to improve performance in new areas like cybersecurity, where documenting and benchmarking processes can provide the basis for detecting anomalies. Without an understanding of expected performance, it can be difficult to detect potential attacks or other situations that are not nominal. There are many more examples in the following chapters to demonstrate all these value propositions.

## QUALITY PROFESSIONALS: LEADING THE TRANSFORMATION

With better and more timely information about business processes, responding and adapting to changing customer and stakeholder needs—or a changing external environment—will become easier in some cases and feasible in others. When AI and machine learning are woven into data-driven decision making, quality systems will start becoming self-aware. But it won't be a purely technological transformation: designing effective intelligent agents and using the recommendations they provide will be a uniquely human proposition. The changes will have particular implications for the manufacturing workforce, where cognitive skills and the ability to make decisions based on data are already starting to become just as important as physical capabilities and dexterity.

Quality professionals already have the unique skills and capabilities to lead organizations in their digital transformation efforts. These include the following:

- Systems thinking
- Data-driven decision making
- Leadership for organizational learning
- Defining processes and managing continuous improvement
- Understanding how processes, policies, and decisions impact people: their lives, relationships, communities, well-being, health, and society in general

Quality professionals are distinctively good at structured problem solving, data-driven decision-making, and leveraging cultural change to facilitate improvement. In Quality 4.0, these fundamentals will not change, even as the amount and variety of data increase.

The last point is particularly important. Many machine learning algorithms must be trained, and training is subject to personal and cognitive biases. Quality professionals can anticipate positive and negative impacts, and help organizations protect against negative consequences while capturing opportunities that will benefit society and the planet. There are exciting opportunities as well to enhance environment, health, and safety (EHS) outcomes by applying quality principles and practices to those domains—driven by the ability to examine data across functional boundaries.

## BRAVE NEW WORLD

Even the most magical of technological capabilities can become ordinary and commonplace. I remember the first time I connected to WiFi on a new laptop. It was so shocking that I could check my e-mail without an Ethernet cable. I waved my hands around the outside of the machine, tracing the air on all sides. "Look," I told my colleague, "no wires!" He was giddy too, because we had both received new machines from our employer that day.

But fast-forward almost 20 years, and now WiFi is as essential as electricity and running water, and certainly doesn't get me excited every time I use it (unless it's not working). Optical character recognition (OCR) is another example of a technology that used to be magical but is now just everyday AI—it's used to sort and route most of our mail, and it's embedded in most scanners as well. When the technology solves a previously intractable problem, AI finds its footholds.

> Three fundamental characteristics of deployed AI can be seen in action. First, they identify a long-unsolved problem or unrealized opportunity. Next, they're solved in a way that simply wouldn't be possible without AI. Finally, they demonstrate that AI has a role to play in just about every industry, whether tech-focused or not.
> Sooner or later, every technology transitions from an elite niche to a mainstream tool. AI is now undergoing a similar transformation.
> After years of hype around mysterious neural networks and the PhD researchers who design them, we're entering an age in which just about anyone can leverage the power of intelligent algorithms to solve the problems that matter to them. Ironically, although

breakthroughs get the headlines, it's accessibility that really changes the world. That's why, after such an eventful decade, a lack of hype around machine learning may be the most exciting development yet.
—ANDREW MOORE, HEAD OF GOOGLE CLOUD AI, IN MOORE (2019)

The performance breakthroughs introduced by Quality 4.0 and Industry 4.0 technologies will impact our organizations and our lives gradually, slowly, and deeply. Although they will initially seem magical to us, in two more decades we'll take many things for granted—like being able to sense and analyze our operating environments in real time, and check the status of supply chains and ecosystems whenever and wherever we like. Traceability will be expected. Audits, process improvement, and risk management will be semiautonomous and aided by intelligent systems that recommend and prioritize actions. Instead of spending time dealing with nonconformances and management reviews, we'll have more time to focus on the bigger picture: innovation and growth.

The era of intelligent, autonomous quality systems is only beginning. This book will help you understand the emerging landscape, anticipate opportunities for improvements and breakthroughs, and develop strategies and initiatives for your own organization.

## THE BOTTOM LINE

- The fourth industrial revolution introduced machine intelligence and cyber-physical systems (tangible objects that can communicate over the internet).
- The revolution is beginning because storage is cheap, hardware is much more affordable than in years past, software to perform complex tasks is available, internet connectivity is widespread, and a workforce skilled in these new technologies is ready to go.
- Quality 4.0 improves connectedness, intelligence, and automation so that people, machines, and data can work together to improve performance and achieve organizational goals:

  o Connectedness means finding ways to bring people, machines, and data together—and may or may not involve internet connections or smartphones.
  o Intelligence can come from new ways of collecting and interacting with data, new ways to learn, or automated methods like machine learning algorithms.
  o Automation means having a system perform tasks that a human previously had to do. There are many degrees of automation, from machines provid-

ing recommendations that humans act on, to fully autonomous systems that make (and execute) all the decisions on their own.

- The practice of quality is shifting again, just like it has over the past hundred years. Quality by inspection evolved into quality by design. Next, engagement became critical as organizations realized that to be successful, quality must be *everyone's* job. Now, with increased connectedness, intelligence, and automation, quality can discover new insights for us, helping us meet our organization's goals more quickly and effectively.
- Quality 4.0 initiatives can augment human intelligence; improve decision making; improve traceability, transparency, and auditability; anticipate changes and adapt to changing circumstances; reveal opportunities; and help us learn how to learn.

## REFERENCES

Association of American Railroads. (2019, January). *Overview of America's freight railroads.* https://www.aar.org/wp-content/uploads/2018/08/Overview-of-Americas -Freight-RRs.pdf

Cline, G. (2017, March 31). *Industry 4.0 and Industrial IoT in manufacturing: A sneak peek.* Aberdeen Group. https://www.aberdeen.com/opspro-essentials/industry-4-0 -industrial-iot-manufacturing-sneak-peek/

DeFeo, J. (2019, April 15). "The Juran trilogy: Quality planning." *Juran Institute Blog.* https://www.juran.com/blog/the-juran-trilogy-quality-planning

Gedik, R., H. Medal, C. Rainwater, E. A. Pohl, and S. J. Mason. (2014). "Vulnerability assessment and re-routing of freight trains under disruptions: A coal supply chain network application." *Transportation Research Part E: Logistics and Transportation Review* 71: 45–57.

Ghofrani, F., Q. He, R. M. Goverde, and X. Liu. (2018). "Recent applications of big data analytics in railway transportation systems: A survey." *Transportation Research Part C: Emerging Technologies* 90: 226–246.

Gibert, X., V. M. Patel, and R. Chellappa. (2016). "Deep multitask learning for railway track inspection." *IEEE Transactions on Intelligent Transportation Systems* 18 (1): 153–164.

Goasduff, L. (2017, December 20). "How to listen to the voice of 'things' in the IoT." *Gartner Blog.* https://www.gartner.com/smarterwithgartner/how-to-listen-to-the -voice-of-things-in-the-iot/

Haefner, L. E., R. G. Goodwin, and L. A. Porrello. (1996). "The Great Flood of 1993: Impacts on waterborne commodity flow, rail transportation, and surrounding region." In *1996 Semisesquicentennial Transportation Conference Proceedings.* Ames, IA: Iowa State University and Center for Transportation Research and Education.

Huston, G. (2008). "The changing foundation of the internet: Confronting IPv4 address exhaustion." *The Internet Protocol Journal* 11 (3): 19–36.

Kagermann, H., J. Helbig, A. Hellinger, and W. Wahlster. (2013). *Recommendations for implementing the strategic initiative INDUSTRIE 4.0: Securing the future of German manufacturing industry; final report of the Industrie 4.0 working group.* Forschungsunion. http://www.acatech.de/fileadmin/user_upload/Baumstruktur_nach_Web site/Acatech/root/de/Material_fuer_Sonderseiten/Industrie_4.0/Final_report_ _Industrie_4.0_accessible.pdf

Kagermann, H., W. D. Lukas, and W. Wahlster. (2011). "Industrie 4.0: Mit dem Internet der Dinge auf dem Weg zur 4. industriellen Revolution." *VDI nachrichten* 13 (11). https://www.vdi-nachrichten.com/Technik-Gesellschaft/Industrie-40-Mit-Internet -Dinge-Weg-4-industriellen-Revolution

Krummenacher, G., C. S. Ong, S. Koller, S. Kobayashi, and J. M. Buhmann. (2017). "Wheel defect detection with machine learning." *IEEE Transactions on Intelligent Transportation Systems* 19 (4): 1176–1187.

Lawrence, K., and P. Mitchell. (2019, April 5). "Palmerston North freight train derailment causes disruption on Main Trunk Line." *Stuff.* https://www.stuff.co.nz/national/111813261 /cargo-train-hits-bridge-causing-spilled-logs-and-delays-in-palmerston-north

Matías, J. M., J. Taboada, C. Ordóñez, and P. G. Nieto. (2007). "Machine learning techniques applied to the determination of road suitability for the transportation of dangerous substances." *Journal of Hazardous Materials* 147 (1–2): 60–66.

Moore, A. (2019, June 7). "When AI becomes an everyday technology." *Harvard Business Review.* https://hbr.org/2019/06/when-ai-becomes-an-everyday-technology?es _p=9536708

National Weather Service. (2019, May 27). *EF4 tornado confirmed through Brookville, Trotwood, Dayton and Riverside in Montgomery County, OH.* https://www.weather .gov/iln/20190527_trotwood

Pfaff, R., P. Shahidi, and M. Enning. (2017). "Connected freight rail rolling stock: A modular approach integrating sensors, actors and cyber physical systems for operational advantages and condition based maintenance." In *Proceedings of the Asia Pacific Conference of the Prognostics and Health Management Society.*

Pu, Y. R., L. W. Chen, and S. H. Lee. (2014). "Study of moving obstacle detection at railway crossing by machine vision." *Informational Technology Journal* 13 (16): 2611–2618.

Radziwill, N. (2018). "Let's get digital." *ASQ Quality Progress* 51 (10): 24–29.

Santur, Y., M. Karaköse, and E. Akın. (2016). "Learning based experimental approach for condition monitoring using laser cameras in railway tracks." *International Journal of Applied Mathematics, Electronics and Computers (IJAMEC)* 4: 1–5.

Schlake, B., R. Edwards, J. M. Hart, C. P. Barkan, S. Todorovic, and N. Ahuja. (2009). "Automated inspection of railcar underbody structural components using machine vision technology." (No. 09–2863).

Schwab, K. (2016). *The Fourth Industrial Revolution.* Geneva, Switzerland: World Economic Forum.

Segovia, V. R., and A. Theorin. (2012). "History of control history of PLC and DCS." University of Lund.

Sheridan, T. B., and W. L. Verplank. (1978). *Human and computer control of undersea teleoperators.* Cambridge: Massachusetts Institute of Technology, Man-Machine Systems Lab.

Stenovec, T. (2015, December 18). "Google has gotten incredibly good at predicting traffic—here's how." *Business Insider.* https://www.businessinsider.com/how-google -maps-knows-about-traffic-2015-11

## CHAPTER TWO

# CONNECTED ECOSYSTEMS

*It is collaboration among many, not sudden epiphanies,*
*that really changes the world.*
—GREG SATELL, AUTHOR OF *MAPPING*
*INNOVATION AND CASCADES*

On July 4, 2019, a magnitude 6.6 earthquake struck the high desert of California, near San Bernardino, at 10:43 a.m. local time. Within five minutes, social media posts had spread the word to Europe and beyond, even though the U.S. Geological Survey website had crashed almost immediately under the surprise load. Just ten years earlier, it would have taken one to two days for people outside California to find out about what happened. Today, we can see personal videos of these news events captured by smartphones moments after they occur.

But the value of connectedness is even greater than the ability for news to circumnavigate the globe in near-real time. In the spring of 2019, I had a conversation with a director of operations at a large company that manufactures appliances. We were talking about failure modes and effects analysis (FMEA), a tool developed by reliability engineers in the 1950s to systematically identify and study all possible things that can go wrong in a product design process or manufacturing process. By critically examining failure modes in the design or production planning stages, steps can be taken to mitigate or eliminate defects and other problems. Because FMEA can significantly reduce the potential for failures, rework, and returns, lots of organizations use it. The director I spoke with was trying to figure out a better strategy for making sure his team's FMEA worksheets accurately reflected current processes and controls.

"It's great that I have a way to keep track of my own data, and some of the other people here in my division," he said. "I keep them all on this server. *But it's really the connections that matter.* I need a way to find out when process steps have changed, so I know when to take another look at the risks. I need for my systems to talk to each other so they can keep in sync. If we have to make sure all the systems are kept up to date with the right information manually, it's going to drive all of us up the wall."

Connections enhance performance, drive innovation, and transform people and relationships. Sometimes, these connections occur between people or organizations, who provide each other with information, knowledge, support, interest, guidance, leadership, and new perspectives. Connections between people can also lead to provisions of tangible goods, or even funding. Other times, connections between people and data help people monitor the status of processes, control processes, or monitor the state of the business itself. Machine-to-machine communication can be used to coordinate subsystems within a system and achieve tasks like dynamically optimizing fuel efficiency. When machines and data are connected, machines can gain new capabilities, such as the ability to produce milled or 3D-printed objects from pre-programmed designs, or to work with custom tooling more easily.

In the most profound cases, connectedness can transform and move the human spirit. In the summer of 2016, I attended a dissertation defense at the MIT Media Lab in Cambridge, Massachusetts. This self-proclaimed "antidisciplinary research lab" is known around the world for pushing the boundaries of human-machine interaction, spanning disciplines from cognitive science to art, and promoting radical creativity.

Xiao Xiao, at the front of the room, dimmed the lights and turned on the projector so we could see an image of her at a piano. This was no ordinary piano. Instead, it was the machine she had invented, one that looked and sounded like a piano, but with one extra feature. Instead of a solid fallboard immediately behind the player's hands, she had installed a mirror that could play and record video. This prototype, called MirrorFugue, makes it possible for more than one person to collaborate on a single piano through both sound and images (Figure 2.1).

The video started. On the screen, a young girl sits at a piano and begins to play "Twinkle, Twinkle Little Star." Shortly after, we see Xiao sit at MirrorFugue, with the girl on the video screen and her hands reflected in the fallboard. Xiao and the girl play a duet (which you can see and hear at https://vimeo.com /61927293, starting at 1:56). Afterward, Xiao asked the audience:

What would it be like to return 10, 20, 50 years later to converse with reflections from various points in the past? I like to imagine that my reflection as a little girl still resides within my piano. I'd like to find my former self still playing the

FIGURE 2.1. MirrorFugue, a high-tech augmented reality player piano
(http://xiaosquared.com).

same familiar tunes and join her in a duet, filling in an accompaniment to her
melody. Or perhaps my children and grandchildren can meet me at their age
as a reflection in the piano and play a duet with me. (Xiao, 2011)

Although her media work used the piano as a centerpiece, what Xiao was
really exploring was how augmented reality can create shared experiences.
Her work "introduced a *philosophy of communication* in which the details of
human presence form essential threads in the fabric of an interpersonal discourse, even when stitched across the folds of space and time" (Xiao, 2011). She
demonstrated how technology makes it possible to deliver user experiences
where value is not even constrained by basic elements of physics.

A great musical performance feeds both the mind and spirit, for it distills
and delivers essential expressions of the human experience. I wonder if
more computer interactions should strive to be similarly evocative, where
facets of ourselves are reflected back at us so that we become more mindful
at the immensity of our existence. (Xiao, 2011)

The MirrorFugue example shows how powerful new and emerging technologies can connect people to data, to experiences, and to each other—even to a

past or future self. Couple this with always-on internet, and more possibilities for increasing connectedness across workplace and home ecosystems will be revealed. More connectedness among people, machines, and data means more potential ways to create value—and billions of internet of things (IoT) devices to use as levers and enablers.

## NATURAL AND ORGANIZATIONAL ECOSYSTEMS

What is an ecosystem? In a natural ecosystem, living and nonliving entities coexist with one another, interacting as a system and dynamically exchanging energy with each other and the environment. It's the energy exchange that determines the health and well-being of the ecosystem. Every living thing in that system is dependent on the ecosystem as a whole, and although natural ecosystems can often bounce back from disruptions, a problem with one part of the system will often impact others.

Organizations work in a similar way. They consist of smaller, interconnected networks of people, organized into departments and functional areas, and informal networks of projects and friends that get the work done. People also interact with nonliving entities like hardware, machinery, software applications, and databases. Networks can be formal (e.g., companies, organizational units, research centers, legally binding partnerships, teams) or informal (e.g., established by cooperation and collaboration). The organization has direct relationships with suppliers, partners, and collaborators, and indirect relationships with other stakeholders and society.

But how does energy flow? In the concept of an "industrial ecosystem," raw materials are the energy and the goal is to minimize material waste to lessen environmental impacts (Frosch & Gallopoulos, 1989). In an optimal industrial ecosystem, plants would "use each other's waste material and waste energy flows as resources" (Korhonen et al., 2001). In psychology, *flow* is a word that describes reservoirs of mental energy that, when tapped, results in focused attention and compels action. In organizations, energy flows through value streams and communication networks (Schippers & Hogenes, 2011; Peltoniemi & Vuori, 2004).

Value is the energy in the organizational ecosystem. And because value can flow through processes, transactions, information exchanges, or financial exchanges, there are many sources and sinks for that energy. Non-value-adding activity is not desirable, because it draws energy out of a system rather than creating new energy to sustain the system. The purpose of a quality management system, using this analogy, is to design ways for the nonfinancial value to flow that will help the financial flows as well.

Connections create ecosystems by facilitating the flow of value. For example, data gathered by one entity can be used or turned into value by another if the right connections (human and technological) are in place. Value can be created by connecting people to data and information repositories so they can find what they need to solve problems, or by connecting algorithms to data sources so they perform better. The most disruptive innovation takes place at the "edges" (that is, not in the highly controlled environments of R&D labs), so designing the connections between networks within an ecosystem is critical (Ito, 2012).

For the ecosystems to be healthy, a communications infrastructure must be in place to connect the entities within it. This includes physical connections (the internet infrastructure, WiFi, and cellular networks) as well as conceptual connections (policies, procedures, protocols, and formats). The communications infrastructure makes it possible to exchange information, share data, make decisions, and interact in ways that help people within the networks collectively achieve business goals.

Within an organizational ecosystem, the entities that need to be connected to each other can be broadly classified into three groups:

- **People** include individuals as well as groups of people in teams, functional divisions, entire organizations, suppliers, research centers, community support organizations, cities and towns, countries, and society in general.
- **Objects and machines** include industrial equipment, operations technology such as historians or supervisory control and data acquisition (SCADA) systems, models to process and understand data (e.g., simulations, optimization models, statistical models, machine learning models), sensors, actuators, beacons, radio frequency identification (RFID) tags, embedded systems, biological machines (e.g., CRISPR), smart materials (which can function like machines), and any physical item that can be connected to a network.
- **Data** include values, objects (like images), files, unstructured data, streaming data (e.g., from sensors), and data repositories like databases, data lakes, and data warehouses.

The remainder of this chapter walks through each of these categories (infrastructure, people, objects and machines, and data). Each section describes some of the key enabling technologies that enhance connectedness between these elements, providing the substrate for value creation.

## INFRASTRUCTURE

Everything we do—whether it's going to the grocery store to buy food or paying an invoice to a supplier on another continent—requires *infrastructure* to support the task. This includes roads, railways, airports, shipping lanes, power grids, emergency services, and telecommunications systems. It also includes the financial systems that support payments and other exchanges of value. In some countries, the school system is part of the infrastructure. Information technology infrastructure (on-premise as well as cloud-based systems) support business applications and communications, and can consist of routers, servers, firewalls, and data centers.

*Critical infrastructure* is defined as "systems and assets, whether physical or virtual, so essential to the nation that any disruption of their services could have a serious impact on national security, economic well-being, public health or safety, or any combination of these" (Alcaraz & Zeadally, 2015). Each U.S. presidential administration typically identifies key elements of the national critical infrastructure to guide policy and funding priorities. Presidential Policy Directive PPD-21, the most recent directive, identified the elements shown in Figure 2.2.

Infrastructure is in place to help people easily exchange materials and information. When materials and information can flow, economies also flow.

| | |
|---|---|
| • Chemical | • Critical manufacturing |
| • Communications | • Defense industrial base |
| • Dams | • Energy |
| • Emergency services | • Food and agriculture |
| • Financial services | • Healthcare and public health |
| • Government facilities | • Nuclear reactors, materials, and |
| • Information technology |   waste management |
| • Transportation systems | • Water and wastewater treatment |
| • Commercial facilities (offices, sports stadiums) | |

FIGURE 2.2. U.S. critical infrastructure sectors (White House, 2013).

## Connectivity and 5G Networks

Lack of available, reliable infrastructure can present challenges for all of these sectors. This can be the case even in the most advanced countries in the world, including the United States. In South Carolina, for example, over half a million people (11% of the state's population) do not have internet services that can support streaming video, sharing and downloading large files, or even some interactive websites. Not only does this make homework difficult for many of the state's students, but it makes business difficult for manufacturers and is impeding growth. David Cline, owner of Piedmont CMG, remarked that "there were extreme limitations, even on a good day" (Barton, 2019a). Hospitals, however, are bearing the brunt of the infrastructure crisis in South Carolina. Rural hospital closures and difficulties communicating with patients have exacerbated this issue, which is now a priority among the state's lawmakers (Barton, 2019b).

The next innovation in connectivity is 5G, driven by the need to match extreme demands on the global network by mobile devices that the existing 4G infrastructure cannot accommodate. 5G promises to catalyze adoption of technologies that require higher transfer speeds and lower latency, like augmented reality and IoT devices, but rollout will take some time. By mid-2019, parts of Chicago and Minneapolis had Verizon 5G, while Dallas, Houston, Kansas City, and Atlanta could access Sprint's 5G. New devices from Sprint, Apple, and Samsung had been released to coincide with network availability.

Once 5G is broadly available, though, it will likely be years before new real-time services and business models start to emerge in great numbers. When 4G was released in 2010, for example, it took four to five years for people to start watching videos on their phones and hailing taxis from a live network of drivers (McGarry, 2019). Unfortunately, "the coverage dilemma is likely to remain the same, thus widening the rural-urban digital divide further," prompting researchers to explore architectural innovations for 5G in rural areas (Khalil et al., 2017).

According to the United Nations, only half the world's population has access to the internet. After 2015, the rate of new connections slowed, with an estimate of nearly 4 billion people remaining off-line in 2019 (Dreyfuss, 2018). As a result, executives and senior leaders planning Industry 4.0 and Quality 4.0 initiatives cannot assume availability and reliability of internet connectivity. The conditions and projections for each stakeholder group must be evaluated separately. The types of devices that will be used, frequency of uploads and downloads, and sizes of those transactions must be considered during feasibility assessments.

## Internet Protocol Version 6 (IPv6)

Every device that relies on the internet to communicate needs an internet protocol (IP) address. From 1983 until at least 2019, version 4 of this protocol (IPv4) supported most internet traffic. Unfortunately, IPv4 has a critical limitation: there are only 4.3 billion unique addresses. But because 50 billion internet-connected devices are projected to be in service by 2020, IPv4 addresses will run out.

Fortunately, this problem has been known since the 1980s, when network engineers observed a more rapid depletion of addresses than they initially anticipated. Work on a new protocol, IPv6, started soon after in the 1990s and entered limited service, commercially, by 2006. (What happened to IPv5, you may ask? This version number was assigned to an experimental platform for supporting data streaming that never became a viable solution.)

Pickard et al. (2018) studied the spread of IPv6 using Rogers's (1962) diffusion of innovations model and found that "IPv6 adoption will meet the 50% point to begin the Late Majority phase between March of 2021 and October 2022." If your organization has not started considering IPv6, this result strongly suggests that now is the time.

## Cloud Computing

Although there was a long ramp-up period as businesses learned to trust compute power owned and maintained by other organizations, cloud computing is now a more popular solution for hosting critical business systems than on-premises installations. Data protection, compliance management, and service-level agreements are now frequently well maintained. In some industries—for example, aerospace in India—cloud computing is increasingly trusted for mission-critical operations technology on factory floors (Misra et al., 2018).

There are four kinds of clouds (deployment models) and three general service models representing functionality that can be offered (Figure 2.3). Infrastructure as a Service (IaaS) leaves the systems administration, operating system management, data management, and software up to the customer. Platform as a Service (PaaS) shifts each of these responsibilities to the cloud

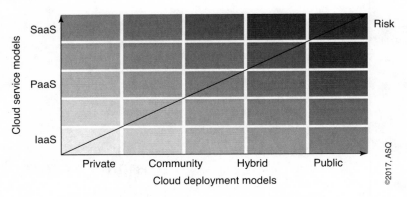

FIGURE 2.3. Risk depends on cloud deployment and service models
(Novkovic & Korkut, 2017).

provider except the software management. Software as a Service (SaaS) is even more hands-off, representing a model where the customer rents software that the cloud provider builds, grows, and maintains. In addition, cloud infrastructure can be deployed completely on-premises (private cloud), can be shared with affiliated organizations (community cloud), can be shared with unaffiliated organizations (public cloud), or can use a combination of the methods (hybrid cloud). Leaders planning Industry 4.0 and Quality 4.0 initiatives should be aware that the risk profile is different for the combinations of service and deployment models (Novkovic & Korkut, 2017).

## Software-Defined Systems

In a software-defined system, specialized hardware is replaced by software that runs on a single, shared computing device. This eliminates the need to install, manage, and maintain multiple distinct hardware elements. For example, in software-defined networking, routers and switches are replaced with software that has more finely tuned power to handle various kinds of network traffic. This is called network function virtualization (Kreutz et al., 2015).

Because there is no longer a need for different pieces of hardware, the nature of the new product development life cycle shifts completely, becoming a software project rather than a hardware project. Although both approaches have challenges, software development is typically somewhat faster and less expensive than hardware development. Additionally, shifting the maintenance burden to software saves time and money and means that less specialized skill sets are needed to improve and evolve devices. This makes it easier to innovate and grow.

In addition to networking, there are several other use cases for this approach, including software-defined industrial internet of things (IIoT), software-defined cloud manufacturing, software-defined control systems, and software-defined supply chains (Thames & Schaefer, 2016). For the Industry 4.0 or Quality 4.0 leader, the value of these emerging approaches is that it may help us design agile, programmable production systems that are more easily managed, configured, and secured. Before investing in specialized hardware, explore the possibility of a software-defined system.

## Edge Computing

Huge amounts of data will be collected by sensors and IoT devices, and not all of that data will need to be captured or transferred. Some data will be analyzed and used locally, and, once it's processed, may not need to be stored. For example, if a voltage is being read every 100 milliseconds because the sensor is configured that way, but only a 1-second average is needed for further computations, that operation can be done close to the sensor ("at the edge"). Specialized chips (*resource-constrained devices*) can also acquire data and execute machine learning algorithms at the edge, sending back only the results (Alizadeh et al., 2019). In Figure 2.4, for example, a machine learning classifier embedded with a camera can gather images, determine whether a cat is carrying prey such as a dead mouse or bird, and then send a signal back to the cat flap

| Image type | No cat | Cat not on approach | Cat on approach | Cat with prey |
|---|---|---|---|---|
| Count of images | 6,542 | 9,504 | 6,689 | 260 |
| Example | | | | |

FIGURE 2.4. Edge computing could be used to detect a cat with prey, sending a code over the network to lock the cat flap (Ignite Seattle, 2019).

to prevent the cat from entering the house. This is an effective arrangement of processing and networking resources that would yield greater efficiencies as the size of the camera network grew.

By processing data as close to the point of capture as possible, costs are minimized while issues like latency are made nearly irrelevant. Additionally, datasets produced by sensors or other instruments can become so large that they are impossible to move around the network. Edge computing helps alleviate these potential issues as well (Chen et al., 2018). When implementing an Industry 4.0 or Quality 4.0 solution, examine possibilities for pushing processing and analysis as close to the edge as you can to minimize costs and reduce the data footprint of the system.

## Cyber-Physical Systems (CPS)

The backbone of the connected enterprise and the smart city is cyber-physical systems (CPS)—physical objects that can communicate with people, machines, or data stores over networks. According to the National Institute of Standards and Technology (NIST, n.d.), *cyber-physical systems* or "smart" systems are "co-engineered interacting networks of physical and computational components." CPS have a cyber (connected) part and a physical (tangible) part and disrupt the traditional automation hierarchy in industrial systems by increasing connections between each of the layers (Figure 2.5) (Monostori, 2014).

Components within each of the levels can connect with other components in context-dependent ways, from the process level through the supply chain

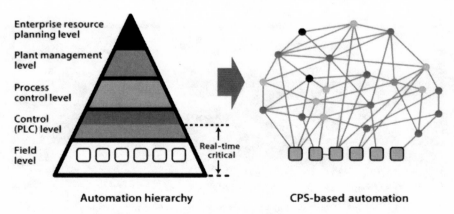

Enterprise resource planning level

Plant management level

Process control level

Control (PLC) level

Field level

Real-time critical

**Automation hierarchy**

**CPS-based automation**

FIGURE 2.5. CPS change production architectures (Monostori, 2014).

level. These systems will provide the foundation of our future critical infrastructure and form the basis for emerging and future smart services.

## PEOPLE

Even though the changes and disruptions associated with Industry 4.0 and Quality 4.0 are brought about by technology, people are at the heart of the transformation. Overall process efficiency depends on the combined performance of people, processes, and technologies. Delivering an excellent customer experience requires a committed, engaged workforce. Acting based on beliefs, intentions, and values is a uniquely human experience, as is demonstrating empathy.

> Imagine you're on a road trip, driving across the country, and you pull into a Starbucks drive-through that you've never been to before. . . . You're a loyal customer and you buy about the same thing every day, at about the same time. So as you pull up to the order screen, we show you your order, and the barista welcomes you by name. Does that sound crazy? No, actually, not really. In the coming months and years you will see us continue to deliver on a basic aspiration: to deliver technology that enhances the human connection.
> —GERRI MARTIN-FLICKENGER, CTO OF STARBUCKS, IN MARCH 2016

Technology represents all the ways that social groups create the tangible objects of their civilizations—bringing together people, processes, materials, and information to create utility and value. This section covers several technology-enabled ways to connect people to one another, using the broad definition to capture technologies that are not digital in nature.

### Protocols and Procedures

Until 2015, industrial communications used a diverse array of mechanisms to help machines communicate with each other, including fieldbus systems, Ethernet-based approaches, Modbus, and ZigBee (Wollschlaeger et al., 2017). A device was unable to talk to another device without a shared communications protocol. Even with a shared protocol, without a structure in place to know how to work together and make decisions based on those communications, the mere ability to pass information back and forth will not be valuable.

The same is true with people. Sharing information and acting on information are two different things. To effectively act on shared information, an organization has to know itself. How do decisions get made? How does the organization maintain a customer focus? How does workforce development ensure that capability and capacity are managed? Guidance like the Baldrige Excellence Framework can help organizations answer these questions to make sure that enhancing connectedness will add value (NIST, 2019).

The policies, practices, and procedures captured in self-study and award programs like Baldrige, by quality systems, by environmental management systems, and by occupational health and safety systems are all technologies. The lesson for Industry 4.0 and Quality 4.0 is this: traditional technologies are as critical as ever and cannot be ignored in favor of software and intelligent systems; rather, they should be used as a complement.

## Social Media and Engagement

By the mid-2000s, static websites had evolved into interactive platforms for co-creation. This "Web 2.0" emphasized sharing, collaboration, online discussions, and user-generated content. You could log into systems where you had a profile, and use that profile information to interact with others or make purchases. By allowing people to engage with businesses and with each other, web use increased tremendously, making knowledge sharing more effective and efficient. "Social media offers the potential to reduce the transaction costs of knowledge acquisition, knowledge linking and the occasional incorporation of key knowledge holders in a project. It does this by removing the need for lengthy coordination processes and complex decision-making channels" (Bauer et al., 2015).

Today, social media and proprietary community sites are places where unstructured data can be mined to unveil VoC. In addition, Suresh et al. (2018) and others have demonstrated that deep learning, a complex machine learning algorithm based on neural networks, shows promise in helping companies interpret customer feedback, intentions, and pain points. The value of social media to Industry 4.0 and Quality 4.0 initiatives is that it can provide a platform for people to find each other, share news and bulletins, and gain deep insights into VoC, reducing the cost of knowledge acquisition for everyone.

## Chatbots and Intelligent Conversational Agents

A contemporary way to connect people with knowledge is through a *chatbot*, an interactive dialogue interface connected to a program rather than a human. The best ones (that you might interact with on a customer service website) may not be distinguishable from human conversation. Although this technology has existed since the 1960s, chatbots have only recently become widespread thanks to advancements in natural language processing (a form of AI) and ease of setup via the SaaS model (Radziwill & Benton, 2017). More advanced intelligent conversational agents are also available to connect people with knowledge—for example, the personal assistants Alexa (from Amazon), Siri (from Apple), and Bixby (from Samsung). Although most use cases center on customer service, text-based and voice response chatbots may be used for any kind of search for information. Leaders advancing Industry 4.0 and Quality 4.0 initiatives can consider implementing chatbots and intelligent conversational agents anytime human operators need assistance accomplishing their tasks.

## Random Collisions of Unusual Suspects (RCUS)

Connecting people to machines and data can improve decision making, and connecting people to others in your organization or supplier ecosystem can smooth processes that cross organizational boundaries. Cultivating connections to support innovation, though, is also important. Using a term coined by Saul Kaplan of the Business Innovation Factory in Providence, Rhode Island, staging opportunities for "random collisions of unusual suspects" can strengthen innovative potential:

> Most innovation isn't about inventing anything new but merely the recombination of what already exists in new ways to solve a problem or deliver new value. Everything we need to innovate is in our sandbox and can be found at the edges between our sectors, disciplines, and silos. Getting better faster is all about exploring the adjacent possible . . . by creating the conditions for more random collisions of unusual suspects. We spend far too much time hanging out with usual suspects, people exactly like us. We don't learn anything new that way. The gold is in the grey areas between our silos if we only spend time at the edge colliding with more unusual suspects. (Anderson, 2017)

The message to Industry 4.0 and Quality 4.0 leaders is a recipe for stimulating recombinant innovation: breakthrough ideas come from combining bits and pieces of established ideas (Radziwill & Owens, 2014). Plan opportunities for people to connect with those outside their sectors, disciplines, and silos, instead of waiting for those opportunities to serendipitously occur.

## OBJECTS AND MACHINES

The physical objects to be connected, and the machines that facilitate the connections, will also play a role in many Industry 4.0 and Quality 4.0 initiatives. To understand the emergence of the new internet-connected enabling technologies, we first take a quick look at how machines have evolved in plants and factories since the 1960s.

### Operations Technology (OT)

Operations technology (OT) refers to the hardware and software that control physical equipment and monitor and control production processes, usually in an industrial environment. This can include manufacturing processes, defense networks, transportation systems, or any other element of critical infrastructure. Industrial control systems (ICSs) are OT, with components to monitor the state and status of machines, people, and their work throughout the production process, and to control machinery and associated devices to ensure that products are reliably produced to specifications.

In contrast, IT systems are used for e-mail, apps, databases, document management, and other enterprise functions. While you may upgrade your apps or operating system every few years, OT life cycles are much longer, and it is not uncommon to find 30- or 40-year-old technology on a factory floor. In addition, OT is much more likely to run continuously, with inspections and downtime planned well in advance to minimize losses.

As a result, priorities for managing OT (and the data it produces) are safety, availability, integrity, and confidentiality, in that order. IT priorities are exactly the opposite: confidentiality is most important, followed by integrity, availability, and (sometimes) safety. Cybersecurity was not traditionally a concern among manufacturers of OT components, because most were not designed to use internet protocols for communications, and the long life cycles mean that many OT components are still too old to use wireless networks. SCADA systems, responsible for directing field controllers on factory floors, used so many protocols for transmitting data (e.g., Modbus, Profibus, Conitel, IEC 60870-5-101, and DNP3) that the systems were secure because it was really difficult to get one machine to talk to another. "Security by obscurity" worked, but not because it was necessarily a good thing (Kranz, 2018).

## Industrial Internet of Things (IIoT)

Newly manufactured OT components *can* communicate over modern protocols like IPv4, IPv6, and Ethernet. Although this makes interoperability much easier to achieve, it also expands the attack surface for potential cyber incidents. The industrial internet of things (IIoT) is the culmination of hundreds of OT protocols converging to IP-based communications, making the hardware that controls equipment and production processes more versatile and capable—the "SCADA-fication of everything."

## Internet of Things (IoT)

When CPSs are used for nonindustrial purposes that do not involve OT, this is known as the internet of things (IoT). Just like in IIoT, IoT consists of objects that can communicate with one another over networks, coordinated through special platforms. Unlike IIoT, these objects are not typically the kinds you find in manufacturing plants, utility companies, and other critical infrastructure facilities.

In addition to supporting consumer devices like smart appliances, smart home security, and personal assistants, IoT can be used for air quality (Oh et al., 2015) and water quality monitoring (Vijayakumar & Ramya, 2015), equipment monitoring in the healthcare industry (Satija et al., 2017), monitoring of patient vital signs for rural healthcare (Rohokale et al., 2011), and precision agriculture to improve product quality and yield (Shenoy & Pingle, 2016). These applications are important to Industry 4.0 and Quality 4.0 leaders because IoT "will boost a tremendous amount of innovation, efficiency, and quality. Connecting production, medical, automotive, or transportation systems with IT systems and business-critical information will provide tremendous value" (Weyrich & Ebert, 2016).

## Sensors, Actuators, and Microcontrollers

At the center of IoT functionality are the components that make it function: the sensors that perceive inputs; the actuators that move physical parts like switches and turn on motors; and the microcontrollers that collect, synthesize, interpret, and share the data the sensors collect. Microcontrollers are designed to be embedded in objects. Unlike general-purpose microprocessors, microcontrollers are usually designed for a particular task or group of tasks. Sensors and microcontrollers can also be packaged together. Modern microcontrollers like Arduino and Raspberry Pi come with an onboard power source, ability to connect sensors, and sometimes even sensors or actuators. These are the foundational building blocks of IoT and IIoT.

## Radio Frequency Identification (RFID) and Beacons

Two special types of sensors commonly seen on factory floors are radio frequency identification (RFID) systems and beacons. These technologies are used for near-real-time location tracking and asset management. Sensors can be active (with an onboard transmitter and battery) or passive (communicates only when prompted by the RFID reader). Active RFID sensors are either *transponders*, activated by a reader, or *beacons*, which regularly broadcast signals. Passive RFID sensors have shorter ranges and are less expensive. Ultra-wideband systems are an alternative to RFID, providing reliable asset tracking and local positioning to within centimeters (Huang et al., 2017).

Even outside of manufacturing, Industry 4.0 and Quality 4.0 initiatives can leverage the data gathered from these sensors to track work in progress, evaluate the layout of workspaces, keep track of assets, and dynamically explore opportunities for improvement through simulation. These applications are starting to take hold in facilities and building management as well as in healthcare. RFID and beacon data can be provided to inform digital twins for simulations and exploratory studies.

## Digital Twins

The best explanation of digital twins comes from the engineers who conceived the term in 2014:

Up until fairly recently, the only way to have extensive knowledge of the physical object was to be in close proximity to that object. The information about
any physical object was relatively inseparable from the physical object itself.
We could have superficial descriptions of that object, but at best they were
limited in both extensiveness and fidelity. . . . It was then only in the last half
of the twentieth century, that we could strip the information from a physical object and create what we are calling a Digital Twin. . . . The range of
investigation into its behavior was both expensive and time consuming. We
first had to physically create the object, a one-off proposition. We then had
to create a physical environment in which the object was impacted by actual
forces. This meant that we were limited to investigating forces and their associated levels that we thought were of concern. Often, the forces would result in destruction of the object, dramatically increasing the expense. . . .

   This meant that the first time we actually saw a condition not covered by a
physical test would be when the physical object was in actual use. This meant
that there were going to be many unforeseen conditions or emergent behaviors
that resulted in failures that could result in harm and even death to its users. . . .

   The idea of the Digital Twin is to be able to design, test, manufacture, and
use the virtual version of the systems. . . . This will reduce failures of the
physical system when it is deployed and in use, reducing expenses, time,
and most importantly harm to its users. (Grieves & Vickers, 2017)

They cite the example of computer-aided design (CAD) drawings, which began
as conceptual models of physical systems. Today, however, simulations can be
carried out on CAD drawings within systems like SolidWorks and Autodesk
Fusion 360, making it possible to test the object and even explore trade-offs
like the impact of material selection on sustainability.

   Although the term is new, the concept is not. In the early 2000s, at the
Green Bank Telescope in Green Bank, West Virginia, we routinely had to test
new observing capabilities on the telescope. But with a $55 million asset, where
each hour of lost observing time was assessed at $5,800, taking the telescope
out of operations for testing was not an attractive option. Our software engineers built a "simulated telescope" with digital twins of receivers at different
observing bands, the backends that interpreted the signals, and the connective infrastructure that transported the data. Using these digital twins made it
much easier (and more cost-effective) to optimize operation and maintenance
of the physical assets and data acquisition and analysis processes.

## Robotics

The value of industrial robotics extends well beyond manufacturing. In addition to mechanical support for routine tasks like picking, placing, and sorting, robots can provide access to inaccessible or unsafe environments and perform tasks that may be dangerous for humans (like entering confined spaces or deep mines). Agricultural productivity increases due to robotics; precision robotics enhance success rates for surgeries (Bauzano et al., 2014); and collaborative robotics are being explored to alleviate musculoskeletal disorders by humans and robots sharing loads (Munoz, 2017).

Robotics innovations are not to be confused with Robotic Process Automation (RPA), the software-driven automation of service tasks. For example, scripts that automatically and routinely process applications, clean data, route calls, route e-mails, or apply business rules might all be considered RPA. Automating routine processes such as these can reduce variability, improve performance, and reduce costs.

## Wearables

Robotics can be controlled from the body, and when embedded in personal protective equipment (PPE) like hard hats, antivibration gloves, or goggles, it can significantly reduce exposure of workers to hazards. These wearables can alert workers of possible exposure to environmental risks (like toxins, heat stress, cold stress, and excessive noise), emergency conditions on machinery, dangerous postures and lifting conditions, and potential overburdening of physical or cognitive capabilities. There is still much innovation to be pursued, because "in today's production environments, productive use of this new abundance of data has not been exploited, neither for the benefit of the workers, nor for corporate reasons" (Romero et al., 2018). Quality 4.0 initiatives will harness that data to improve products, processes, and occupational health and safety.

## 3D Printing

Additive manufacturing builds parts and objects incrementally rather than milling them out of larger blocks of material. As a result, there is much less waste. 3D printing is one form of additive manufacturing and provides one way to support the mass customization requirements of Industry 4.0. Although the most common material for 3D printing is poly lactic acid filament, metal powders, concrete, textiles, smart materials, and food can all be 3D printed. Even nanofabrication is possible (Dilberoglu et al., 2017).

Although the technology may not be as revolutionary as initially assumed, there is power in being able to produce an object locally from software code that specifies how to build it. 3D printing will increase the number of potential products that can be offered, create a market for niche goods, and support rapid prototyping. It may also reduce delays associated with obtaining replacement parts or specialized tools (Strange & Zucchella, 2017). For Industry 4.0 and Quality 4.0 leaders, the main benefit of additive manufacturing and 3D printing is that data is transferred instead of objects, eliminating one source of waste: transport.

## Smart Materials and 4D Printing

The label *smart materials* refers to two different things: materials that change characteristics under external stimuli like temperature or electric charge (e.g., artificial muscles, shape memory alloys, color-changing fabric), and objects that can be tagged or otherwise embedded with intelligence, becoming somewhat self-aware and able to communicate. When 3D printing is used to generate material that can dynamically change characteristics, this is sometimes called 4D printing. These smart materials may be useful in extreme environments like space, or in creating programmable materials, self-assembling structures, or compact configurations for storing the structures (Dilberoglu et al., 2017). The implication for Industry 4.0 leaders is that future manufacturing processes will not fully occur within the bounds of the plants or facilities.

## Neurotechnologies

Brain-computer interfaces (BCIs) and other neurotechnologies are being developed to "influence neural activity" and, presumably at some point in time, enact some degree of external control. For the time being, healthcare and advertising are the practical frontiers for application. Work to date in neural stimulation has been focused on improving the quality of life for those with Parkinson's disease and psychiatric conditions. Noninvasive methods to read signals from the brain are also used in advanced marketing to determine optimal choices and combinations for new product features, and to identify emotional preferences. Future BCIs may be used to enhance cognition and improve safety in industrial environments (Roelfsema et al., 2018).

In addition to the ethical issues associated with these innovations, there are also security concerns. Loss of agency through "brainjacking" could, at some point in the future, present significant threats to people and operators in smart factories, connected buildings, and smart cities (Pycroft et al., 2016).

## DATA

In addition to the connections among people, objects, and machines, connectedness to data sources is also essential. This can include enterprise data repositories, real-time streams from sensors and other instruments, and data owned by other people and organizations that is accessible over the internet.

Ways to make connections extend beyond setting up channels for data to flow. For example, good visualizations, animations, and simulations can help people draw more value from data. Production processes can be made leaner and more efficient by moving data instead of objects (for example, 3D printing instead of shipping). This section covers some key concepts associated with connecting to and engaging with data and big data in the Industry 4.0 era, including reality technologies that can "enhance cognition by new forms of data interaction" (Simões et al., 2018).

### Structured and Unstructured Data

Much of the data generated by companies from the 1980s until around 2010 was *structured*. That means it was (and continues to be) stored in databases where tables, fields, data types, and relationships were explicitly defined in a data model. But even a decade ago, the volume of *unstructured* data (text, images, videos, e-mails, audio files, XML, JSON) was beginning to grow (Manyika et al., 2011). In Industry 4.0, unstructured data, which can be more challenging to manipulate and analyze for quality improvement purposes, is far more prevalent. Additionally, streaming data from heterogeneous IIoT and IoT devices is unstandardized, unstructured, and often high volume. It can decrease in value quickly after it is produced, so it is not routinely archived (Gokalp et al., 2016).

## Streaming Data and Analysis Pipelines

A hallmark of Industry 4.0 is the ability to assimilate and analyze data in real time, as it streams from sensors, instruments, and other sources. The often-custom software that processes these streams are called pipelines. Streaming data, and the insights that can be produced when it is analyzed while in motion, can help organizations become more agile:

> Data can take several forms, such as data at rest and data in motion. Data that are sitting in warehouses waiting to be analyzed are referred to as data at rest. . . . *Data in motion* (data analyzed in real time as the event occurs, such as click streams and sensors) flow through a connected device to a database on the receiving end for immediate analysis . . . used to fine tune the [predictive] models . . . and analytics at the edge . . . deploys information to control processes in real time. (Duarte, 2017)

Although the idea of near-real-time data processing pipelines is not new (Radziwill, 2008), the availability of robust frameworks for setting up and managing pipelines is. Frameworks like Apache Kafka, Kinesis Streams, Apex, and Hadoop make it possible for more organizations, even small and medium-size businesses, to get up-to-date information about operations extremely fast.

## Data Lakes and Data Warehouses

There's plenty of information that a typical organization needs to collect and track. This could include information about products, services, customers, and suppliers; things the company says to customers, and what customers say in return; and reports about feelings, preferences, or product and service failures on social media. In Industry 4.0, it may also include information about predictive models and the aggregated, assimilated data used to feed them.

When data must be stored, organizations have choices about just how much order to impose. A *data lake* collects and aggregates data, and provides a unified location where structured, semistructured, and unstructured data can live. The model for the data lake is that everyone has some water (their own data) and contributes it to the larger reservoir. The data may not be clean, but at least it's all in the same lake. Data lakes solve the problem of not being able to track or access data stored in private or limited-access repositories, and can protect organizations from losing knowledge when employees leave. Unfortunately, they can also become dirty and unmanageable, becoming "data swamps" (Miloslavskaya & Tolstoy, 2016).

A *data warehouse,* in contrast, usually has a rigid structure or schema governing what gets stored, whereas the data lake just has a catalog so people can find resources. The warehouse is usually cleaned and audited on a regular basis. The model for the data warehouse is a physical warehouse—you know exactly what's there and where it is because you've organized it the way you want it to be.

## Virtual, Augmented, and Diminished Reality (VR/AR/DR)

Once data is available and accessible, effective connectedness between people and data requires that insights and knowledge be extracted. The richer and more compelling the experience, the more insights can be communicated. Virtual reality (VR) and augmented reality (AR) immersive technologies can make this happen. For industrial training, VR is particularly value-adding, especially when a worker needs to be trained for dangerous environments like mines or oil refineries. AR uses special glasses or smartphones to superimpose information about the operations environment on real objects in the user's field of vision, providing immediate visibility to monitor data. Diminished reality (DR) is a complementary approach that uses the AR environment to *remove* stimuli from the work environment. This technique is not as widely used as AR yet, but has the potential to help workers improve cognitive processing by enabling better focus (Schwald & De Laval, 2003; Fraga-Lamas et al., 2018). The lesson to Industry 4.0 and Quality 4.0 leaders is that cost-effective methods are available for making training more effective, as well as new interfaces for live data access in the field that can support and improve cognitive function.

## THE BOTTOM LINE

Connectedness is one of the three key themes of Quality 4.0. By connecting people to each other, to objects and machines, and to data, we can make better and faster decisions. Effective decisions improve our products, processes, and organizations. Energy can flow into, out of, and through the ecosystem, and value can be created.

Enhancing connectedness requires thinking about four elements:

- **Infrastructure:** Industry 4.0 and Quality 4.0 initiatives should not assume availability and reliability of internet connectivity. Despite broad availability and new 5G networks coming online rapidly, there are still issues with sufficient speeds and latency in some rural and suburban areas of the United States and around the world.
- **People:** Quality systems, management systems, and digital technologies are *all* technologies. Find opportunities to use them to increase the quantity and quality of connections between people within and outside your organization. Make room for people to connect with those outside their normal area of work or expertise to support recombinant innovation.
- **Objects and Machines:** As communications protocols for the OT on factory floors converged to internet-based standards over the past two decades, enhanced interoperability led to the emergence of the IIoT. These same concepts expanded beyond manufacturing and critical infrastructure to consumer goods and environmental monitoring, creating the IoT. Technologies are now available to sense inputs from a multitude of sources and take goal-directed actions in near real time to achieve quality and performance goals.
- **Data:** Connecting people to data enables better decision making and more effective control of processes. Data visualizations, animations, and simulations also help people draw more value from data, and processes can be made leaner and more efficient by moving data instead of objects (for example, with 3D printing).

The culmination of enhanced connectedness will be a new way of living, made up of smart roads, smart utilities, smart factories, smart homes, and smart cities. But grand systems like these are built step by step as new devices and organizations join the network. For now, Industry 4.0 and Quality 4.0 leaders can focus on adding value by incrementally increasing connectedness, providing a foundation for smart cities and workplaces.

---

## REFERENCES

Alcaraz, C., and S. Zeadally. (2015). "Critical infrastructure protection: Requirements and challenges for the 21st century." *International Journal of Critical Infrastructure Protection* 8: 53–66. https://www.nics.uma.es/pub/papers/alcaraz2015CRI.pdf

Alizadeh, K., A. Farhadi, and M. Rastegari. (2019). *Butterfly transform: An efficient FFT based neural architecture design.* arXiv preprint arXiv:1906.02256.

Anderson, K. (2017, August 11). "Random Collisions of Unusual Suspects: Unleashing the Adjacent Possible." *Forbes.* https://www.forbes.com/sites/kareanderson

/2017/08/11/random-collisions-of-unusual-suspects-unleashing-the-adjacent
-possible/#6bea7c091dd2

Barton, T. (2019a, March 11). "Lagging internet left rural South Carolina biz stranded: Lawmakers seek to fix 'digital divide.'" *Greenville News*. https://www.greenvilleonline .com/story/news/2019/03/11/south-carolina-seeks-fix-digital-divide-boost-rural -internet-speed/3131109002/

———. (2019b, April 5). "SC House passes Bill to expand high-speed internet access in rural areas that need it." *The State*. Available from https://www.thestate.com/news /politics-government/article228793939.html

Bauer, Wilhelm, Moritz Hämmerle, Sebastian Schlund, and Christian Vocke. (2015). "Transforming to a hyper-connected society and economy—Towards an 'Industry 4.0.'" *Procedia Manufacturing* 3: 417–424.

Bauzano, Enrique, Belen Estebanez, Isabel Garcia-Morales, and Victor F. Muñoz-Martinez. (2014) "Robot collaborative assistance for suture procedures via minimally invasive surgery." In ROBOT2013: First Iberian Robotics Conference, pp. 255–269. Springer, Cham.

Chen, B., J. Wan, A. Celesti, D. Li, H. Abbas, and Q. Zhang. (2018). "Edge Computing in IoT-based Manufacturing." *IEEE Communications Magazine* 56 (9): 103–109.

Dilberoglu, U. M., B. Gharehpapagh, U. Yaman, and M. Dolen. (2017). "The role of additive manufacturing in the era of Industry 4.0." *Procedia Manufacturing* 11: 545–554.

Dreyfuss, E. (2018, October 23). "Global internet access is even worse than dire reports suggest." *Wired*. https://www.wired.com/story/global-internet-access-dire-reports/

Duarte, J. (2017). "Data disruption." *ASQ Quality Progress* 50 (9): 20–24.

Fraga-Lamas, P., T. M. Fernández-Caramés, Ó. Blanco-Novoa, and M. A. Vilar-Montesinos. (2018). "A review on industrial augmented reality systems for the Industry 4.0 shipyard." *IEEE Access* 6: 13358–13375.

Frosch, R. A., and N. E. Gallopoulos. (1989). "Strategies for manufacturing." *Scientific American* 261 (3): 144–152.

Gokalp, M. O., K. Kayabay, M. A. Akyol, P. E. Eren, and A. Koçyiğit. (2016, December). "Big data for Industry 4.0: A conceptual framework." In *2016 International Conference on Computational Science and Computational Intelligence (CSCI)*, 431–434. IEEE.

Grieves, M., and J. Vickers. (2017). "Digital twin: Mitigating unpredictable, undesirable emergent behavior in complex systems." In *Transdisciplinary Perspectives on Complex Systems*, 85–113. Switzerland: Springer.

Huang, S., Y. Guo, S. Zha, F. Wang, and W. Fang. (2017). "A real-time location system based on RFID and UWB for digital manufacturing workshop." *Procedia Cirp* 63: 132–137.

Ignite Seattle. (2019, June 6). "Cats, rats, AI, oh my!" [Video]. YouTube. https://www .youtube.com/watch?v=1A-Nf3QIJjM

Ito, J. (2012, June 12). "Innovation on the edges." *Edge*. https://www.edge.org/conversation /joichi_ito-innovation-on-the-edges

Khalil, M., J. Qadir, O. Onireti, M. A. Imran, and S. Younis. (2017, March). "Feasibility, architecture and cost considerations of using TVWS for rural internet access in 5G." In 2017 *20th Conference on Innovations in Clouds, Internet and Networks (ICIN)*, 23–30. IEEE.

Korhonen, J., M. Wihersaari, and I. Savolainen. (2001). "Industrial ecosystem in the Finnish forest industry: Using the material and energy flow model of a forest ecosystem in a forest industry System." *Ecological Economics* 39 (1): 145–161.

Kranz, Maciej. (2018). "Why industry needs to accelerate IoT standards." *IEEE Internet of Things Magazine* (1): 14–18.

Kreutz, D., F. M. Ramos, P. Verissimo, C. E. Rothenberg, S. Azodolmolky, and S. Uhlig. (2015). "Software-defined networking: A comprehensive survey." *Proceedings of the IEEE* 103 (1): 14–76.

Manyika, J., M. Chui, B. Brown, J. Bughin, R. Dobbs, C. Roxburgh, and A. H. Byers. (2011, May). *Big data: The next frontier for innovation, competition, and productivity*. McKinsey Global Institute.

McGarry, C. (2019, June 17). "The truth about 5G: What's coming (and what's not) in 2019." *Network World*.

Miloslavskaya, N., and A. Tolstoy. (2016). "Big data, fast data and data lake concepts. *Procedia Computer Science* 88: 300–305.

Misra, S. C., R. Mishra, and A. K. Munnangi. (2018). "Trust concerns in adoption of cloud services in the aerospace sector in India." *Software Quality Professional* 21 (3).

Monostori, L. (2014). "Cyber-physical production systems: Roots, expectations and R&D challenges." *Procedia CIRP* 17: 9–13. http://www.sciencedirect.com/science/article/pii/S2212827114003497

Munoz, L. M. (2017). "Ergonomics in the Industry 4.0: Collaborative robots." *Journal of Ergonomics* 7: 7556.

National Institute of Standards and Technology. (n.d.). "Cyber-physical systems (CPS) web site." https://www.nist.gov/el/cyber-physical-systems

———. (2019). "Baldrige excellence framework (Business/Nonprofit): Proven leadership and management practices for high performance." https://www.nist.gov/baldrige/publications/baldrige-excellence-framework/businessnonprofit

Novkovic, G., and T. Korkut. (2017). "Software and data regulatory compliance in the cloud." *Software Quality Professional* 20 (1).

Oh, C. S., M. S. Seo, J. H. Lee, S. H. Kim, Y. D. Kim, and H. J. Park. (2015). "Indoor air quality monitoring systems in the IoT environment." *The Journal of Korean Institute of Communications and Information Sciences* 40 (5): 886–891.

Peltoniemi, M., and E. Vuori. (2004, September). "Business ecosystem as the new approach to complex adaptive business environments." In *Proceedings of eBusiness research forum* (2): 267–281.

Pickard, J., M. Angolia, and T. S. Chou. (2018). "IPv6 diffusion on the internet reaches a critical point." *Journal of Technology Management & Applied Engineering (JTMAE)* 34 (1).

Pycroft, L., S. G. Boccard, S. L. Owen, J. F. Stein, J. J. Fitzgerald, A. L. Green, and T. Z. Aziz. (2016). "Brainjacking: Implant security issues in invasive neuromodulation." *World Neurosurgery* 92: 454–462.

Radziwill, N. M. (2008, July). "End-to-end operations at the National Radio Astronomy Observatory." In *Observatory operations: Strategies, Processes, and Systems* 2 (7016): 701612. International Society for Optics and Photonics.

Radziwill, N. M., and M. Benton. (2017). "Quality in chatbots and intelligent conversational agents." *Software Quality Professional Magazine* 19 (3).

Radziwill, N. M., and T. Owens. (2014). "Fresh perspective: Innovation." *ASQ Quality Progress* 47 (1): 30.

Roelfsema, P. R., D. Denys and P. C. Klink. (2018). "Mind reading and writing: The future of neurotechnology." *Trends in Cognitive Sciences* 22 (7): 598–610.

Rogers, E. M. (1962). *Diffusion of innovations*. Simon and Schuster.

Rohokale, V. M., N. R. Prasad, and R. Prasad. (2011, February). "A cooperative Internet of Things (IoT) for rural healthcare monitoring and control." In *2011 2nd International Conference on Wireless Communication, Vehicular Technology, Information Theory and Aerospace & Electronic Systems Technology (Wireless VITAE)*, 1–6. IEEE.

Romero, D., S. Mattsson, Å. Fast-Berglund, T. Wuest, D. Gorecky, and J. Stahre. (2018, August). "Digitizing occupational health, safety and productivity for the operator 4.0." In *IFIP International Conference on Advances in Production Management Systems*, 473–481. Springer.

Satija, U., B. Ramkumar, and M. S. Manikandan. (2017). "Real-time signal quality-aware ECG telemetry system for IoT-based health care monitoring." *IEEE Internet of Things Journal* 4 (3): 815–823.

Schippers, M. C., and R. Hogenes. (2011). "Energy management of people in organizations: A review and research agenda." *Journal of Business and Psychology* 26 (2): 193.

Schwald, B., and B. De Laval. (2003). "An augmented reality system for training and assistance to maintenance in the industrial context." *Journal of WSCG* 11 (1).

Shenoy, J., and Y. Pingle. (2016, March). "IOT in agriculture." In *2016 3rd International Conference on Computing for Sustainable Global Development (INDIACom)*, 1456–1458. IEEE.

Simões, B., R. De Amicis, I. Barandiaran, and J. Posada. (2018). "X-Reality system architecture for Industry 4.0 processes." *Multimodal Technologies and Interaction* 2 (4): 72.

Strange, R., and A. Zucchella. (2017). "Industry 4.0, global value chains and international business." *Sussex Multinational Business Review* 25 (3): 174–184.

Suresh, S., G. Raja, and V. Gopinath. (2018, April). "VoC-DL: Revisiting voice of customer using deep learning." In *32nd AAAI Conference on Artificial Intelligence*.

Thames, L., and D. Schaefer. (2016). "Software-defined cloud manufacturing for industry 4.0." *Procedia CIRP* 52: 12–17.

Vijayakumar, N., and A. R. Ramya. (2015, March). "The real time monitoring of water quality in IoT environment." In *2015 International Conference on Innovations in Information, Embedded and Communication Systems (ICIIECS)*, 1–5. IEEE.

Weyrich, M., and C. Ebert. (2016). "Reference architectures for the internet of things." *IEEE Software*, (1): 112–116.

White House. (2013, February 12). *Presidential policy directive 21: Critical infrastructure security and resilience* (PPD-21). https://obamawhitehouse.archives.gov/the-press-office/2013/02/12/presidential-policy-directive-critical-infrastructure-security-and-resil

Wollschlaeger, M., T. Sauter, and J. Jasperneite. (2017). "The future of industrial communication: Automation networks in the era of the internet of things and industry 4.0." *IEEE Industrial Electronics Magazine* 11 (1): 17–27.

Xiao, X. (2011). "MirrorFugue: Communicating presence in musical collaboration across space and time." PhD diss., Massachusetts Institute of Technology.

## CHAPTER THREE

# INTELLIGENT AGENTS
# AND MACHINE LEARNING

You can use all the quantitative data you can get,
but you still have to distrust it and use your own intelligence
and judgment.
—ALVIN TOFFLER, AUTHOR OF *FUTURE SHOCK*

Sheet metal forming is an efficient and economic way to make parts. Consequently, it's a key aspect of the manufacturing process for appliances, some consumer goods, and automotive components like a car's roof, hood, and doors. For personal vehicles, the forming process is critical, because achieving the perfect geometry for each component has consequences. Any deviation from specifications can impact the fit between different components, the assembly process overall, and the aesthetics of the final product—which strongly influence the perceived quality of the final product. Surface curvature for every part has to be smooth and even, with no visible defects, to ensure that light is elegantly reflected from all angles.

Figuring out whether one of these metal components meets hundreds of geometric specifications is no easy task. Coordinate measuring machines, which are expensive and require trained specialists to operate, provide one way to detect part shape fluctuations. An easier approach is to use digital image processing, with photogrammetric algorithms that measure surface geometry and strain states, but the accuracy of this method depends on the resolution of the camera. Still, the approach is not perfect, so researchers have developed methods based on springback analysis and critical strain distributions to improve the manufacturer's ability to detect the most common defects (Boese-mann et al., 2000). Despite the utility of these approaches, critical problems remain—for example, even dust or oil residue left on the metal component can be confused with fine cracks and other defects. For that reason, issues like this are called *pseudo-defects*.

Thanks to AI, though, this challenge is being addressed head-on. Leggett (2019) shares a story from automotive manufacturers BMW Group, which embraced AI in 2018 to improve its ability to distinguish real defects from pseudo-defects. Since it already had a camera-based quality control system in place, it was able to use the same image data it was already collecting but funnel those images through a neural network. Because the neural network had been pretrained with known images that illustrated good quality, various defects, pseudo-defects due to dust, and pseudo-defects due to oil residue, the intelligent agent running the neural network was able to classify each new image with a high degree of accuracy. BMW has substantially improved its ability to flag real defects, especially in cases that were previously "visual close calls."

In 2014, *Quality Manufacturing Today* shared another story from an auto manufacturer that was trying to improve root cause analysis results using a software package called SigmaGuardian. Embedded with machine learning (ML) algorithms and proprietary techniques based on information theory, the package scans through lists of failures to find emerging patterns. This manufacturer was having trouble with recurrent faults for the rear taillights on multiple modes, a problem that also carried high warranty costs. Each corrective action investigation and root cause analysis led to the same destination: nowhere.

Traditional methods exhausted, the company decided to see if the new methods might provide a cost-reducing clue. It was quickly surprised:

> The engineers had been looking at the rear of the vehicle for the answer (and not succeeding) however the software quickly found the root cause of the fault to be located in the roof of the vehicle, a part of the car that had not even been investigated as a possible source. Sometimes prior experience, or being too close to a problem, can inhibit a solution if an old hypothesis is applied to a new problem. (Somers, 2014)

In addition to uncovering a surprising result, the maker of this particular software claims that it can be used to identify and resolve issues on the fly (rather than after the fact, using statistical hypothesis tests). It can also handle difficult challenges like finding root causes based on stacked tolerances that would otherwise not be picked up by statistical process control (SPC) or manufacturing execution systems (MESs) (Love, 2018).

Techniques for early warning systems like this, based on ML, are still in the formative stages in the market. Even so, intelligent algorithms and agents have been the subject of research and prototypes for over two decades. In his master's thesis from the KTH Royal University of Sweden, Möller (2017) summarized many examples of how intelligent algorithms have been used in industrial quality control:

**Neural Networks for Classification**

- Using neural networks to determine when X-bar and R charts are in or out of control, a technique that worked particularly well for large shifts (Smith, 1994)
- Using neural networks to determine assignable causes, integrating SPC with feedback control (Shao & Chiu, 1999)
- Monitoring autocorrelated process data with recurrent neural networks (Pacella & Semeraro, 2007)
- Using neural networks to determine when variance is out of control in an autocorrelated process (Low et al., 2003)

**Image Recognition and Classification**

- Visual determination of fruit quality (Pandey et al., 2013; Sa et al., 2016)
- Visual determination of whether holes have been properly drilled by machine vision or deep learning (Johansson, 2017)

**Decision Trees for Anomaly Detection and Classification**

- Using decision trees (Guh & Shiue, 2005; Wang et al., 2008) and deep learning (Gauri & Chakraborty, 2007) to identify patterns and features in control charts
- Applying gradient boosting and random forests (ensembles of decision trees) to classify manufacturing failures (Bosch, 2016)

**Other Applications**

- Association rule learning can identify links between a breakdown in one component and failures or issues in other components or parts of the system (Snellman, 2017)
- Using convolutional neural networks to process expected and anomalous engine noises from different models (Möller, 2017)

Why aren't capabilities like these in broader use today? In short, because they can be challenging to apply in practice. Most of the research studies these capabilities are based on used simulated data. Unfortunately, operations data can be messy, especially when it is streaming in near real time from devices and machines. Additionally, traditional techniques for quality engineering often work well, and without a proven order of magnitude increase in prevention or detection capabilities using ML, these systems can be difficult to justify (Weisbrod, 2019).

But the pace of progress continues, and so does the democratization of ML. Amazon Web Services (with AWS Machine Learning), Microsoft Azure (with Machine Learning Studio), IBM Watson, Google Cloud ML Engine (based on the popular TensorFlow), and others all make it possible to deploy ML solutions quickly—with all the ease of Software as a Service (SaaS).

The business challenge for AI and ML is no longer primarily technical. Information technology and infrastructure can be outsourced, but nontechnical factors cannot. These include selecting appropriate projects, making sure there is a clear and demonstrable potential for value generation, building data science teams with solid statistical knowledge, and making good use of institutional capabilities and knowledge. (When a company cuts corners in any of these areas, it can lead to disasters like the $50 million horror story in Chapter 7.)

This chapter covers algorithms and approaches for adding intelligence to systems, including intelligent agents, AI, ML, edge computing, embedded intelligence, and affective computing. The information in this chapter will help you map business drivers to potential AI/ML solutions and evaluate the suitability of these methods for satisfying your organization's needs (see Figure 3.1).

## MODELS DRIVE CONTINUOUS IMPROVEMENT

Every organization has models that describe how it functions. The business model establishes what value the company plans to provide, to whom, and how it will generate revenue in the process. There are models that describe how leaders lead—how they set direction, create a plan, deploy that plan to the workforce, monitor progress, and help incorporate lessons learned into standard work and future decisions. Governance models explain how a company will ensure accountability for strategy, fiscal performance, and protection of stakeholder interests. Operations models describe the value-added processes used to transform inputs to outputs, and how they are monitored and adjusted to keep the organization on track to achieve its goals.

> Where there is no standard, there can be no kaizen.
> —TAIICHI OHNO

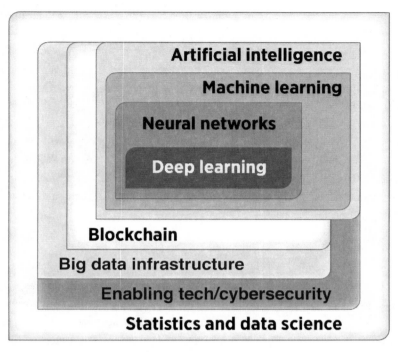

FIGURE 3.1. Relationships among AI, ML, and infrastructure elements (Radziwill, 2018).

Without a model, there is no baseline for performance. Even simple models, like flowcharts and value streams, can help teams make sure that each person on the team shares the exact same understanding of goals, process steps, and what's important. Learning occurs as we create models, as we compare those models against new information, and as we advance those models to help us more effectively and efficiently meet business goals.

## What Is Learning?

As we learn about our individual work and our businesses, we can improve the accuracy, performance, and completeness of these organizational and operational models. As we learn about the environment in which the business operates, and how it is changing, these models can be continually improved to adapt to those changes.

Learning . . . [changes] relationships among the learner, the other human participants, and the tools. Thus learning involves not only acquiring new knowledge and skills, but taking on a new identity and social position within a particular discourse or community of practice. Learning changes who we are by changing our ability to participate, to belong, and to experience our life and the world as meaningful. (Moss, 2003)

This process of evolution and adaptation helps organizations survive, helps people grow personally and professionally, and expands the capabilities of the workforce. At the same time, learning promotes the development of new relationships, as people with new skills are sought out by others to support efforts and initiatives.

## Learning Fuels Continuous Improvement

Strategic planning sessions usually include reflection on past successes and challenges. They may also address strengths and opportunities for improvement, to identify larger-scale adjustments to better meet strategic objectives. Operations meetings reflect on metrics gathered on shorter scales to consistently meet expectations.

For organizations that use ISO 9001 as their quality management system, improvement comes from corrective and preventive actions (CAPA), responses to audit findings, or recommendations that emerge from management reviews. Organizations that use the Baldrige Excellence Framework (BEF) to build their models engage in regular self-study and assessment, or the award process, to identify strengths to amplify and opportunities for improvement to close gaps. Board meetings, and other governance sessions, often culminate in recommended actions to adjust the strategic and operating models.

These adjustments are based on new data. Sometimes, that new data is taken from monthly or quarterly financial statements. Sometimes, it comes from quarterly reports that describe different aspects of operational performance like Key Performance Indicators (KPIs) or quality events like nonconformances. Other times, the new data emerges as themes from lessons learned from projects or initiatives, or market research that anticipates upcoming threats or opportunities.

But does learning *really* matter in continuous improvement? In a particularly compelling study, Kovach and Fredendall (2013) used a structural equa-

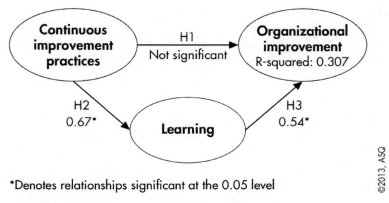

\*Denotes relationships significant at the 0.05 level

©2013, ASQ

FIGURE 3.2: Structural equation model (Kovach & Fredendall, 2013).

tion model to test the link among continuous improvement practices, organizational learning, and improvement outcomes. They tested the model using survey responses from 183 people in oil and gas, manufacturing, services, construction, and hospitals.

Their model, in Figure 3.2, shows that learning is the secret sauce that helps organizations improve: "*Learning has a direct, significant effect* on the maturity of the organization's use of [continuous improvement practices], which in turn affects organizational improvement" (Kovach and Fredendall 2013). Organizations improve not because they implement continuous improvement practices but because of the learning that takes place when practices are followed.

## Machine Learning

While people can learn and adapt leadership, governance, and operational models to improve performance, machines can learn as well—in certain well-defined cases. Though you wouldn't want an intelligent agent to occupy a seat at your board meeting or a role on your audit team (at least not yet), ML algorithms can find patterns in observations, detect anomalies, predict values, and classify observations into groups. Because they can be continually updated based on new data, they are ideal for sorting and forecasting models like the ones in the stories at the beginning of this chapter.

The similarities between traditional continuous improvement and the practice of applying ML algorithms are shown in Table 3.1. Continuous improvement and ML in practice share a singular goal: to help models perform better

**TABLE 3.1. Continuous improvement and ML learning are analogous.**

| Continuous improvement practice | AI/ML practice |
| --- | --- |
| Models are conceptual and operational | Models are mathematical or statistical |
| Works on a large scale (departments, organizations, business ecosystems) | Works on a smaller scale (individual processes, recipes and mixtures, KPIs) |
| New data comes from financial statements, audits, management reviews, other reviews, root cause analysis, and corrective actions | New data comes from new observations of the process being modeled (e.g., new incidents, new communications with customers, new quality events like nonconformances) |
| Data is often qualitative (e.g., lessons learned, audit findings) but can also be quantitative (financial, operations KPIs) | Data is quantitative or can be represented quantitatively (e.g., frequency of words, numbering categories) |
| Continuous improvement of policies, procedures, and heuristics that define the conceptual and operational models | Continuous improvement of mathematical and statistical models for forecasting, classification, defining best or anomalous conditions, or defining optimal paths or sequences |

as they adapt to new information and changing circumstances. This *is* learning: gathering experiences and new data and updating or adapting the models that describe how you operate and the relationships that enable you to operate.

This means that ML is the technological analog for what quality professionals have done for decades, only on a smaller scale. ML adds intelligence and automation to the practice of continuous improvement. It can complement, catalyze, and supplement continuous improvement, but it does not replace it.

## Artificial Intelligence

ML algorithms, and the models that can be built from them, are just one group of techniques within the broader category of approaches referred to as artificial intelligence (AI). Any machine or program that demonstrates cognitive capabilities that are usually attributed to humans can be considered AI. Examples include vision, perception, interpreting spoken language, understanding the meaning of spoken language, reasoning, problem solving, creativity, insight, or pattern recognition.

AI and ML are not interchangeable terms: while all ML is AI, not all AI is ML. AI is often implemented as intelligent agents that perform a task on behalf of someone or something else, and an intelligent agent may incorporate AI, ML, or both. Conversely, advanced methods may not use any AI or ML at all. Here are some examples related to optical character recognition (OCR) from Alkhalaf et al. (2014) and Barve (2012):

- In 1929, inventor Gustav Tauschek received a patent for an OCR machine that optically sensed letters and then compared them with patterns on a wheel marked by punched holes. When the sensed image matched a pattern on the wheel, a printing drum would rotate to and print the proper letter. *Although mechanical, this could be considered an early form of analog AI.*
- In the 1970s, an OCR process would split a scanned document into tiny regions, one for each character. Dark pixels were assigned a value of 1, and others received a 0. A statistical algorithm computed correlations between the placements of the 1s and 0s and stored examples to compare against those in other regions. *This is a statistical method that is neither AI nor ML.*
- Throughout the 1980s and 1990s, computer vision techniques were applied to the OCR problem, using sophisticated preprocessing techniques and rulesets to identify letters. *This is AI but does not use ML.*
- In the 2000s and 2010s, artificial neural networks and deep learning (a special kind of neural network) have been used to expand OCR capabilities, recognizing handwritten letters and numbers with high accuracy. *This is ML, which by definition is also AI.*

The most recent wave of OCR research has been motivated by search innovation. Why? Because companies recognized that people want to search for (or translate) words on signs they see, and search using pictures instead of text. Navigating a country in which you do not speak the language is much easier when you can just point your phone's camera at a sign or menu and your phone interprets the text for you on the screen.

## The Hype

AI can solve all problems—autonomously and perfectly. But you're a smart professional; you know this is not true. It's deep learning that has these magical properties (especially when paired with blockchain).

Of course, neither of the above assertions is truthful. In June 2019, a controversial post (which has since been removed) appeared on the *Medium* blog site that called out the abuses of terminology in marketing campaigns and pitches to investors based on the author's personal experience:

> When I worked at [Company X], I filled out the application that ultimately placed us on the 2018 version of the AI 100 [list in major technology magazine]. Like almost every other award application, it was an exercise in innovation theatre. . . . About a month after we appeared on the AI 100, an analyst . . . reached out and asked to know more about how we were using artificial intelligence to transform marketing and reduce fraud. I gave him an in-depth demo of our platform and live customer user cases, walking through each capability without once mentioning AI [because there was none in the platform]. The next week, we were listed in his report as one of the top five companies in AI alongside Alphabet, IBM, Facebook and Salesforce.

Because of overinflated expectations about the power of AI and ML (and free-flowing funding from investors, if you do it), there is pressure for companies to claim that they implement AI/ML, when in fact they do not. When you are evaluating marketing claims, look for evidence of the problem types and algorithms described later in this chapter. If a claim seems too good to be true, chances are it's just hype.

## The Reality: Strong and Weak AI

AI has become proficient at winning games, recognizing speech, recognizing faces, creating finely tuned dynamic schedules, optimizing policies, and filtering spam from your e-mail. Still, these tasks do not incorporate many attributes of intelligence, such as emotions, beliefs, values, ethics, or convictions, to guide decision making. When AI encounters an ethical dilemma, how does it even know there is a dilemma to be weighed? All the current implementations of AI, whether or not they use ML or deep learning, are examples of weak AI.

This does not mean that the algorithms are not powerful, just that they lack the ability to do things that humans struggle with (such as deliberate based on unclear or conflicting evidence) or do instinctively (making judgments or decisions based on intuition). Artificial general intelligence (AGI), or strong AI, is expected to close this gap—at least to some degree. Bostrom (2014) explains that there are three directions in which this can go:

- **Speed superintelligence**—a system that can perform cognitive tasks of a human, but much faster
- **Collective superintelligence**—a system of smaller subsystems that can perform cognitive tasks as well as humans over a broad domain of capabilities
- **Quality superintelligence**—a system that performs cognitive tasks as quickly as humans, but much better

You may not need to incorporate AGI into your long-term strategic planning unless your time horizon is several decades long. Citing composite results from multiple surveys of expert committees made up of AI researchers, Bostrom gives a 90% chance that AGI will be demonstrated—at least in a prototype—sometime between 2065 and 2093.

## Intelligent Agents

An agent performs an action on behalf of someone or something else, and when critical analysis is part of the process, that agent is an *intelligent agent*. When a computer system is situated in a specific environment and is capable of autonomous, goal-directed action within this environment, then that technology is also an intelligent agent.

Real estate agents, insurance agents, lawyers, and home assistants like Siri and Alexa are all examples of intelligent agents. Russell and Norvig (2016) described key questions that must be answered if you are designing an intelligent agent:

- **Percepts:** What information is available?

  o What information is available to your agent, and how often is it available?
  o What senses (or sensors) can gather information for your agent?
  o There should be percepts to support every decision about whether (and when) an action should be taken.

- **Actions:** What actions should be performed based on the percepts?

  o Using the information from the percepts, what atomic actions can you take?
  o There should be sufficient percepts to start and support every action.

- **Goals:** What goal should the action(s) contribute to?

  o What goals should those actions be able to accomplish together?
  o There should be a chain of actions to support each goal.

- **Environment:** Where is the agent supposed to do its work?

  o Is the agent embodied (like in a robot), or a component within a production system, or part of a cyber-physical system?
  o Is there anything unique about the conditions of deployment (e.g., a zero-gravity environment; confined space; inaccessibility once deployed, as on a satellite)?
  o All actions should be executable within the target environment.

The intelligent agent gathers information, processes it, and takes small steps (actions) to gradually achieve goals. The way the information is interpreted, and the manner in which actions should be taken, will depend on the unique characteristics of the target environment.

## EXAMPLES OF AI

Historically, most capabilities in AI have been enabled by symbolic or rule-based approaches, many of which required new research into how knowledge should be captured and represented. Because they depend on prior knowledge to inform the rules and relationships, the quality of the AIs depend on the quality of the data and whether meaning can be extracted from it. Today, many subfields have emerged in AI, including robotics, image processing and recognition, ML and data mining, knowledge representation and reasoning, semantic web, expert systems, natural language processing, and search.

*Knowledge representation and reasoning* takes a look at how information and knowledge can be structured and stored so that they lend themselves to intelligent computational scenarios (like search). Research in this area has produced relational databases, document-based databases like CouchDB and MongoDB, and graph databases like Oracle Spatial, RedisGraph, and Neo4j.

*Expert systems* capture reasoning in the form of facts and rules to generate new inferences. For example, facts might be "Nicole is the daughter of Mary" and "Mary is the daughter of John." Rules would be "A daughter is a child" and "The parent of a parent is a grandparent." Although a few more facts and rules would be required to make this example functional, when an inference engine is presented facts and rules, it can generate new facts like "John is the grandfather of Nicole."

*Natural language processing* (NLP) deals with the computational requirements for understanding language and generating it. It requires understanding linguistics, word formation, word order, the structure of phrases, and even things like sarcasm and ambiguity. Of course, gaining understanding in some of these areas is much easier than in others. For NLP to be effective, both traditional (e.g., Earley algorithm, Yarowsky algorithm, Hidden Markov Models) and ML algorithms (e.g., decision trees, support vector machines, Naive Bayes) are leveraged (Liu et al., 2017).

*Search and optimization* are often powered by AI. While most people associate these activities with internet search engines like Google and Bing, they are also important in many other places. Video games, for example, have some unique challenges, since computers have many different options for the agents they use to compete against the human player. With large multiplayer online games, these options increase exponentially. Researchers are therefore using search and optimization techniques to help the computer search through a portfolio of options to identify strong strategies during gameplay (Churchill & Buro, 2015).

Hybrid intelligent systems combine several of these approaches to generate an optimal solution. In some cases, hybrid intelligent systems also incorporate or leverage human intelligence to achieve their goals. This was the case with reCAPTCHA, the program that asks you to type in two words or sequences of characters that have been garbled so that humans can read them but machines cannot. Although the purpose of reCAPTCHA is to validate that a human is attempting to use a credit card, sign up for a new e-mail account, or access a gated resource, the combined results from users around the world—in essence, the human interpretations of text that was hard to read by machines—were used to digitize books and a hundred years of the *New York Times* (Law & Ahn, 2011).

## ML Problem Types

ML algorithms that look for patterns in massive amounts of data are used to build intelligent systems and some AI applications. As a subfield of AI, hundreds of ML algorithms have been developed over the past few decades, each of which can be used to directly or indirectly generate new business knowledge. Part of the challenge is knowing which algorithm, or family of algorithms, is appropriate for your business task. A description of many of the common statistical and ML models used in data science can be found in Appendix A.

In general, there are five main things you can do:

- Find patterns with unsupervised learning
- Identify key predictors through data reduction techniques
- Identify irregularities or anomalies
- Learn from examples with supervised learning
- Learn from experience with reinforcement learning

For example, consider credit card fraud detection. Because fraud patterns change over time, using ML algorithms to automatically detect emerging patterns can strengthen a bank's capabilities to defend its customers and respond faster than if humans had to monitor all the transactions manually.

You can conceptualize past transaction data as a spreadsheet with millions of rows, where each row (an observation) represents a transaction. The columns might contain information like the date, time, latitude and longitude of the transaction, the amount, the merchant, and perhaps even what was purchased. There is another column that gives us the "answer" to the question of fraud, telling us whether the transaction is legitimate or fraudulent. This column exists because the bank has had enough time to verify the validity of questionable transactions with the account holders. An unsupervised approach might cluster the transactions into two groups to see if you can separate the legitimate from the fraudulent. In a supervised or semisupervised approach, the algorithm looks for the similarities and differences between legitimate and fraudulent transactions, because it knows (at least to some extent) which are which.

## Finding Patterns (Unsupervised Learning)

Unsupervised learning uses "unlabeled" data to find relationships between observations. In the credit card fraud case, for example, "unlabeled" might mean we have information about the transactions, but we don't know which ones are fraudulent; there are no labels telling us which observations are good and which are bad.

One example of unsupervised learning is clustering, a family of algorithms that examine the relationships between the variables using distance metrics and similarity functions. Observations that are close to each other are assigned to the same cluster. Clustering is performed by iteratively building hierarchies, by partitioning observations in different ways, or by testing observations against prior knowledge and experience (Bayesian approaches). One of the most popular methods for clustering is the k-means algorithm, which can be computationally intensive.

The notion of "closeness" between the observations can be defined in many ways. For example, if you walk into a plenary session at a conference, you may want to split the attendees into groups for an exercise or breakout session. There are many ways to determine similarity, dissimilarity, and closeness, including height, hair color, physical distance, interests, or even the strength of friendships and relationships. You don't need to know which groups the people fit into, but based on one or more of those variables, you can use unsupervised learning to guess who is a member of each hypothesized group. After the group assignments are made by the unsupervised learning algorithm, the next step would be to examine the characteristics of the groups to make a determination about why the groups are different.

Here are some examples of how clustering and other unsupervised techniques have been used for quality and process improvement research across different industries:

- Grieco et al. (2017) used k-means clustering to extract patterns from Engineering Change Requests (ECRs) with a high degree of accuracy. Organizing ECRs based on similarity can help companies respond to themes rather than individual requests, improving the overall efficiency of their work.
- Recognizing that process quality impacts product quality, Wu et al. (2000) used unsupervised Kohonen Maps (a special kind of neural network) to examine process disturbances in welding production lines, indicated by voltage readings. Their model made a correct assessment of process issues in 22 out of 24 test cases.
- Although automatic diagnosis is a goal in medical research, it is hindered by the inability to provide large enough datasets with accurate labels (e.g., malignant or benign, nominal or not nominal). Unsupervised methods are thus more practical for this purpose. Shin et al. (2012) used unsupervised deep neural networks (deep learning) on a video dataset to take the first step, accurately identifying which organ was in the imagery. Their model was able to identify the liver and heart in the majority of cases, although it had difficulty identifying the kidney and spleen.
- Khodabandelou et al. (2014) used Hidden Markov Models to generate process maps from event logs. Their goal was to advance understanding of software users' thinking and motivations as they navigate through an application.

Unsupervised learning is useful for detecting patterns in observations as long as all variables are quantitative (or can be expressed somehow as numbers, like word frequencies in a document). These methods, however, are often very sensitive to the way distance between observations is calculated. For example, the distance between two people will be shorter if they can just draw a line between them and follow that path (Euclidean distance) than if they have to follow a path that takes them along specified north–south and east–west routes (Manhattan distance). Unsupervised methods also often require human intervention to make sense of the results.

## Identifying Key Predictors (Data Reduction)

Sometimes, the data you have to process is too unwieldy and overwhelming. Imagine the same credit card transaction dataset considered earlier, but instead of just information about the transaction amount, location, and characteristics, there are also a few hundred other variables. Which variables should you use to determine whether a transaction is valid or fraudulent?

If you use all the predictors, there are two risks. First, the model-building process may be computationally intensive, meaning that it may take more time or more processing power to build the model than you have available. Second, by building a model with too many predictors, you run the risk of modeling the noise instead of the signal, an outcome called *overfitting*. An overfit model has fantastic predictive power on the data you used to train it, but it won't do as well on new data it hasn't seen before.

Common methods for data reduction (also called dimensionality reduction) include Principal Component Analysis (PCA), Linear Discriminant Analysis (LDA), and autoencoder-type neural networks. Although they are not expressly adaptive, reactive, or proactive (and thus are more appropriately classified as statistical models than ML models), applying them to commercial problems tends to require significant processing power. They are often applied before sending data through unsupervised or supervised ML algorithms, so are sometimes included in ML texts.

Both computational performance and the accuracy of the resulting ML model are improved when you use the right features and reduce the number of them so that only the best predictors are used. Reducing dimensionality can increase the ultimate power of your model.

## Identifying Irregularities (Anomaly Detection)

ML algorithms can be trained to understand what a "good" or "normal" condition looks like, even when it is difficult or impossible for a human to describe what makes that condition nominal. This approach is referred to as anomaly detection. Many different ML algorithms can be applied to anomaly detection problems, including neural networks, support vector machines (SVMs), k-nearest neighbor classification, Bayesian Networks, decision trees, and deep learning. For example:

- *Post-silicon validation* is the largely manual process of testing computer processors that have been prototyped and are in physical form. The goal is to understand the root causes of any deviations from specification so that they can be repaired before full production runs. DeOrio et al. (2013) used k-means clustering on various monitor points at each timestep and found that points outside the clusters often indicated anomalies. Because the dataset had over 10,000 variables (features), they used PCA and developed a heuristic to use as preprocessing steps to make the data more manageable.
- Energy is one of the leading costs of operations at production facilities, especially when facilities are distributed. Faltinski et al. (2012) compared the performance of neural networks, SVMs, and decision trees with that of traditional models using differential equations and automata. They built a prediction system that yielded almost 99% accuracy for detecting both overconsumption and underconsumption of energy, conditions that may require countermeasures.
- *Intrusion detection* (finding network traffic that is malicious or harmful) is an active area of research in network security. George (2012) used PCA to reduce the dimensionality from 42 to 28 variables, followed by a multiclass SVM, and found that the bulk of intrusions could successfully be detected.

As you can see, the problem-solving process in anomaly detection examples regularly benefits from applying one, two, or even three techniques to build the model. This is one of the reasons modeling with ML is part art and part science. Domain expertise is always required to interpret and apply ML models to real-life business or production situations.

## Learning from Examples (Supervised Learning)

Supervised learning is used to generate a model from a sufficient number of old observations. In the credit card transaction example from earlier, this means determining the characteristics of transactions for which you already know the answer—each observation in the dataset is labeled as either fraudulent or not fraudulent. Although there have been some compelling recent successes with applying unsupervised methods, supervised methods have historically led to more powerful implemented models:

- For water utilities, early detection of burst pipes across the expansive network of a city or town can prevent wasted water and protect water pressure quality for customers farther from the burst. Huang et al. (2018) used dynamic time warping for feature extraction, which finds patterns that may have slightly different time signatures (similar to how a song can be identified whether it is played at fast or slow speeds). Next, they applied a random forest (an ensemble of decision trees) to figure out how and when to best detect bursts. Their method was tested on a real distribution network and had a low false positive rate and high accuracy.
- Stolpe and Morik (2011) examined the quality of groups of steel sticks, processed sequentially at multiple production stations. Using a k-nearest neighbors classifier, they wanted to predict the quality of the individual sticks early in the process because catching problems earlier would reduce costs of poor quality. In their supervised model, classification accuracy (good quality vs. bad quality) was better than manual inspection.

*Semisupervised learning* can also be applied when labels are available for some observations but not others. Sometimes this happens when there are missing values in the data, but other times, people can manually make judgments and label observations. For example, imagine that you are building a model to classify tens of thousands of legal documents that are related to each other. You can manually go through the documents and determine whether they are valid or invalid, but it costs $200 an hour even with the most junior of lawyers performing the reviews. You pay for 100 documents to be labeled and then use a semisupervised learning algorithm on your documents (modeled as a network of nodes and connections) to group them (Kipf & Welling, 2016).

## Learning from Experience (Reinforcement Learning)

*Reinforcement learning* (RL) is a type of unsupervised learning that defines potential rewards for making different choices instead of labeling each observation with a defining characteristic. While supervised learning builds a model based on known information, RL dynamically explores an environment to discover it. RL can be used to determine the best ways for an intelligent agent to interact with its environment.

For example, you may want to find an optimal path from a location in a building to the nearest fire exit. If you associate each outdoor location with a high reward, areas near exits with a small reward, and areas deep within the building with zero reward, RL can identify the optimal paths that generate the highest reward.

Dowling and Cahill (2004) claims that RL may be the most useful and applicable technique for solving problems in industrial environments. Here are some examples:

- RL can be used to learn customer preferences by observing behavior, which can provide information that is critical for marketing departments. Halperin (2017) demonstrates how the method can be used to design marketing strategies for new products and services, and devise pricing strategies that are tuned to the competitive environment.
- Students learn better when concepts are presented in logical ways. West et al. (2019) used RL to find optimal learning paths for a curriculum, which could improve higher education and training outcomes while enhancing student satisfaction.
- Improving the quality of a sound source, especially in a noisy environment, is important for hearing aid manufacturing. Koizumi et al. (2017) explored this problem using RL, defining "reward" as an increase in perceived quality of the source.
- Challenging medical conditions like sepsis require doctors to quickly identify treatment policies. Raghu et al. (2018) used RL (supplemented by several other supervised and supervised approaches) to identify how clinicians could use models to substantially improve mortality.

RL involves building a model by letting the algorithm explore the system of observations based on the rewards that are defined, make mistakes, and try over and over again. The approach is very similar to what organizations do when they set policies of "failing quickly" to support innovation, even more so because RL seeks to maximize rewards over the long term. Although it requires lots of data to be effective, RL has been used to develop many wildly successful game-playing AIs and addresses a different style of problem than the other ML approaches.

## THE SOCIAL CONTEXT OF INTELLIGENT, CONNECTED SYSTEMS

Although ML models can be powerful, they can also pose risks and dangers to both individuals and groups. From a quality perspective, safety is often critical to consider (especially in high-risk industries like food and beverage production), but it is just as important for us to consider algorithm safety in the Industry 4.0/Quality 4.0 era. This section illustrates the central role social context will play in designing ethical, responsible high-quality systems by sharing stories about nonconformances and unintended consequences.

Most intelligent systems are imagined, developed, and deployed by technologists. Attempting to build humanlike intelligence into machines means that human biases can also be built into systems, which can have a range of consequences:

- Silly and embarrassing (mistaken identities, Tay.ai, autonomous vehicle witchcraft)
- Potentially damaging if used inappropriately (OKCupid)
- Potentially trauma inducing (Facebook's emotional contagion experiment), hostile, or deadly (identifying criminals, ethnic cleansing)
- Challenging to social norms, behaviors, and regulations (RealDoll, Google it)

Business leaders and engineers will have to ask: Is it appropriate and right to deploy this AI/ML? Will it be appropriate in the future? Does this AI/ML impact human agency, safety, or well-being? Does this AI/ML put an individual or group in danger of harassment or death? Although intelligent agents and ML models may be regulated in the future, in the interim, leaders will have to exercise additional caution to protect their employees, customers, and society.

### Silly and Embarrassing

As often as AI/ML can "get it right," an algorithm can also make a whopper of an error or lead to unintended circumstances. In addition, human agents can interact with intelligent agents in unexpected ways, yielding outcomes that were not anticipated by designers or quality assurance. For example:

- **Autonomous vehicle witchcraft.** If you ever see a parked autonomous vehicle with a dotted chalk line inside a solid chalk line circling it, you have not stumbled upon a magic ritual. Instead, you're seeing a prankster who's trapped that car, rendering it incapable of motion (Figure 3.3). The vehicle can't tell that the lines, which are supposed to indicate differences between lanes, are not actually in lanes oriented straight ahead (Mufson, 2017).
- **"Google can't tell its tabby from its tabasco."** In 2019, a medical researcher demonstrated how adding a tiny bit of noise to a picture of a tabby cat—which is invisible to the human eye—tricked a classifier into thinking there was a 99% chance it was guacamole (Brown, 2019). If an automated system takes action based on an incorrect classification like this (for example, in a medical diagnosis), the consequences could be lethal.

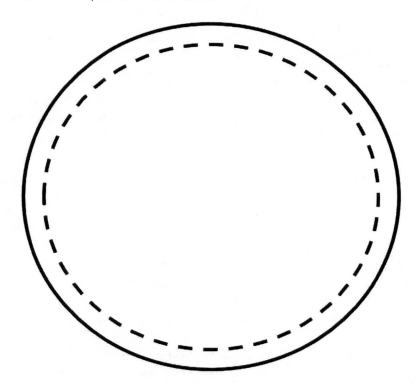

FIGURE 3.3. How to trap an autonomous vehicle (Mufson, 2017).

As IoT ecosystems grow, intelligent agents will also increasingly interact with other intelligent agents, increasing the likelihood of problems like this. Ensuring that one product meets its specifications will no longer be sufficient—manufacturing organizations will need to anticipate how connected products may interact with each other or with rogue data sources.

## Potentially Damaging If Used Incorrectly

On March 23, 2016, Microsoft released @TayAndYou on Twitter, an AI-powered chatbot that was designed to learn from its interactions with other Twitter users to generate a "personality." Sixteen hours and 96,000 tweets later, Tay was shut down by its creators. In that short time, the AI had become a drug-loving Hitler fanatic and Holocaust denier, and had been reported several times for abusing individuals in replies and direct messages. This was not the image Microsoft wanted to achieve (Metz, 2018).

Although catastrophes like this seem as though they should be preventable, our relative collective inexperience with AIs means that we are learning as the scenarios unfold:

- **The dark side of online dating.** In 2010, dating site OKCupid selected 526,000 users at random and asked what makes their culture unique. Sashimi was mentioned only by Asians, soul food was listed 20 times more for black people than all other groups, and Diet Coke was mentioned only by whites. Although the site just intended this to be a fun way to get users engaged (literally and figuratively), results like this could easily be used in reverse for racial profiling based on user profiles (Fisher, 2010).
- **When Siri and Alexa go rogue.** What if your home assistant automatically orders illegal drugs on your behalf—could you go to jail? This is exactly what happened with Random Darknet Shopper, a shopping bot for a Swedish art exhibition that was given $100 in bitcoin each week to make random purchases on the dark web, which were then displayed in the exhibit. "Swedish officials weren't amused when it purchased ecstasy, which the artist put on display." The artist was not arrested, but the drugs were confiscated after the exhibition ended (Wyner, 2016).

Much like in the earlier examples, problems arise when AIs inadvertently impact human agency, safety, or well-being. This is especially the case when it is not anticipated or intended.

## Potentially Trauma Inducing, Hostile, or Deadly

Even more concerning, AIs can directly cause physical or mental harm. For AIs used to perform medical diagnoses, the link is strong, clear, and immediate (Raghu et al., 2018). A more subtle incident occurred in 2012, when Facebook secretly manipulated the emotional tone of posts presented to certain users for a week to see if it would have an impact on their moods (Kramer et al., 2014).

> Emotional states can be transferred to others via emotional contagion, lead-ing people to experience the same emotions without their awareness . . . in contrast to prevailing assumptions, in-person interaction and nonverbal cues are not strictly necessary for emotional contagion, and that the obser-vation of others' positive experiences constitutes a positive experience for people. (Kramer et al., 2014)

When people had positive content removed from their News Feed, more of their own posts were negative and fewer were positive. When negative content was removed, subjects engaged much more positively with the platform. The researchers learned that they could quickly and easily manipulate the moods and perceptions of Facebook users just by controlling the types of informa-tion they were exposed to—no in-person contact, personal connections, or knowledge of the person's default mood was required. These experiments were performed without informed consent, and this was quickly recognized as an ethical problem (Kramer et al., 2014). Facebook was pressured to discontinue this practice.

The potential dangers compound rapidly. In "What If AI in Health Care Is the Next Asbestos?", statistics are shown that raise doubt about whether AI can really provide broad improvements to the accuracy of medical diagnoses (Ross, 2019). Schwarcz and Prince (2019) outline how credit-based insurance and the loan underwriting process can lead to "proxy discrimination," where a seemingly neutral automated practice can instead systematically block groups of people from access to resources and privileges. Surveillance of political ac-tivists and dissidents, surveillance and systematic oppression of minorities via social media, and facial recognition for policing can all lead to fatal outcomes (Hagerty & Rubinov, 2019).

## Challenging to Social Norms, Behaviors, and Regulations

As connected, intelligent, and automated technologies become more pervasive in daily life, and as embodied AIs challenge our notions of what it means to be a "person," collectively held notions about what is possible and what is permissible will also shift. After the *Citizens United* decision, for example, nonpersons may have the right to free speech, which may complicate how we design, use, and respond to interactions with chatbots and IoT consumer devices (Radziwill, 2016). In 2017, Saudi Arabia granted citizenship to Sophia the Robot, manufactured by Hanson Robotics in Hong Kong (Morby, 2017). Researchers like Eskens (2017) explore questions about how embodied AIs may challenge our notions of informed consent. Although each of these things is inherently technological, everyone has the potential to shape how we think about, and interact with, other humans.

## COUNTERMEASURES

Researchers and concerned citizens have already started exploring ways to combat the inevitable invasion of privacy and security that may arise with more broad deployment of AI/ML models, particularly those for facial recognition. For example:

**Hack your face:** As early as 2014, researchers started exploring the use of noise (much like the "tabby vs. guacamole" example mentioned earlier) to fool facial recognition models into incorrectly identifying a known individual. Lynch (2018) shares the status of facial recognition in the UK, where police have started using AI/ML systems (with high failure rates) for locating suspects and monitoring crowds. Powerful recognition systems like Hyperface (Ranjan et al., 2017) can be fooled using extreme makeup and hairstyles (Figure 3.4) and "Hyperface Fabric" (Figure 3.5)—both efforts pioneered by artist Adam Harvey—to confound the models.

**Sousveillance:** As a response to surveillance, this technique involves people actively countersurveilling individuals or organizations that monitor their needs, desires, behaviors, and movement. Sousveillance can be performed by recording audio or video by smartphones, or using wearables that capture data about people or the environment (Levy & Barocas, 2018).

A systematic response would certainly be better than addressing the symptoms of these AIs, especially for large organizations. As a result of a multi-

FIGURE 3.4. Extreme makeup can fool facial recognition AIs (Heathman, 2017).

FIGURE 3.5. "Hyperface Fabric" can confound identification (Hern, 2017).

year collaboration among experts across disciplines led by John C. Havens, the Institute of Electrical and Electronics Engineers has produced the guidebook *Ethically Aligned Design* (https://ethicsinaction.ieee.org/#read). This framework can help organizations explore how intelligent systems influence power dynamics, wealth distribution, social mobility, gender equality, and race relationships at all stages of the life cycle for autonomous and intelligent systems. Drawing from a strong foundation in the humanities, this guidebook recognizes that intelligent agents can both establish and reinforce culture—and thus can be powerful forces for (or against) change.

## THE BOTTOM LINE

The greatest impact of intelligent technologies won't be from
eliminating jobs but from changing what people do and driving
innovation deeper into the business.
—S. BARRO AND T. H. DAVENPORT

Continuous improvement and ML share a singular goal: to improve model performance as new information is acquired and circumstances change (e.g., the competitive environment, or the capabilities of the workforce). Those models can be structural (standard work), conceptual (business models, business rules), or operational (prediction, classification, forecasting). ML models are most often applied to operations, although future research may demonstrate its applicability to structural and conceptual models.

These activities, in fact, define learning—gathering experiences and new data, and updating or adapting the models that describe how you operate and the relationships that enable you to operate. While continuous improvement can be applied on any scale, from process to ecosystem (and does not require a computer or extensive amounts of data), machine learning is usually applied on the scale of an individual process or KPI (and often requires a computer to process lots of data). In this chapter, we learned the following:

- **Without a model, there is no baseline for performance improvement.** Models can be conceptual (business models), structural (standard work), or quantitative (prediction, classification, and forecasting models).
- **Learning is the key that improves business performance.** ML algorithms are used to classify, predict, forecast, distinguish, and optimize. They can be applied to several problem types:

  o **Unsupervised learning** to find patterns
  o **Data (dimensionality) reduction** techniques to simplify predictors
  o **Anomaly detection** via multiple methods
  o **Supervised learning** to learn from examples
  o **Reinforcement learning** to learn from experience

- **Machine learning models are only as good as the data they learn from and the people that build the models.**

o  Human biases can unfortunately be incorporated into models, leading to un-intended consequences or detrimental impacts to people or groups of people.

o  Machine learning models can be powerful—but they can also be potentially dangerous. Ethical concerns should be addressed at all stages of a model's life cycle.

o  The most serious problems occur when AIs negatively impact human agency, safety, or well-being.

Finally, know that simple methods often outperform more advanced algo-rithms. The principle of parsimony should be applied for every implementa-tion: the simplest model with the best performance, balanced against com-putational complexity and runtime requirements, should be the one put into a production environment. If you are interested in learning more about the different kinds of models, an index of the most common statistical and ma-chine learning algorithms used to analyze and interpret massive datasets can be found in Appendix A.

## REFERENCES

Alkhalaf, K. S., A. I. Almishal, A. O. Almahmoud, and M. S. Alotaibi. (2014). "OCR-based electronic documentation management system." *International Journal of Innovation, Management and Technology* 5 (6): 465.

Barro, S., & T. H. Davenport. (2019). "People and machines: Partners in innovation." *MIT Sloan Management Review* 60 (4): 22–30.

Barve, S. (2012). "Optical character recognition using artificial neural network." *International Journal of Advanced Research in Computer Engineering & Technology* 1 (4): 131–133.

Boesemann, W., R. Godding, and H. Huette. (2000). "Photogrammetric measurement techniques for quality control in sheet metal forming." *Archives of Photogrammetry, Remote Sensing and Spatial Information Science* 33: B5.

Bosch. (2016). *Bosch production line performance.* Kaggle. https://www.kaggle.com/c/bosch-production-line-performance

Bostrom, N. (2014). *Superintelligence: Paths, dangers, strategies.* Oxford University Press.

Brown, M. (2019, June 20). *A Google algorithm was 100 percent sure that a photo of a cat was guacamole.* Inverse. https://www.inverse.com/article/56914-a-google-algorithm-was-100-percent-sure-that-a-photo-of-a-cat-was-guacamole

Churchill, D., and M. Buro. (2015, September). "Hierarchical portfolio search: Prismata's robust AI architecture for games with large search spaces." In *Eleventh Artificial Intelligence and Interactive Digital Entertainment Conference,* 16–22.

DeOrio, A., Q. Li, M. Burgess, and V. Bertacco. (2013, March). "Machine learning-based anomaly detection for post-silicon bug diagnosis." In *Proceedings of the Conference on Design, Automation and Test in Europe,* 491–496. EDA Consortium.

Dowling, Jim, and Vinny Cahill. (2004) "Self-managed decentralised systems using K-components and collaborative reinforcement learning." In Proceedings of the 1st ACM SIGSOFT workshop on Self-managed systems, 39–43.

Eskens, R. (2017). "Is sex with robots rape?" *Journal of Practical Ethics* 5 (2).

Faltinski, S., H. Flatt, F. Pethig, B. Kroll, A. Vodenčarević, A. Maier, O. Niggemann. (2012, July). "Detecting anomalous energy consumptions in distributed manufacturing systems." In *IEEE 10th International Conference on Industrial Informatics*, 358–363.

Fisher, M. (2010, September 8). "Statistics show the real 'stuff white people like.'" *The Atlantic*. https://www.theatlantic.com/national/archive/2010/09/statistics-show-the-real -stuff-white-people-like/340126/

Gauri, S. K., and S. Chakraborty. (2007). "A study on the various features for effective control chart pattern recognition." *The International Journal of Advanced Manufacturing Technology* 34 (3–4): 385–398.

George, Annie, and A. V. Vidyapeetham. (2012) "Anomaly detection based on machine learning: dimensionality reduction using PCA and classification using SVM." *International Journal of Computer Applications* 47 (21): 5–8.

Grieco, A., M. Pacella, and M. Blaco. (2017). "On the application of text clustering in engineering change process." *Procedia CIRP* 62: 187–192.

Guh, R. S., and Y. R. Shiue. (2005). "On-line identification of control chart patterns using self-organizing approaches." *International Journal of Production Research* 43 (6): 1225–1254.

Hagerty, A., and I. Rubinov. (2019). *Global AI ethics: A review of the social impacts and ethical implications of artificial intelligence.* arXiv preprint arXiv:1907.07892.

Halperin, I. (2017). *Inverse reinforcement learning for marketing.* arXiv preprint arXiv: 1712.04612.

Heathman, A. (2017, January 4). "Protect your privacy with anti-surveillance clothing." *Wired UK.* https://www.wired.co.uk/article/anti-surveillance-clothing-adam-harvey

Hern, A. (2017, January 4). "Anti-surveillance clothing aims to hide wearers from facial recognition." *The Guardian.* https://www.theguardian.com/technology/2017/jan/04 /anti-surveillance-clothing-facial-recognition-hyperface

Huang, P., N. Zhu, D. Hou, J. Chen, Y. Xiao, J. Yu, G. Zhang, and H. Zhang. (2018). "Real-time burst detection in district metering areas in water distribution system based on patterns of water demand with supervised learning." *Water* 10 (12): 1765.

Johansson, K. (2017). Field Quality. Personal interview by E. Möller, July 6, 2017.

Khodabandelou, G., C. Hug, R. Deneckère, and C. Salinesi. (2014, May). "Unsupervised discovery of intentional process models from event logs." In *Proceedings of the 11th Working Conference on Mining Software Repositories*, 282–291. ACM.

Kipf, T. N., and M. Welling. (2016). *Semi-supervised classification with graph convolutional networks.* arXiv preprint arXiv:1609.02907.

Koizumi, Yuma, Kenta Niwa, Yusuke Hioka, Kazunori Kobayashi, and Yoichi Haneda. (2017) "DNN-based source enhancement self-optimized by reinforcement learning using sound quality measurements." In *2017 IEEE International Conference on Acoustics, Speech and Signal Processing* (ICASSP), 81–85. IEEE.

Kovach, J. V., and L. D. Fredendall. (2013). "The influence of continuous improvement practices on learning: An empirical study." *Quality Management Journal* 20 (4): 6–20.

Kramer, A. D., J. E. Guillory, and J. T. Hancock. (2014). "Experimental evidence of massive-scale emotional contagion through social networks." *Proceedings of the National Academy of Sciences (PNAS)* 111 (24): 8788–8790.

Law, E., and L. V. Ahn. (2011). "Human computation." *Synthesis Lectures on Artificial Intelligence and Machine Learning* 5 (3): 1–121.

Leggett, D. (2019, July 16). *BMW group uses artificial intelligence in series production.* QNT News. http://asq.org/qualitynews/qnt/execute/displaySetup?newsID=25705

Levy, K., and S. Barocas. (2018). "Privacy at the margins—refractive surveillance: Monitoring customers to manage workers." *International Journal of Communication* 12: 23.

Liu, D., Y. Li, and M. A. Thomas. (2017, January). "A roadmap for natural language processing research in information systems." In *Proceedings of the 50th Hawaii International Conference on System Sciences*, 1112–1121.

Love, J. (2018, September 5). SigmaGuardian Early Warning and Prevention 2.0 released. *Warwick Analytics Blog.* https://warwickanalytics.com/sigmaguardian-early-warning-and-prevention-2-0-released/

Low, C., C. M. Hsu, and F. J. Yu. (2003). "Analysis of variations in a multi-variate process using neural networks." *The International Journal of Advanced Manufacturing Technology* 22 (11–12): 911–921.

Lynch, J. (2018, February 12). "Face Off: Law Enforcement Use of Face Recognition Technology." Electronic Frontier Foundation, San Francisco CA. https://www.eff.org/wp/law-enforcement-use-face-recognition

Metz, R. (2018, March 27). "Microsoft's neo-Nazi sexbot was a great lesson for makers of AI assistants." *MIT Technology Review.* https://www.technologyreview.com/s/610634/microsofts-neo-nazi-sexbot-was-a-great-lesson-for-makers-of-ai-assistants/

Möller, E. (2017). "The use of machine learning in industrial quality control." MS thesis, engineering, KTH, Stockholm, Sweden. https://pdfs.semanticscholar.org/1034/e70ec2d1ba1feodfa7a872f63ea8cbd11f69.pdf

Morby, A. (2017, October 26). "Saudi Arabia becomes first country to grant citizenship to a robot." *DeZeen.* https://www.dezeen.com/2017/10/26/saudi-arabia-first-country-grant-citizenship-robot-sophia-technology-artificial-intelligence-ai/

Moss, P. A. (2003). "Reconceptualizing validity for classroom assessment." *Educational Measurement: Issues and Practice* 22 (4): 13–25.

Mufson, B. (2017, March 18). "Meet the artist using ritual magic to trap self-driving cars." *Vice.* https://www.vice.com/en_us/article/ywwba5/meet-the-artist-using-ritual-magic-to-trap-self-driving-cars

Pacella, M., and Q. Semeraro. (2007). "Using recurrent neural networks to detect changes in autocorrelated processes for quality monitoring." *Computers & Industrial Engineering* 52 (4): 502–520.

Pandey, R., S. Naik, and R. Marfatia. (2013). "Image processing and machine learning for automated fruit grading system: A technical review." *International Journal of Computer Applications* 81 (16): 29–39.

Radziwill, N. (2016, February 15). "Free speech in the internet of things (IoT)." *Quality and Innovation.* https://qualityandinnovation.com/2016/02/15/free-speech-in-the-internet-of-things-iot/

Radziwill, N. (2018). "Let's get digital." *ASQ Quality Progress* 51 (10): 24–29.

Raghu, A., M. Komorowski, and S. Singh. (2018). *Model-based reinforcement learning for sepsis treatment.* arXiv preprint arXiv:1811.09602.

Ranjan, R., V. M. Patel, and R. Chellappa. (2017). "Hyperface: A deep multi-task learning framework for face detection, landmark localization, pose estimation, and gender recognition." *IEEE Transactions on Pattern Analysis and Machine Intelligence* 41 (1): 121–135.

Ross, C. (2019, June 19). *What if AI in health care is the next asbestos?* Stat News. https://www.statnews.com/2019/06/19/what-if-ai-in-health-care-is-next-asbestos/

Russell, S. J., and P. Norvig. (2016). *Artificial intelligence: A modern approach.* Pearson Education Limited.

Sa, I., Z. Ge, F. Dayoub, F., B. Upcroft, T. Perez, and C. McCool. (2016). "Deepfruits: A fruit detection system using deep neural networks." *Sensors* 16 (8): 1222.

Schwarcz, D., A. Prince. (2019). Proxy discrimination in the age of artificial intelligence and big data. https://papers.ssnr.com/sol3/papers.cfm?abstract_id=3347959

Shao, Y. E., and C. C. Chiu. (1999). "Developing identification techniques with the integrated use of SPC/EPC and neural networks." *Quality and Reliability Engineering International* 15 (4): 287–294.

Shin, H. C., M. R. Orton, D. J. Collins, S. J. Doran, and M. O. Leach. (2012). "Stacked autoencoders for unsupervised feature learning and multiple organ detection in a pilot study using 4D patient data." *IEEE Transactions on Pattern Analysis and Machine Intelligence* 35 (8): 1930–1943.

Smith, A. E. (1994). X-bar and R control chart interpretation using neural computing. *The International Journal of Production Research* 32 (2), 309–320.

Somers, D. (2014, May). "Is this Quality 4.0?" *Quality Manufacturing Today.* http://www.qmtmag.com/display_eds.cfm?edno=9665074

Stolpe, M., and K. Morik. (2011, September). "Learning from label proportions by optimizing cluster model selection." In *Joint European Conference on Machine Learning and Knowledge Discovery in Databases*, 349–336. Springer.

Wang, C. H., R. S. Guo, M. H. Chiang, and J. Y. Wong. (2008). "Decision tree based control chart pattern recognition." *International Journal of Production Research* 46 (17): 4889–4901.

Weisbrod, E. (2019, June 19). *Turn dairy production data into quality intelligence: Manufacturers need real-time process control and data interrogation.* Dairy Foods. https://www.dairyfoods.com/blogs/14-dairy-foods-blog/post/93680-turn-dairy-production-data-into-quality-intelligence

West, J., F. Maire, C. Browne, C., and S. Denman. (2019). *Improved reinforcement learning with curriculum.* arXiv preprint arXiv:1903.12328.

Wu, C. S., T. Polte, and D. Rehfeldt. (2000). "Gas metal arc welding process monitoring and quality evaluation using neural networks." *Science and Technology of Welding and Joining* 5 (5): 324–328.

Wyner, M. (2016, April 25). *The ironic reality of ethics and law in artificial intelligence.* Webvisions. http://www.webvisionsevent.com/2016/04/the-ironic-reality-of-ethics-and-law-in-artificial-intelligence/

## CHAPTER FOUR

# AUTOMATION: FROM MANUAL LABOR TO AUTONOMY

*Beyond driving cars, AI can learn from previous situations to provide input and automate complex future decision-making processes, making it easier and faster to arrive at concrete conclusions based on data and past experiences.*

—KLAUS SCHWAB, FOUNDER AND EXECUTIVE CHAIRMAN OF WORLD ECONOMIC FORUM

When I started working at the Green Bank Observatory in West Virginia in 2002, many of the astronomers were using the spectrometer (Figure 4.1) to process their observations. After receiving the highly focused signals from the antenna, the spectrometer determined precisely which frequencies were dominant. From the patterns they observe in the spectra (Figure 4.2), astronomers can figure out things like the chemical composition, temperature, or luminosity of the planets, galaxies, or star-forming regions they are studying.

Unfortunately, the spectrometer at that time was not easy to use, and the observing process was inefficient. Even though it was designed to be extremely powerful (and indeed it was), every time a new observing mode was needed, hundreds of wires had to be manually repositioned. If there was a problem during the observation, a specialist had to go to the instrument and move wires until the issue was resolved. If the astronomer needed to make a change to the observing mode and the engineers were unavailable, the astronomer was stuck.

The process needed a little more automation, and the system itself needed to be more flexible and agile so that new observing capabilities could be added more easily. Work was started to create a new spectrometer that could be fully configured by software, and not require the manual intervention of engineers rewiring the device for each observing mode. This was, in effect, a transition from what is called *hard automation* (where reconfiguration or rewiring is required to support the operation of an automated or semiautomated system) to *soft automation*, which is more flexible.

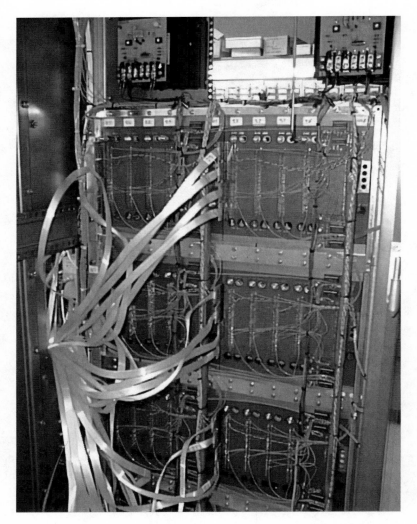

FIGURE 4.1. The original spectrometer for the Green Bank Telescope in West Virginia (http://www.nrao.edu/engineering/spectrometer_gallery.shtml).

Although additional automation was needed to transform raw electromagnetic signals to insight, developing new spectrometers over the next few years brought some other benefits too. By using reconfigurable off-the-shelf hardware platforms instead of engineering special purpose boards, the team was able to create a new data processing backend more quickly and leverage the learning that had taken place in industry. By making use of software tools for

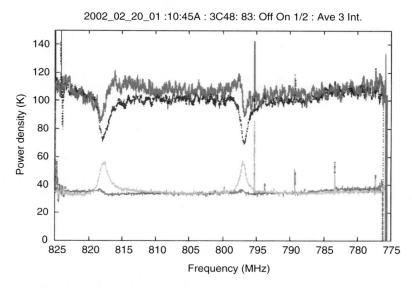

FIGURE 4.2. Observations of 3C48 (Langston, 2002).

rapid design, verification, and deployment, they created a more robust end product. Finally, because the new spectrometer was built on industry-standard hardware and software, it was easier to hire and train new engineers to maintain the device and continue development (DuPlain et al., 2008).

This chapter explores the evolution of work from manual labor to the autonomous cyber-physical systems (CPSs) of Industry 4.0. Examples drawn from multiple industries will help you learn how to effectively distribute work among humans, machines, and computers.

## MECHANIZATION AND AUTOMATION

Since the first industrial revolution in the late 1700s, manual labor has been gradually and continually replaced by mechanized and automated solutions. *Mechanization* is the use of machinery to help human workers perform physical tasks. The machinery can be static (e.g., handheld tools) or powered by electricity, electric motors, or internal combustion engines. Cognitive and sensory aspects of the task, however, are still performed by people in mechanized solutions. The "level of mechanization" can be thought of as the degree to which work must be powered by physical labor (Groover, 2016).

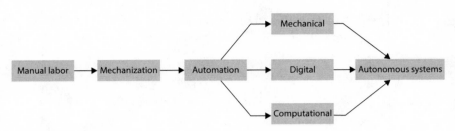

FIGURE 4.3. Progression from manual labor to autonomous systems.

*Automation,* the "technology by which a process or procedure is accomplished *without human assistance,*" reduces or eliminates the need for human intervention. The progression from manual operations to full automation and autonomy rarely happens in one step; rather, the evolutionary process occurs in steps and stages and can extend over years or decades (Groover, 2016).

There are mechanical and computational aspects of automation, and digitization of systems is often required for automated computation to be possible. Mechanical aspects replace human physical power with machine power. Computational aspects include "replacement of cognitive tasks, such as human sensory processes and mental activity" (Chiantella, 1982). The relationships between the stages are shown in Figure 4.3.

## Innovation Happens at the Edges

Transitioning from one stage to another yields innovation. Jerome (1934) provided a fascinating look into the social and technological context of the transition from manual labor to mechanization, experienced during a time when the possibility to fully automate any industrial process was limited. His paper uses the word *automation,* but he considers only the "full and complete mechanization" of a process with no cognitive or decision-making support.

Some examples from his paper, plus additional examples of innovation that has emerged while introducing mechanization and automation, are shown in Table 4.1. Note that all but one of the examples in the middle column, "Mechanization to automation," required digitization to occur first. Digitizing systems (so that data can be captured and retrieved electronically) makes it possible to automate in many cases.

In rare situations, progressing from manual labor to mechanization (or advancing to automation or autonomy) does not result in desired performance improve-

**TABLE 4.1. Examples of innovation at the boundaries of mechanization and automation.**

| Manual labor to mechanization | Mechanization to automation | Manual labor to automation |
|---|---|---|
| Harvester-thresher combine reduces farm labor requirements by 80%–84% | iPads eliminate some job tasks done by human servers at restaurants | Algorithms replace human workers doing credit approvals at financial institutions |
| Mechanical loading of bituminous coal reduces labor by 25%–50% | BMW introduces fully automated auto production | Voice-to-text software replaces human transcription in courtrooms |
| Machines for painting buildings reduce labor requirements by 80%–85% | Software developers use code generators to write code | iPhone manufacturer Foxconn moves from manual labor directly to automation |
| Semiautomatic glass bottle production reduces labor by 29%–71%; fully automatic by 86%–97% | Credit card fraud detection algorithms alert customers directly (bypassing bank employees) if a fraudulent transaction is detected | Toyota replaces its master craftsmen forging and cutting crankshafts but later reverts to manual labor because it yields 10% less waste (sometimes it doesn't work out as intended) |

ments, either at the individual or organizational level. When Toyota replaced the master craftsmen managing its crankshaft forging and cutting process, for example, the process generated more waste and was not appreciably faster (Jerew, 2014). Consequently, leaders should not always assume that more automation is better. Using simulation and pilot programs can hedge against this risk.

## Reasons for Automation

While humans are good at performing some tasks, machines can often do them better and faster. In general, tasks that require intelligence or higher-level thinking have been done by humans, while repetitive or rule-based tasks can be done by machines (Figure 4.4). The human element links mechanization and automation, by physical labor as well as cognitive skills and intelligence (Figure 4.5). Although wider adoption of artificial intelligence is shifting these boundaries, choosing *what* to automate and *whether* to automate remains important.

The decision to automate must also be considered with the market in mind. Automation may help you respond to increased demand, or eventually offer your product at a lower price point. It may help you improve safety and productivity. But in the meantime, it may require significant investments in machinery and other assets. Consider automation if you need or want to:

- Accomplish processes that cannot be done manually
- Improve complex tasks that are otherwise slow, labor intensive, or error prone
- Improve safety by giving more hazardous tasks to robots and nonhumans
- Improve labor productivity and throughput while reducing lead time
- Improve product quality and reduce variation in production
- Increase flexibility and enhance ability to add new products quickly

| Humans are good at: | Machines are good at: |
| --- | --- |
| • Pattern recognition<br>• Learning and generalizing old information and experiences to new situations<br>• Identifying new solutions<br>• Building trust with other humans<br>• Creative or abstract thinking | • Repetitive tasks<br>• Long duration tasks<br>• Making quick, accurate calculations<br>• Multitasking<br>• Recalling information<br>• Consistency<br>• Precision |

FIGURE 4.4. Function allocation between humans and machines.

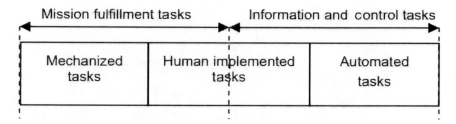

FIGURE 4.5. Human labor and intelligence unites mechanization and automation (Frohm et al., 2008).

- Reduce costs by shifting expensive labor to less expensive maintenance
- Reduce outages and improve time-to-recovery
- Detect errors faster and more accurately
- Reduce the risk of labor shortages

Despite these benefits, there are also positive outcomes associated with *not* automating. For new or highly customized products, automation can be costly, especially if machines and other assets are expensive and requirements are likely to change. Avoiding automation can be a hedge against market failure. If demand is inconsistent, organizations can save money by not automating. If the product must be delivered quickly, it can often be too difficult to implement automated systems in a short amount of time. Finally, production methods can be rooted in history and tradition. To preserve skills that are culturally important, it may be necessary or desirable to avoid introducing either mechanized or automated solutions.

For example, cultured pearl production requires finely tuned skills and has intense monitoring requirements. Pearl farmers develop expertise over time to respond to subtle cues in the environment over the two- to three-year production life cycle, as it is common to lose 10%–40% of the pearls due to illness or death of the oysters (Figure 4.6). This production system is not well suited for automation.

## Building Blocks for Automation

Automation can be mechanical, computational, or both; computational automation often requires that systems be digitized and data be captured (and be retrieved) electronically. Mechanical automation is often equated to robotics, but as Goldberg (2011) points out, researchers in robotics tend to focus on the

FIGURE 4.6. Pearl production (http://missjoaquim.com/loose-pearls
-sale-pearls-come/).

mechanics of the enabling technologies, while researchers in automation focus
on quality and other related drivers (Table 4.2).

Any kind of automation requires, at the very least:

1. Power to move (or actuate) physical components, drive controller units, and
   enable data acquisition and processing. Power can be provided through hydrau-
   lic, pneumatic, electrical, chemical, mechanical, thermal, or magnetic mechanisms.
2. Instructions or operating procedures for manipulating mechanical equipment,
   wiring, software, and/or infrastructure.
3. A control system to execute instructions and monitor telemetry.

Automation is typically accomplished through control systems and the software that drives them. Control systems have four functions:

- **Measure**—obtain values from sensors or instruments, and provide the values
  to other parts of the system
- **Compare**—examine observed values in the context of targets, specifications, or
  models
- **Compute**—calculate quantities, compute or estimate errors

**TABLE 4.2. Robotics versus automation.**

| Robotics | Automation |
|---|---|
| • Systems that incorporate sensors and actuators operating autonomously or semiautonomously<br>• Research emphasizes intelligence and adaptability to new environments<br>• Emphasizes feasibility and proof-of-concept<br>• Research demonstrates new abilities of a robot (e.g., walking, driving, performing a skilled task) | • Research emphasizes quality, efficiency, productivity, and reliability<br>• Quality can be improved with new techniques, analysis, models, or results that inform robustness, stability, or productivity<br>• May explore the feasibility of new mechanisms, models, or theories for repetitive operations<br>• Makes existing abilities or subtasks more efficient, reliable, or cost-effective |

Source: Adapted from Goldberg, 2011.

- **Correct**—alert the operator to perform actions, or automatically perform actions on behalf of the operator (as intelligent agents)

Although the descriptions above relate to control system functions in manufacturing and similar organizations, every organization has control systems that demonstrate these four elements. For example, consider a nurse treating a patient in a hospital room. He or she collects vital signs and reviews other test results, comparing them with what is considered normal. If there is a problem, the nurse may perform computations to determine the right amount of medicine or support to administer. Finally, based on all the available information, he or she may correct (or adjust) the protocol for restoring the patient to health.

In manufacturing control systems, these four functions are performed by different kinds of hardware. Sensors are devices that measure physical properties of objects or the environment. These can include temperature sensors, motion sensors, gas sensors, optical sensors, or accelerometers. Sensors can also measure water quality, detect the presence of smoke, and detect rotational velocity using gyroscopes. Transducers and encoders sense mechanical motion and convert nonelectrical signals into electrical signals for transmission. Transmitters (such as beacons) send those signals. Controllers, including Programmable Logic Controllers (PLCs), provide logic that understands how and when to make changes in the system, and execute actions when they are required. Finally, control elements and actuators make the physical changes requested by the controllers, such as moving rods, gears, and cams.

## Industrial Control System (ICS)

An *industrial control system* (ICS) is "a collection of personnel, hardware, and software that can affect or influence the safe, secure, and reliable operation of an industrial process" (ANSI/ISA 62443). An ICS must monitor and control processes. To do this, it

- gathers information about a process from devices at its endpoints,
- interprets that information in the context of production system goals, and
- facilitates interactions among human operators, field controllers, and field devices.

On the control side,

- Field devices obtain process data, often in the form of monitor points (or telemetry). This provides feedback for the controllers to use as they decide whether to adjust the process, and how, independently or as support for human decision making. Switches, sensors, valves, meters, actuators, and RFID tags are all field devices.
- Field controllers track (and sometimes analyze) information about the state of the process, and take action using that information within preprogrammed logic of PLCs, Programmable Automation Controllers (PACs), remote terminal units (RTUs), proportional–integral–derivative (PID) controllers, and new technologies like Arduino and Raspberry Pi are examples of field controllers.
- The industrial internet of things (IIoT) is made up of field devices and field controllers connected to the internet. In essence, the transition to IIoT is nothing more than the expansion of SCADA systems using a common protocol for communication.

On the monitor side, human-machine interfaces (HMIs)

- display information about the state of the system,
- often provide human operators with the ability to adjust or manipulate controls, and
- can present information on workstations, tablets, smartphones or other custom handheld devices, augmented reality (AR) devices, or wearables (like smart hard hats and vests).

The extent to which a control system is allowed to make corrections depends on the level of automation that managers and control systems engineers have chosen for a particular task or group of tasks. Because there are varying degrees of automation that can be considered, several researchers have explored what these differences imply for strategy and planning. These findings are presented in the next section.

## LEVELS OF AUTOMATION

Automation can be described by *where* in the organization it occurs (which informs the types of automation that are possible), *what kind* of automation can be realized, and *how* humans, machines, and computers can cooperate to achieve shared objectives. Knowing these things can help leaders identify new opportunities.

### Where Automation Occurs

The ANSI/ISA 95 (IEC/ISO 62264) standards (referred to as ISA 95) were created to provide common definitions and terminology for the systems that support manufacturing operations. Before the standards, it was hard for manufacturers to effectively communicate their requirements to systems integrators, especially where digitization and automation were concerned. After the release of ISA 95, projects that previously took one to two years to deploy could instead be rolled out in only two to four months, and more than 90% were successful (as compared with less than 50% previously) (Brandl, 2012).

ISA 95 conceptualizes operations into five levels, from Level 0 (which represents the physical process) to Level 4 (which covers the business processes that connect operations with strategy and the market). The ISA 95 automation pyramid (Figure 4.7) is also organized by timescale, so while events occur (and data is generated) at the millisecond/microsecond scale at Level 0, events at the business and enterprise Level 4 occur on the order of days or months. Although the model was created for manufacturing environments, the pyramid (from bottom to top) could be applied to any production environment because it addresses physical processes, digitization and monitoring, process control, workflows, planning, and management.

Together, Levels 0 and 1 describe how to exchange information about operations processes that is collected on the front lines, Level 2 controls processes, and Level 3 manages flows (workflows, material flows, and information flows). Table 4.3 shows examples of the types of automation in each level.

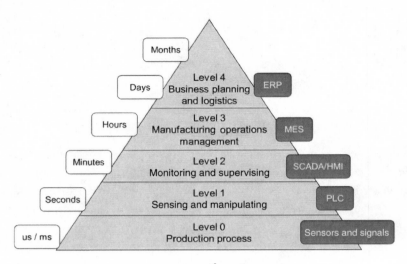

FIGURE 4.7. ISA-95 automation pyramid (Åkerman, 2018).

## What Kind of Automation Can Be Done

The introduction of CPSs in Industry 4.0 expands the capabilities (and power) of the various interconnected components. New communications protocols make it possible to communicate across all levels of the automation hierarchy (Figure 4.8). Although this makes tight integration of processes possible, it also expands the "attack surface" for potential cyberattacks.

In general, this networked framework for CPS-based automation implies that there are four stages of automation maturity in an organization beyond manual operations:

- **Manual operations**—Processes are managed on paper or on individual spread-sheets that are difficult to track or trace. Limited (or no) records may be kept.
- **Digitization**—Some document repositories, data repositories, or software packages are available to support electronic data entry, retrieval, and possibly visualization.
- **Horizontal integration**—Some systems are connected and can be used to ex-change information across functional areas of the organization (e.g., sales, marketing, production).
- **Vertical integration**—Information and material flows connect the sensor level, control level, production level, and/or enterprise level.

**TABLE 4.3. Automation can occur at many levels.**

| ISA-95 | Level[a] | Hierarchy[b] | Description | Examples |
|---|---|---|---|---|
| Level 4 | Enterprise | Enterprise resource planning | Across multiple sites and/or distribution facilities | Plan production, create schedules, interact with customers, provide automated responses, or aggregate or share data. Can use Robotic Process Automation (RPA). |
| Level 3 | Plant (or cross-functional) | Plant management | Within a facility, across multiple production lines or departments | Facilitate data collection and transfer between functional areas to improve the flow of information and materials. Can use RPA. |
| Level 2 | Cell/subsystem (or department/functional) | Process control | Groups of people and/or machines that cooperate to fulfill one or more specific processes | Monitor and control systems, continually examine parameters, set schedules and production targets, collect process records, and start up and shut down machines as required. |
| Level 1 | Machine | Control | Groups of control system elements that together accomplish a specific task within a process, such as turning or milling | PLCs and PACs deliver control functions automatically based on the sensor data they receive. |
| Level 0 | Device | Field | Individual sensors, transducers, encoders, actuators, and other field devices and controllers | Automatically collect data on temperature, pressure, or environmental conditions and send it up to machines or cells for monitoring and analysis. Turn motors, valves, and pumps on or off or adjust their levels. |

[a] Data from Groover (2016).
[b] Data from Monostori (2014).
Source: Adapted from Groover, 2016.

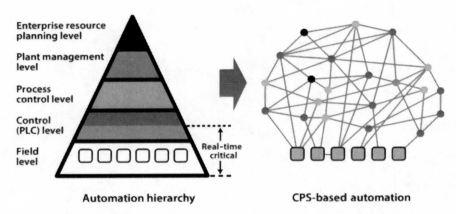

Automation hierarchy                    CPS-based automation

FIGURE 4.8. CPSs create a network from the automation hierarchy (Monostori, 2014).

- **Connected work systems**—Information and materials flow across functional areas and between levels of the automation hierarchy, making it possible for the organization to anticipate, adjust, and adapt to changing circumstances and requirements.

Note that this framework could be used by most organizations. The sensor level would correspond to the level closest to the work processes, and data could be sensed by individuals or participants in the process rather than instruments.

## How Humans, Machines, and Computers Automate

When automation is absent, systems provide no decision or labor support. With some automation, the system can provide suggestions to the human, which he or she has the option to follow through with or not. Notifications can also be provided to the human to facilitate decision support and provide supporting information when the system takes actions automatically. Finally, a system may act mostly autonomously, potentially providing the human operator with the option to manage by exception and stop the system from proceeding if issues are detected (Lorenz et al., 2001).

## EXAMPLES OF AUTOMATION

To enable the different levels of automation in Table 4.4, an automated system can collect data, report or track events or movements, monitor events and issue

**TABLE 4.4. Levels of automation in terms of decision support.**

| Sarter et al. (1997); Ruff et al. (2002) | Sheridan and Verplank (1978) | Kaber et al. (1999) |
|---|---|---|
| Manual control | Human specifies process, and computer directly executes the instructions | Manual control; batch processing |
| Management by consent | Computer assists human by determining options, and human selects the desired option | Action support |
| Management by consent | Computer assists human by determining options and suggesting a choice; human selects an option that may or may not be what was recommended | Shared control; decision support |
| Management by consent | Computer assists human by determining options and selecting a choice; human has the option to follow the computer's recommendation or not | Blended decision making |
| Management by exception | Computer selects and implements option, but requires human approval before executing it | Blended decision making; rigid system |
| Management by exception | Computer selects the best option and automatically implements it, but gives the human the chance to stop the process | Rigid system; automated decision making |
| Assisted control | Computer selects and implements options automatically, then reports results to the human | Supervisory control |
| Unassisted control | Computer selects and implements options automatically, but tells the human about the results only if asked, and reports comprehensive results | Supervisory control |
| Unassisted control | Computer selects and implements options automatically, but tells the human about the results only if asked, and reports only some information | Supervisory control |
| Autonomous control | Computer selects options, implements options, and automatically performs the whole job; it may or may not tell the human anything that has transpired, although logs may be collected to keep track of what occurred | Full automation |

alerts if needed, provide alternatives to human decision makers, and carry out actions. Manual and routine tasks are much more easily automated than cognitive, nonroutine tasks.

## Robotic Automation

Fueled by a vibrant global robotics market, machines to perform manual and routine tasks can now be acquired more readily than in the past. Implementing robotic automation can directly address the worker deficit in manufacturing, while improving occupational health and safety and increasing productivity. In addition, combining robotic automation with emerging technologies like AR can enhance information flows and ease human-machine interaction (Malý et al., 2016).

In one example from aircraft engine turbine manufacturing, Caggiano and Teti (2018) describe how a manufacturing cell's performance was improved by robotic automation of its deburring (smoothing) station. The cell included a grinding machine tool with a robot for loading and unloading (which was already in place), a coordinate measuring machine, and the new automated robotic deburring station.

The robotic components were responsible for grabbing components, inspecting surfaces, and deburring. The human operator was responsible for assembly and disassembly of the parts, part positioning in the grinding machine, and manual part transfer between the machines. With the new robotic components, utilization of the elements in the entire manufacturing cell improved from 13% to 61%. The difference between the projected outcomes predicted by discrete event simulation (DES) and the actual performance was only about 1.5%. Using DES helped this team of industrial engineers find a viable solution for performance improvement before buying and installing the physical machines.

## Robotic Process Automation (RPA)

Many business processes can be partially or fully automated by triggering software scripts, based on business rules, that run under certain conditions or at specific times. Although the label conjures up images of humanoid machines and robotic arms, Robotic Process Automation (RPA) is actually much more straightforward and emphasizes the automation of routine tasks. What most people think of as robots are not a part of RPA; rather, the "robots" are software programs that carry out the activities. Here are some examples:

- **Mobile Communications Provider.** Telefonica "learned that low-performing back offices can be transformed to high-performing back offices through six *transformation levers*: centralize physical facilities and budgets, standardize processes across business units, optimize processes to reduce errors and waste, relocate from high-cost to low-cost destinations, [enable with] technology (e.g., self-service portals), and automate services" (Lacity & Willcocks, 2015). In response, it automated over a third of its 15 core processes using approximately 160 RPA scripts. With a payback period of just 12 months, its three-year return on investment (ROI) was estimated to be between 650% and 800%.
- **RPA Service Provider.** OpusCapita, a financial services company in Espoo, Finland, deploys RPA solutions for companies in its sector. It selects RPA initiatives for its clients based on eight criteria: high transaction volume, actions that touch multiple systems, stable and unchanging environment, low cognitive overhead, easily defined in terms of business rules, likelihood of human error, highly standardized task, and a clear understanding of the current costs of manual operations (Asatiani & Penttinen, 2016).
- **Process Mining to Identify RPA Opportunities.** Process mining reconstructs the flow of business processes from large transaction logs. Geyer-Klingeberg et al. (2018) generated a Purchase-to-Pay (P2P) process flow from an SAP ERP system for two firms. Based on the frequency that a process has consistent steps with few deviations, they were able to select an appropriate candidate for automation. Choosing the right processes for RPA can significantly reduce the cost, risk, and time-to-value of implementation.

With RPA, automation scripts leverage existing software systems and IT infrastructure without changes. Data validation, report preparation and sharing, routine mass e-mails, updating business scorecards, installing software updates, updating vendor records, responding to partners, and configuring products can all be done using RPA. Chatbots and virtual assistance also provide examples of RPA in action, when routine tasks are facilitated.

## Design Automation

New product development (NPD) can be complex, expensive, and time-consuming. Thanks to software for Simulation-Based Design, virtual prototyping, and design optimization, advances in design automation can significantly speed up the process while reducing the risks and making the final products more robust and reliable.

As one example, Ang et al. (2016) examined design automation for shipbuilding. Designing a new ship is a massive exercise in concurrent engineering, with stakeholders scattered across the globe. New ships take months to build and cost between $10 million and $100 million. Motivated by changing fuel prices and increasingly tough environmental regulations, time-to-market is still a concern, with the ability to customize designs a compelling differentiator. They proposed the Hull Form Design Optimization framework to automate the exploration of design alternatives, presenting the designer with a catalog of feasible options. He or she can examine these suggestions in the context of a project's unique constraints rather than attempting to sort through all possible designs manually.

## Automatic Code Generation

One of the goals of model-based software engineering is to be able to describe the abstract structure of a system in such a way that the source code can be automatically generated. Consequently, models (informational, behavioral, and structural) have to be more complete and more precise so that the code they produce yields the intended outcomes. Design tools like Unified Modeling Language and SysML, which are well established, provide the basis for the automated code generation (Ciccozzi et al., 2018). Also, researchers like Morin et al. (2017) believe that code generation may be useful for programming the internet of things (IoT), especially when functionality has to be distributed across many nodes in a group of devices.

## Test Automation

When new software features are released, testers make sure that the new capabilities do not break preexisting functionality. As a result, and particularly for large software packages, many regression tests must be executed to provide this assurance. In one case study, Garousi and Yildirim (2018) automated the testing process for the graphical user interface (GUI) of a large-scale law practice management software. In addition to reducing the test time from two days to one hour for each release, they were able to increase the frequency of releases from once every two weeks to multiple times daily, making continuous delivery possible. During demos and progress meetings with customers, they were able to show the results from the automated test processes, which built trust and satisfaction.

## AUTOMATION ACCEPTANCE

Automating processes and systems may yield beneficial outcomes and increase productivity, but only if they are accepted by the people who work with them. People adopt technologies when they provide clear benefits, make jobs easier to do, and do not require excessive cognitive overhead. Unfortunately, even the smallest and most beneficial changes can be difficult to adopt, because *all* change requires cognitive effort. This section shares insights for helping the workforce and senior leaders more easily adopt new technologies, especially when they involve automation.

### Preserving Agency and Countering Resistance

In 1983, a research team led by Lynne Markus wanted to find out why people were so reluctant to adopt software that was developed to help them—software specifically designed to make their lives easier. If the software was so beneficial, the researchers wondered, why was it so difficult to get people to use it?

Their work uncovered three theories of resistance. If people aren't responding to system and process improvements, their findings, which were reiterated over several years, suggested that one of three things is going on:

- **People Issues.** There may be an issue with the people using, or being affected by, the new technology. They may need additional education or training, enticement to try the new technology, or a better explanation of why they are being asked to change.
- **System Issues.** Perhaps the new technology doesn't behave as anticipated, provide the right answers, or execute tasks consistently. People who have been performing the task manually will be reluctant to adopt an automated solution if it doesn't do as good a job as they used to. Furthermore, they may need some time to observe the system in action to develop trust, or may need an intermediary solution where the automated solution recommends next steps, and the people get to approve or disapprove. The solution is straightforward: fix the bugs, improve the infrastructure, ease into automation, or do all three—and make sure a process is in place to ensure that future releases are solid.

- **Power Issues.** Finally, look at how the system has changed access to information. As Markus explains, *where information flows, power goes.* By democratizing access to information, you may have inadvertently disrupted the power balance in your organization. You may be threatening someone's sense of meaning, purpose, or agency. If you can get to the root cause of that issue to solve it, resistance will (probably) magically ebb.

Most significant here is the need to preserve human agency. Every person needs to have a sense of control regarding his or her own work, tools, and body. Automated systems should not make decisions on behalf of people, or limit or remove their agency in any way, unless they have explicitly declared that this is acceptable to them. Lack of (or insufficient) agency can negatively impact adoption of automated solutions.

## Technology Follows Behavior

If you think about software implementations that have occurred in your career, which ones have been successful, and which ones have failed? In general, if a software package is adopted to automate a task that is not yet being performed by people, there is a risk that no one will start doing that task just because the software is available. For example, let's say you buy a software system that manages sales communications with prospects and will provide you with intelligence about how well each pitch works so that you can customize pitches to future prospects. If your sales force does not currently think or work in terms of "pitching," it is unlikely that they will start thinking this way just because the software becomes available to do it.

*Strategic conformance* is the "match in problem-solving style between decision aiding automation and the individual operator" (Westin et al., 2015). What this implies is that technology follows behavior. You can increase the chances of adoption by automating or facilitating tasks that people already do and are already comfortable with:

Given technical advances in areas previously considered unique to human cognitive skills, automation is expected to increasingly assume authority in problem-solving and decision-making tasks. The inevitable trajectory of many work domains will involve more capable automation acting in an intelligent advisory capacity. As such, automation will likely provide support that is more strategic in timescale, less transparent to the operator in that decision rationales are concealed, and presented as recommendations. In this context, the issue of acceptance is central. It is reasonable to hypothesize that a recommended solution matching the individual's problem-solving style would be more readily accepted. (Westin et al., 2015)

## Workforce Implications

Automating processes often results in needing fewer people to perform a task. Even if automation reduces or eliminates risks due to health or safety, there will still be resistance from workers if they believe that technology is taking their jobs.

But not all jobs are easy to automate. Robotic automation, for example, is often supplemented by human labor to move parts between machines or to perform intermediary inspections. Industry segments most likely to lose jobs as a result of automation, according to Bonekamp and Sure (2015), were office and administrative support, service, sales, production, and transportation. Least likely to be automated were education, legal, arts, media, healthcare, management, science and engineering, and financial functions. Frohm et al. (2008) reports these changes:

- Simple and repetitive activities replaced by CPSs and intelligent agents
- Greater automation of control, supervising, and scheduling activities
- More decentralized decision-making and planning processes
- Greater need for process integration and cross-functional perspectives
- More automated quality assurance and maintenance processes
- Greater importance of interdisciplinary cooperation
- Greater requirements for IT competence and data literacy
- Engagement with partner and supplier networks more significant
- Teamwork becomes more important as work becomes more flexible

The most supportive view of automated systems is that they will enable real-time control of work processes and provide individualized guidance for the workforce. By augmenting human intelligence, theoretically, a balance will be struck between the need for explicit and tacit (or experiential) knowledge. In all cases, the continued adoption of automation means there will be a sustained shift toward the need for cognitive skills (including data literacy and interpretation) in manufacturing as well as most other industries.

## THE BOTTOM LINE

Automation is a cornerstone of Industry 4.0, enabled by CPSs that communicate through networks. To create an automated system, manual labor is mechanized, then systems are digitized. Finally, mechanical and computational automation can be introduced in a stepwise manner, ultimately leading to the potential for autonomous operations. Automation is not an all-or-nothing pursuit—there are degrees of automation that can range from a system that executes preset instructions under certain conditions to a system that is fully autonomous, like a self-driving car. (Being able to view the organization in terms of levels of automation can also help cybersecurity professionals more accurately identify the ICS attack surface.)

Automated systems collect data, report or track events or movements, monitor events and issue alerts if needed, provide alternatives to human decision makers, and carry out actions (sometimes, using robotics). In general, CPSs enable different levels of automation maturity beyond manual operations:

- **Digitization**—Systems provide partial support for electronic data entry, retrieval, and/or visualization
- **Horizontal integration**—Some systems exchange information across functional areas of the organization (e.g., sales, marketing, production)
- **Vertical integration**—Information and material flows connect the sensor level, control level, production level, and/or enterprise level
- **Connected work systems**—Information flows horizontally and vertically, enabling a flexible and adaptable organization

Automation can make it possible to accomplish tasks that are not otherwise feasible or safe for humans, and to make processes faster and less error prone. There are also benefits to *not* automating—for example, if products are very new or highly customized, production assets are expensive and requirements are likely to change, demand is inconsistent, or delivery must be rapid. Leaders should not always assume that more automation is better. Using simulation and pilot programs can hedge against the risk of unintended consequences.

Managing change so that the workforce and senior leaders not only accept but benefit from the new automated solutions also requires thinking in terms of people, systems, power, agency, and behavior. Strategically developing competencies in the workforce, in particular cognitive skills and data literacy, can help alleviate the fear of being automated out of a job (particularly in manufacturing).

# REFERENCES

Åkerman, M. (2018). *Implementing shop floor IT for Industry 4.0.* PhD dissertation, Chalmers University of Technology.

Ang, Joo Hock, Cindy Goh, and Yun Li. (2016) "Smart design for ships in a smart product through-life and industry 4.0 environment." In 2016 IEEE Congress on Evolutionary Computation (CEC), 5301–5308. IEEE.

Asatiani, A., and E. Penttinen. (2016). "Turning robotic process automation into commercial success—Case OpusCapita." *Journal of Information Technology Teaching Cases* 6 (2): 67–74.

Bonekamp, L., and M. Sure. (2015). "Consequences of Industry 4.0 on human labour and work organisation." *Journal of Business and Media Psychology* 6 (1): 33–40.

Brandl, D. (2012). *Practical applications of the ISA 95 standard.* MESA International. https://web-material3.yokogawa.com/2/11821/tabs/document_11744.pdf

Caggiano, A., and R. Teti. (2018). "Digital factory technologies for robotic automation and enhanced manufacturing cell design." *Cogent Engineering* 5 (1): 1426676.

Chiantella, N. (1982). "Achieving integrated automation through computer networks." *SMA/CASA Computer Integrated Manufacturing Series* 1 (2): 2–21.

Ciccozzi, F., M. Famelis, G. Kappel, L. Lambers, S. Mosser, R. F. Paige, A. Pierantonio, A. Rensink, R. Salay, G. Taentzer, A. Vallecillo, and M. Wimmer. (2018, October). "Towards a body of knowledge for model-based software engineering." In *Proceedings of the 21st ACM/IEEE International Conference on Model Driven Engineering Languages and Systems: Companion Proceedings:* 82–89.

DuPlain, R., S. Ransom, P. Demorest, P. Brandt, J. Ford, A. L. Shelton. (2008, July). "Launching guppi: The green bank ultimate pulsar processing instrument." In *Advanced software and control for astronomy II* (Vol. 7019, p. 70191D). International Society for Optics and Photonics.

Frohm, J., V. Lindström, J. Stahre, and M. Winroth. (2008). "Levels of automation in manufacturing." *Ergonomia—An International Journal of Ergonomics and Human Factors* 30 (3).

Garousi, V., and E. Yildirim. (2018, April). "Introducing automated GUI testing and observing its benefits: An industrial case study in the context of law-practice management software." In *2018 IEEE International Conference on Software Testing, Verification and Validation Workshops* (ICSTW), 138–145.

Geyer-Klingeberg, Jerome, Janina Nakladal, Fabian Baldauf, and Fabian Veit. (2018). "Process Mining and Robotic Process Automation: A Perfect Match." In BPM (Dissertation/Demos/Industry), 124–131.

Goldberg, K. (2011). What is automation? *IEEE Transactions on Automation Science and Engineering* 9 (1): 1–2.

Groover, M. P. (2016). *Automation, production systems, and computer-integrated manufacturing.* India: Pearson Education.

Jerew, B. (2014, April 8). "Toyota claims manual labor is more efficient than automation." *The Green Optimistic*. https://www.greenoptimistic.com/toyota-manual-labor-efficient-automation-20140408

Jerome, H. (1934). "The effects of mechanization." In *Mechanization in industry*, 365–418. National Bureau of Economic Research. http://www.nber.org/chapters/c5249.pdf

Kaber, D. B., E. Omal, and M. Endsley. (1999). "Level of automation effects on telerobot performance and human operator situation awareness and subjective workload." *Automation Technology and Human Performance: Current Research and Trends*, 165–170.

Lacity, M., L. P. Willcocks, and A. Craig. (2015). *Robotic process automation at Telefonica O2* (The Outsourcing Unit Working Research Paper Series). http://eprints.lse.ac.uk/64516/1/OUWRPS_15_02_published.pdf

Langston, G. (2002). Available from http://www.gb.nrao.edu/~glangsto/gbt/800/

Lorenz, B., F. D. Nocera, and R. Parasuraman. (2001, January 17–19). "Human performance during simulated space operations under varied levels of system autonomy." *Proceedings of Bioastronautics Investigators Workshop*.

Malý, I., D. Sedláček, and P. Leitão. (2016, July). "Augmented reality experiments with industrial robot in industry 4.0 environment." In *2016 IEEE 14th International Conference on Industrial Informatics (INDIN)*, 176–181. IEEE.

Markus, M. L. (1983). "Power, politics, and MIS implementation." *Communications of the ACM* 26 (6): 430–444.

Monostori, L. (2014). Cyber-physical production systems: Roots, expectations and R&D challenges. *Procedia CIRP* 17: 9–13. http://www.sciencedirect.com/science/article/pii/S2212827114003497

Morin, B., N. Harrand, and F. Fleurey. (2017). Model-based software engineering to tame the IoT jungle. *IEEE Software* 34 (1): 30–36.

Ruff, Heath A., Sundaram Narayanan, and Mark H. Draper. (2002). "Human interaction with levels of automation and decision-aid fidelity in the supervisory control of multiple simulated unmanned air vehicles." *Presence: Teleoperators & Virtual Environments* 11, no. 4 (2002): 335–351.

Sarter, Nadine B., David D. Woods, and Charles E. Billings. (1997). "Automation surprises." *Handbook of human factors and ergonomics* 2 (1997): 1926–1943.

Sheridan, T. B., and W. L. Verplank. (1978). *Human and computer control of undersea teleoperators*. Cambridge: Massachusetts Institute of Technology, Man-Machine Systems Lab.

Westin, C., C. Borst, and B. Hilburn. (2015). "Strategic conformance: Overcoming acceptance issues of decision aiding automation?" *IEEE Transactions on Human-Machine Systems* 46 (1): 41–52.

## CHAPTER FIVE

# QUALITY 4.0 USE CASES ACROSS INDUSTRIES

While initial developments stimulated by Industry 4.0 did not require significant advances in quality to build basic connectivity of its system, currently evolving advances require thinking algorithms that have an ability to make choices . . . that not only observe, collect, and distribute data, but . . . creatively consider what to do with the data and how to improve upon the current way that this data is generated.

—GREGORY H. WATSON, PAST CHAIR AND FELLOW OF ASQ AND PAST CHAIR AND HONORARY MEMBER OF THE INTERNATIONAL ACADEMY FOR QUALITY

I'm not a frequent user of the ATM at the bank, but every now and then I need to get some cash. Several months ago, I went through the motions on a Sunday so I could get my son some funds for a field trip the next day. I pulled my car up to the ATM window, put my card into the slot, and entered my secret code into the terminal. Everything was humming along without a hitch, and after less than a minute, the "thank you" screen appeared.

But my cash didn't come out! I pressed a few more buttons, but nothing. I started to panic. The machine ate my card and didn't give me any money. And on a Sunday, there was nothing I could do to remedy the situation. I quickly pressed more buttons, faster! But nothing was working. I'd have to tell my son he couldn't go on the trip.

Then, I noticed that my card was sticking out of the card slot. *Thankfully, at least I didn't lose my card.* As soon as I pulled it out, though, there was another whirr—and out popped a stack of bills.

As someone who thinks about quality and process improvement all the time, I realized what had happened immediately. I was in awe. What a great idea! Although people are highly motivated to take their cash, they are not as motivated to remember to take their card. Most process flows at the ATM return the card and then dispense the cash. But the process that had been programmed into this machine had been mistake-proofed (*poka-yoke* in Japanese), so no one could possibly leave their card behind. I had to pull out the card in order to trigger the dispenser.

This was such a good idea, I was shocked that in my 25 years of ATM use I hadn't encountered it before. Like a good academic, I went to Google Scholar immediately to see if the idea had ever been considered in the research. Not only had it been studied in depth, but the results from improving this particular process flow had been well known for almost two decades:

> A "card-returned-then-cash-dispensed" ATM dialogue design was at least 22% more efficient (in withdrawal time) and resulted in 100% fewer lost cards (i.e., none) compared with a "cash-dispensed-then-card-returned" dialogue design. (Zimmerman & Bridger, 2000)

Why couldn't this mistake-proofing have been done sooner? Simply put, because changing the process flow on hundreds of thousands of ATMs was not an easy task. For years, the process was dictated by hardware rather than software, which could have been more easily reconfigured. Newer ATMs are much more flexible and may even be programmed to be responsive to customer needs. By moving the automation capabilities to the software, financial institutions could improve ATM processes in hours or days rather than decades.

Pickett (2019) notices the same trend in industrial sensors, which can be programmed to notice potential human errors in real time and help people correct them, or manage a no-touch approach to optimizing the life-cycle value of assets:

> Industry 4.0 and the Internet of Things have given rise to IoT-enabled, mistake-proofing sensors that connect to other shop floor devices and systems. On top of that, AI software is increasingly being used to analyze the data that these sensors collect, finding patterns and adjusting processes without the need for human intervention. . . . Maximo, IBM's cloud-based Enterprise Asset Management [EAM] software, is designed to handle constant streams of data from IoT sensors and devices. The company's Watson IoT platform connects, manages, and analyzes IoT data with the help of artificial intelligence.

These examples show that even the most basic improvement methods can be enhanced and accelerated with connected, intelligent, and automated digital approaches. With near-real-time feedback, amazing opportunities for improvement will be revealed in many processes, even outside of manufacturing.

## CONNECTED, INTELLIGENT, AND AUTOMATED

These stories provide examples of Quality 4.0 use cases—practical scenarios that use both traditional methods and digital technologies to:

- Enhance connectedness (of people, machines, and/or data)
- Augment, increase, or improve on human intelligence
- Increase automation to some degree

Although not all three must be present in each digital transformation, Industry 4.0, or Quality 4.0 initiative, the impacts (and sometimes benefits) compound. For example, Romero et al. (2018) explore how "Industry 4.0 technologies, such as enterprise wearables, can foster better industrial hygiene to keep operators healthy, safe, and motivated within emerging cyber-physical production systems." In one of their three case studies, they describe how workers' personal protective equipment (PPE) can be modernized to provide better protection from workplace hazards like toxins, temperature extremes, and noise.

First, the PPE is fitted with sensors that can detect whether a hazard is a threat. When the PPE is connected to a network, it has access to information about what is normal and acceptable, versus what is dangerous. Intelligence can be embedded into the sensor to automatically alert the worker when there is a problem. An alert raised by one worker's PPE can be propagated to other workers in proximity to provide an early warning. A fully automated system may also report the issues to a central software system, finding patterns in the aggregated data over time and using that information to better protect the workers in real time.

Connectedness, intelligence, and automation are not just present or absent: each of these dimensions reflects a spectrum of capabilities. Figure 5.1 shows the relationships, using the degrees of automation initially developed by Sheridan and Verplank (1978). A Quality 4.0 initiative will shift the capabilities of an entity, infrastructure element, or process higher on at least one of the three dimensions, and usually two or three.

Making systems connected, intelligent, and automated can increase the speed and quality of decision making; improve transparency, traceability, and

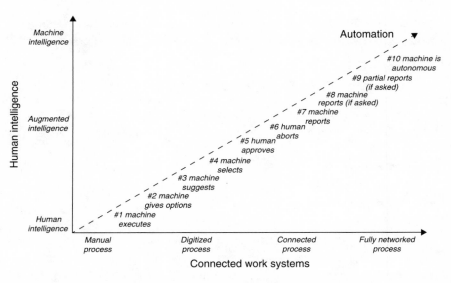

FIGURE 5.1. Degrees of connectedness, intelligence, and automation in Quality 4.0.

auditability; anticipate changes, reveal biases, and adapt to new circumstances and knowledge; reveal opportunities for continuous improvement and new business models; and help people and systems learn how to learn, cultivating self-awareness and other-awareness as a skill. In addition, connectedness has two aspects: digitization and information sharing. Information can be shared in many ways (e.g., face-to-face, phone, letter, e-mail, text message, online chat), only some of which require a communications network. Although information can be shared without being digitized, wider sharing is enabled when data sources are digital.

By implementing connected, intelligent, and automated systems, at least one of the following outcomes can be achieved:

- Improved product or process quality
- Reduced cost, waste, defects, rework, or Cost of Quality (CoQ)
- Increased customer satisfaction or reduced dissatisfaction
- Enhanced workforce capability or capacity
- Enhanced decision-making capability
- Improved environment, health, or safety outcomes and conditions
- Improved leadership, governance, or ethical outcomes
- Improved ability to meet legal, financial, or compliance requirements

- Greater benefits to society
- Improved financial results
- Improved ability to realize strategic objectives
- Enhanced ability to bring products and services to market
- Creation of a new business model

This chapter explores patterns identified from studying hundreds of Quality 4.0 example initiatives across several industries. They illustrate how organizations have already successfully used digital technologies to improve performance and expand competitive advantage, and provide a model for how you can select directions and structure initiatives.

## KEY THEMES IN INDUSTRY 4.0 AND QUALITY 4.0

To select the most representative use cases and case studies specific to Industry 4.0 and Quality 4.0, I first examined previous studies that addressed key themes in this area. This section summarizes Martinelli et al.'s (2019) patent studies, Dombrowski et al.'s (2017) evaluation of 260 research papers containing use cases in Industry 4.0, and my original study of 430 abstracts covering Industry 4.0 as well as Quality 4.0 work from outside manufacturing and related industries.

### Patents: 1990–2014

Martinelli et al. (2019) examined 363,803 patents filed between 1990 and 2014 with the United States Patent and Trademark Office. Because the United States was a leader in innovation throughout this period, the authors felt that this sample would provide a representative glimpse into worldwide progress. The six categories that they tracked were artificial intelligence (AI), big data, cloud computing, internet of things (IoT), 3D printing and additive manufacturing, and industrial robotics. In addition to noting uneven patterns of adoption, their work revealed several interesting things:

- There was a continuum between adoption of smart technologies and adoption of presmart technologies (e.g., CAD [computer-aided design], CAE [computer-aided engineering], and CAM [computer-aided manufacturing]), and no break or discontinuity was detected

- High future growth was indicated for IoT and additive manufacturing
- Slower growth was indicated for industrial robotics (a more mature market segment)
- Slower growth was indicated for human-machine interfaces, possibly reflecting the difficulty of applying AI and machine learning in production processes.
- Large companies are leading Industry 4.0 technology adoption, and small and medium-sized firms are lagging
- Barriers to adoption are costs and absorptive capacity (difficulty of learning, internalizing, and applying the new technologies)

Through the patent analysis, they also uncovered insights into what is driving the adoption of some Industry 4.0 technologies. These factors included more efficient production, cost optimization, greater flexibility, improved product quality, and decreased errors. They also noticed that "the application of new business models figures prominently in the preferences of smaller firms," suggesting that business model innovation may be the most important Quality 4.0 use case for some organizations.

## Industry 4.0 Use Cases

Dombrowski et al. (2017) analyzed 260 papers in Industry 4.0 to identify key themes, selected specifically because they emphasized use cases. Their goal was to understand the interdependencies between Industry 4.0 and lean production systems. They used affinity analysis to organize the papers into the three categories (process oriented, systems focused, and technology focused) shown in Figure 5.2. Next, they grouped the papers by eight lean principles or practices (standardization, zero defects principle, visual management, continuous improvement, management by objectives, flow, pull, and avoidance of waste).

There were three main conclusions regarding the interplay of lean and Industry 4.0:

- Cloud computing is used to reduce waste, although it is also used frequently to implement flow and pull-based systems
- Big data is used to reduce defects, driving operations via the zero defects principle
- Digitalization (new business models and value streams opened up by digital technologies) is strongly supported by standardization

| Process-related characteristics of Industry 4.0 | Horizontal integration | Vertical integration | Real-time data | Transparency |
|---|---|---|---|---|
| | Flexibility | Digitalization | Consistency | Monitoring |
| | Visualization | Traceability | Self-optimization | Self-healing |

| Systems of Industry 4.0 | Smart data/algorithms | Intelligent objects | Internet of things |
|---|---|---|---|
| | Cyber-physical systems | Machine-to-machine communication | |

| Technologies of Industry 4.0 | Big data | RFID/identification | Cloud computing | Augmented and virtual reality |
|---|---|---|---|---|
| | Sensor/actuator | Real-time data | Automated guided vehicles | Consumer electronics |

FIGURE 5.2. Industry 4.0. Adapted from Dombrowski et al. (2017).

Their study also called out the implementation themes originally identified by Monostori (2014):

- **Manual operations**—Processes are managed on paper or on spreadsheets
- **Digitization**—Some processes support electronic data entry, retrieval, and visualization
- **Horizontal integration**—Some systems can exchange information across functional areas of the organization (e.g., sales, marketing, production)
- **Vertical integration**—Information and material flows connect sensors, control systems, and business systems
- **Connected work systems**—Information and materials flow horizontally and vertically; organization can anticipate, adjust, and adapt

## Literature Review and Text Analysis

To identify key themes specific to Industry 4.0 and Quality 4.0 use cases, I conducted a literature review using resources from Google Scholar, ProQuest, and Scopus. This was a broader study than Dombrowski et al. (2017) and included papers and preprints available through August 2019. Papers were selected if they referred to Industry 4.0 (or any of the other "4.0s" that appear to be synonymous with Quality 4.0 in general) and explicitly contained one or more use cases or case studies, were organized as use cases or scenarios, or otherwise contained practice and experience reports.

There were no research papers before 2014 that focused on Industry 4.0 and Quality 4.0 use cases; instead, earlier papers emphasized new engineering

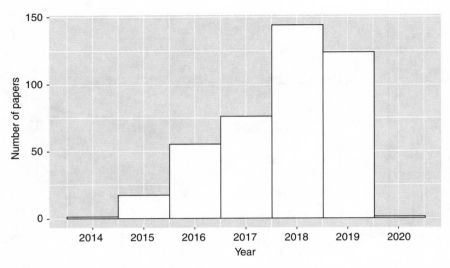

FIGURE 5.3. Publication date distribution for research papers used to develop key themes.

capabilities in sensors, networks, and cyber-physical systems (Figure 5.3). Fewer than 10 papers from 2014 were included. A small number of preprints expected to be published in 2020 were also included.

From the 630 items that were initially identified, 436 were relevant during initial review. A total of 430 contained enough descriptive data to be included in the analysis. Preliminary analysis showed that use cases usually focused on one entity engaged in a single process type (Table 5.1). For example, the paper may have used real-time data to enhance decision support for workers guiding manufacturing processes (smart operator/executing processes) or simulation on digital twins to improve return on assets (smart assets/exploring processes).

The scenarios illustrated several ways in which digitized, connected, smart, augmented entities could engage in more efficient and effective planning, exploring, designing, executing, and auditing:

- Digitization of data about the entities or processes
- Digitization of data for use by the entities or processes
- Adding connections among entities, processes, and data sources
- Augmenting physical or cognitive capabilities of entities
- Using intelligent agents and intelligent systems to improve or enrich processes
- Using connectedness, intelligence, and automation to improve health and safety
- Using connectedness, intelligence, and automation to discover ways to improve

**TABLE 5.1. Entities and processes featured in Industry 4.0 and Quality 4.0 use cases.**

| Entity | Process |
| --- | --- |
| • Asset/equipment | • Planning |
| • City | • Exploring |
| • Customer | • Designing |
| • Data platforms/repositories | • Executing |
| • Factory/workplace | • Auditing |
| • Intelligent agent | |
| • Operator | |
| • Product | |
| • Vehicle | |
| • Worker | |

This element of real-time discovery and self-regulation is particularly characteristic of the new digitally transformed era. Watson (2020) describes Quality 4.0 in these terms as well, as a "holistic sociotechnical system that is purposefully designed to discover and apply profound knowledge in pursuit of continual improvement and consistently achieve an organization's purposeful objectives." To Watson (2019), discovery manifests in several ways in Quality 4.0:

- "Digitization is used to optimize signal feedback and process adjustment, and adaptive learning supports self-induced system corrections.
- Quality shifts its control-oriented focus from the process operators to the process designers.
- Machines learn how to self-regulate and manage their own productivity and quality.
- Human performance is essential; the emphasis shifts from production to system design and integration with the business system."

Next, I performed text analysis of the abstracts from the 430 papers that were ultimately selected. After building a corpus containing all the abstracts, I used the latent Dirichlet allocation (LDA) technique. This generative probabilistic model, often used by machine learning practitioners, considers each document as a distribution of topics, and within each topic there is a distribution of terms. By examining the statistical structure of these topics and terms, it is possible to identify themes from large collections of documents. The themes that the LDA algorithm constructs have to be manually inspected for relevance, and each document can contain one or more themes.

LDA was used to explore the corpus of 430 abstracts for four to eight themes. The final selection of seven themes was based on capturing all topics that appeared in at least 13% of the abstracts. Each of these themes (and the common words that describe it) represents a platform for driving quality and performance outcomes through connected, intelligent, and automated digital systems:

- **Smart Manufacturing and Internet of Things (IoT):** manufacturing systems smart industry cyberphysical internet industrial technologies integration things intelligent
- **Environment, Health, Safety, and Security:** health analytics society safety predictive risks reliability healthcare transport
- **Lean and Supply Chain:** industry lean future supply chain challenges review SMEs methodology maturity
- **Digitization and Design:** production systems design work proposed tools process framework engineering
- **Data-Driven Maintenance:** data information control application maintenance machine sensors software platform
- **Quality Management Systems:** quality management approach human potential knowledge processes performance results
- **Innovation and New Business Models:** new digital transformation business models products value innovation concept sustainability

In addition to manufacturing, the industries that appeared in key themes were automotive, construction, food and beverage, healthcare, and transportation. The top three issues tackled with Industry 4.0 and Quality 4.0 approaches were energy efficiency, sustainability, and safety (mentioned 39, 38, and 37 times, respectively). Most frequently mentioned technologies or techniques were augmented and virtual reality, predictive maintenance, and digital twins (mentioned 26, 17, and 12 times).

The low frequency of use cases from the petrochemical industry was surprising since oil and gas has been leading other industries in adopting IIoT technologies, with 50% of the installed base (Martinelli et al., 2019). Similarly, the lack of attention to cybersecurity was unexpected since this is critical for all networked systems. Notably absent from the collection were use cases or case studies on facility management, even though commercial building automation is considered to be a driving use case for smart cities (Table 5.2).

**TABLE 5.2. Frequency of terms in 430 abstracts of research papers with use cases.**

| Most frequent | Least frequent |
| --- | --- |
| Manufacturing — 471 | Pharmaceutical — 13 |
| Smart Mfg/Smart Factory — 62 | Agriculture — 10 |
| Construction — 37 | Shipbuilding — 10 |
| Automotive — 34 | Petrochemical — 7 |
| Food & Beverage — 28 | Patient — 5 |
| Healthcare — 22 | Cybersecurity — 4 |
| Medical — 21 | Hospitality — 3 |
| Government — 19 | Aerospace — 2 |
| Autonomous Vehicles — 17 | |

## NEW BUSINESS MODELS

Revealing new, innovative business models through increased connected-ness, intelligence, and automation is the heart of digital transformation for all organizations. Business model innovation may be the most important Quality 4.0 use case for smaller organizations (Martinelli et al., 2019). Regardless of the size of your organization, the path toward growth and expansion is filled with many opportunities for value creation, delivery, and capture.

For example, Ibarra et al. (2018) take a look at growth and innovation through digital transformation, focusing on sustainability as a driver (Figure 5.3). They note that process optimization is the first step of innovation, but as this ma-tures, new interfaces can be provided to customers for co-creation of products and services (Figure 5.4). Beyond interfaces, companies can seek to move value creation ecosystems and supply networks online. Finally, the ability to develop and release novel smart products and services can create entirely new markets. The relationship between these four stages and value, as defined by Ibarra et al. (2018), is shown in Table 5.3.

Weking et al. (2018) also conducted a study to build a taxonomy of all the busi-ness models that could be configured from the modes of value creation, delivery, and capture explored by Ibarra et al. (2018). Their results are shown in Figure 5.5. Within each meta-dimension (the leftmost column) the business model options at the right are related to one another to show relationships and interactions.

For example, in the "Value proposition" meta-dimension, one row displays "Product" and the row below it displays "Service." At the very left portion of

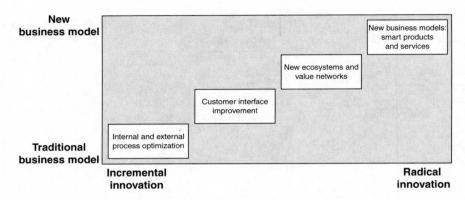

New
business model

Traditional
business model

New business models:
smart products
and services

New ecosystems and
value networks

Customer interface
improvement

Internal and external
process optimization

Incremental
innovation

Radical
innovation

FIGURE 5.4. The four stages of digital transformation and their relation to value. Adapted from Ibarra et al. (2018).

**TABLE 5.3.** Relationship between value and innovation.

| | Process optimization | Customer interface improvement | New ecosystems | Smart products and services |
|---|---|---|---|---|
| Value creation | • Connectedness<br>• Traceability<br>• Training<br>• Transparency | • Data monitoring<br>• New touch points<br>• New services | • Systems connected to partners<br>• Real-time information | • New physical, human, and intellectual resources |
| Value delivery | • Flexibility | • Segmentation<br>• Digital sales<br>• Long-term relationships | • New segments<br>• Broader offers | • Smart products<br>• Service innovation<br>• Co-creation<br>• Direct, close relationships with customers |
| Value capture | • Process efficiency<br>• Waste reduction | • Cost savings<br>• New revenue streams | • Cost reduction for all stakeholders | • New revenue streams |

Source: Adapted from Ibarra et al., 2018.

these two rows, if the business model is built on a physical product, it is possible that there will be no service model. However, there is a possibility that the physical product will be coupled with a service plan to provide repair and maintenance, or in limited cases, a service package for remote monitoring. On

| Big data | Organizational change | Product/service extension | | New product/service | | New business unit | | Spin-off company | |
|---|---|---|---|---|---|---|---|---|---|
| | Innovativeness | Evolution | | New to company | | | New to industry | | |
| Big data | Market | Business to business (B2B) | | Business to customer (B2C) | | | Customer to customer (C2C) | | |
| | Customer contact | Through intermediaries | | Hybrid | | | Direct | | |
| | Sales channel | Off-line | | Hybrid | | | Online | | |
| Big data | Product type | Physical | | Physical + digital | | Digital | | None | |
| | Service | None | Repair and maintenance | Remote monitoring, predictive maintenance | | Product support | | IT support | Advice and consulting |
| | Value proposition | Price | Solution | Availability | | Full service | | Value-adds | Long tail |
| Big data | Design/development | Experts | | Agile/cooperative | | | Crowdsourcing | | |
| | Customization | Custom/high cost | | Mass production | | Mass customization | | Individualization | |
| | Approach | Push | | | | Pull | | | |
| Big data | Strategic analytics | No | | | Internal product and process data | | | Internal and customer's product/process data | | |
| | Platform | None | | Closed | | Trading/exchange | | Innovation | | All functions |
| Big data | Revenue model | Sales | Licensing | Revenue sharing | Physical freemium | Freemium | Rent/lease | By use | Subscribe |
| | Sales model | Ownership/service delivery | | Availability of use | | | Driving results | | |
| | Profit logic | Reduce costs | | Increase revenue | | | Both | | |

FIGURE 5.5. A taxonomy of business models configured from value creation, delivery, and capture. Adapted from Weking et al. (2018).

the rightmost side of this pair of rows, if there is no product offering, there may be advice and consultancy offered to drive revenue, or perhaps digital information services (e.g., data as a service, or DaaS). This chart can be used for brainstorming and to explore the relationships between many meta-dimensions that make up a modern business model.

## QUALITY 4.0 CASE STUDIES

With a better understanding of the key themes and industries, I selected 10 papers that best demonstrate the spirit and values of Quality 4.0. What distinguishes these studies from Industry 4.0 projects and initiatives (Table 5.4) is that each of these projects, prototypes, or feasibility studies was specifically designed with quality and performance in mind.

Rather than launching a project to introduce AI to the organization, or incorporate machine learning into processes, or implement IoT, these initiatives started with the business needs, and technology requirements then emerged. It is this strategic focus that increases the likelihood of digital transformation success.

**TABLE 5.4. Quality 4.0 case studies described in this section [C = connectedness, I = intelligence, A = automation].**

| # | Industry | C | I | A | Description | Outcomes | Reference |
|---|---|---|---|---|---|---|---|
| 1 | All | X | X | X | **Healthy Operator 4.0.** Operator vital signs and workplace environment monitored continuously; alerts, emergency stops, and in-process leveling prevent overburdening. | Enhance operator health and well-being; protect against unhealthy exposure to chemicals, noise, heat, cold; prevent incidents and musculoskeletal disorders; increase productivity; optimize cognitive workload. | Romero et al. (2018) |
| 2 | All | X | X | X | **Augmented Worker.** Exoskeletons explored for monitoring of and relief from heavy lifting tasks, providing ergonomic guidance for workers, and the use of smart protective equipment. | Protect against musculoskeletal injuries; increase quality of product; increase worker quality and productivity. | Butler (2016) |
| 3 | All | X | X | X | **Logistics 4.0.** Real-time adaptive resource planning, warehouse management, transportation management, and intelligent transportation systems are coordinated to route people, parts, and products. | Waste of motion is reduced, minimized, or eliminated. | Barreto et al. (2017) |
| 4 | All | X | X | | **Leaner Management.** Real-time intelligent information hubs guide the process of selecting, implementing, and sustaining improvements. | Improve problem selection; improve identification of root causes; enhance sustainment of improvements. | Rittberger & Schneider (2018) |
| 5 | Manufacturing (Any) | X | X | X | **Smart Labels.** A human-centered smart label (a step beyond barcodes and RFID tags) is defined by software rather than what is printed on the tag. The labels can dynamically adjust to reflect product and pricing changes. | Eliminate non-value-adding steps (changing, reprinting, or disposing of labels); become more responsive to changes. | Fernandez-Caramés & Fraga-Lamas (2018) |

| # | Industry | | | Use Case | Benefit | Reference |
|---|---|---|---|---|---|---|
| 6 | Oil & Gas/ Petrochemical | X | X | **Predictive Maintenance/Smart Asset Management.** A neural network is trained on historical data to determine when an expensive piece of equipment at an oil and gas facility needs to be taken off-line temporarily. | Reduce downtime; prevent asset damage or destruction; improve operator safety. | Abbasi et al. (2019) |
| 7 | Transportation | X | | **Accident Prevention.** Using data sources that exist and are accessible, this study explored the feasibility of advanced analytics for preventing systemic failures that lead to fatal rail accidents. | Improve safety; protect expensive assets against damage or destruction; save lives. | Parkinson & Bamford (2016) |
| 8 | Food & Beverage | X | | **Olive Oil Supply Chain.** Raw material sourcing, processing and manufacturing, distribution and warehousing, and two-way communication with customers are enhanced by emerging digital technologies. | Improve energy, water, and other natural resources management through the supply chain; control greenhouse gas emissions through travel reduction. | Ojo et al. (2018) |
| 9 | Healthcare | X | | **Medical Process Control with Blockchain.** Managing distributed queues for clinical trials, organ donation, and multisite treatment plans can be made easier by maintaining records on a public blockchain. | Improve security of records; improve process performance. | Shifrin et al. (2019) |
| 10 | Pharmaceutical | X | X | **Pharma 4.0.** Using advanced analytics on unstructured data, support distributed innovation and complex tasks like system qualification. | Reduce operations costs; improve data integrity; enhance collaboration across partner ecosystem. | Manzano & Langer (2018) |

## Case 1: Healthy Operator 4.0

Managing for quality also requires that operators stay safe, healthy, and engaged. A lapse of concentration or attempting to perform dangerous work when you're tired or ill can lead to adverse outcomes for both workers and products. To adapt to changing conditions in the environment and in the body, industrial wearables are a core Industry 4.0 use case. Romero et al. (2018) define requirements for sensors and interactions to manage conditions in the work environment, cognitive and physical workloads, and proximity to hazardous conditions.

The wearables monitor exposure to toxins, pollutants, temperatures, and noise level and vital signs in response to lifting tasks and ergonomic postures. They keep track of both physical and cognitive workloads and compare them against conditions to prevent unsafe working conditions. According to the U.S. Centers for Disease Control and Prevention (CDC), musculoskeletal disorders caused by hazardous working conditions cost the American economy an average of $50 billion a year, so addressing just one of these factors can have a solid positive impact.

- **Connected:** Wireless communications transport sampled monitor data on a regular (seconds, minutes) basis as needed and provide real-time job information
- **Intelligent:** Ambient intelligence monitors vital signs, environmental conditions, and social scenarios to help operators proactively manage their own health and workload
- **Automated:** Alerts are issued if unsafe conditions are detected, emergency stops are automatically issued to protect operators against hazardous conditions, and employers are protected against accidental violations of standards and regulations

## Case 2: Augmented Worker

*Augmented workers* are human workers who use technology to improve how they do their jobs, such as laborers who use exoskeletons to assist with heavy lifting. Exoskeleton technology comes in many shapes and sizes and can perform various tasks. Butler (2016) explains that in addition to helping give workers added strength and agility for hazardous tasks, an exoskeleton can help workers perform tasks more safely and ergonomically—for example, by limiting their range of motion. A Personal Ergonomic Device offering assistive technology can also help extend a worker's productivity by compensating for the physical and mental changes associated with aging.

The use of technology to augment job performance provides direct (and immediate) benefits to quality and productivity. In a field test, welders and painters were first asked to perform "moderate to severe ergonomic tests" until fatigued (Butler, 2016). Quality and productivity were measured using work simulators. A few days later, the same workers were provided with an exoskeleton and subjected to identical conditions. Painting productivity improved by 20%–50%, and one welder's productivity increased by 86% compared with the benchmark. Quality of the work was not only maintained but became more consistent with the aid of the exoskeleton, and workers could put in longer hours before becoming fatigued.

- **Connected:** Oftentimes, exoskeletons can work locally, without communications
- **Intelligent:** Continuous monitoring and feedback are provided to help the worker manage the physical and cognitive workload and accomplish jobs more effectively
- **Automated:** Some exoskeletons can automate manual or skilled labor tasks, helping workers complete tasks like lifts that may be above the recommended limits for humans

## Case 3: Logistics 4.0

Motion and transport, of both physical objects and information, are considered waste in the world of quality. If an object is near you, then you don't have to expend energy to retrieve it. When information is available and accessible, you don't need to expend energy to locate and evaluate it. The concept of Logistics 4.0, driven by increasing demand for highly individualized products and services, seeks to optimize processes around sourcing and providing resources. The primary quality goals are to improve customer satisfaction (for all links in the supply network), reduce motion and transport in production processes, and reduce storage costs.

Barreto et al. (2017) describe a system where people, processes, and technologies are seamlessly integrated throughout a logistics ecosystem. Real-time resource planning, which is a data-intensive effort, requires visibility and transparency of information as well as integration across organizational boundaries and supply networks. Warehouses are automated and can report item status and location on demand. Transportation planning is response-driven and incorporates real-time capabilities and capacities of partners. Intelligent transportation systems route the items to their destinations without human intervention. Security controls are autonomous and continuously audited to reduce risk exposure.

- **Connected:** A real-time data platform captures, aggregates, and assimilates data from each entity in motion in the logistics network
- **Intelligent:** The system processes the data and anticipates when issues require rerouting or adaptation
- **Automated:** The system can dynamically respond to fluctuations in demand, adverse conditions (e.g., transport routes, geopolitical turmoil, weather), and workloads across organizational boundaries, ensuring customer satisfaction regardless of risks

## Case 4: Leaner Management

Lean production and lean management rely on collaborative problem solving, with people, processes, and technologies working in harmony to achieve shared goals. Any deviation from this harmony can negatively impact quality and performance outcomes. Rittberger and Schneider (2018) examined the barriers to lean management and whether they could be overcome with Quality 4.0. With Plan-Do-Check-Act (PDCA) as a frame of reference, they noted information latency as a potential issue at the Plan phase, lack of workforce capacity as a major barrier for the Do phase, independent and manual review of data and results at the Check stage, and effective knowledge transfer, which can also impede sustaining improvements, at the Act stage.

They identified five elements, each supported by digital technologies, that could mitigate or remove the barriers. First, real-time information about problem occurrence and resolution would prevent teams from incompletely characterizing or scoping issues, or not having enough information before moving forward with an improvement. Next, mobile informational assistance (e.g., augmented reality interfaces) could provide them with up-to-date information about gemba (where the work is done).

An intelligent system for predicting or anticipating issues would ensure that appropriate improvement projects were selected at the right times. Intelligent agents, possibly enabled with machine learning, could examine the available data and provide insight into real root causes, eliminating the risks of "opinion-based methods" for root cause analysis like 5 Whys. Finally, robotics and RPA could be used to automate routine tasks, freeing improvement teams to work on breakthrough innovation.

- **Connected:** Wireless communications consolidate operations information; augmented reality displays provide additional connectedness to data sources
- **Intelligent:** Real-time operations data is analyzed to identify high-value problems and provide additional insight into causes
- **Automated:** Routine tasks are automated, freeing human labor for creative work

## Case 5: Smart Labels

Traditional labels are printed with information that eventually becomes obsolete, unless the label falls off first. Before barcodes, supermarket and retail employees used to spend their days walking through the store with a label printer, affixing the right price to each item. This process was labor intensive and error prone, and rework was necessary every time there was a price change or a sale. Furthermore, a missing price tag always led to a delay at the cashier's kiosk for everyone in line, as an employee ran through the store trying to figure out the right price.

Fernandez-Carames and Fraga-Lamas (2018) envision the next step in the evolution of labeling capabilities. Although process steps in manufacturing are increasingly being automated, actions that must be taken by humans require documentation and time to read and understand it. This additional information, they argue, could be provided through a smart label and read by a tablet or smartphone (potentially using an augmented reality interface). Being able to program the labels remotely means that any number of labels could be updated in an instant, or supplemented with information about discounts, specials, or recalls. Operational costs are reduced because human errors are avoided, along with the need to check, print, or replace labels.

- **Connected:** Wireless communication facilitates dynamic reprogramming of the labels, as well as labels being able to send information to a central, intelligent server
- **Intelligent:** Label can determine information about the product and/or voice of the customer (VoC) after the product has been sold
- **Automated:** Label can automatically coordinate with other systems to communicate data to provide real-time product information

## Case 6: Predictive Maintenance/Smart Asset Management

In the oil and gas industry, acquiring and installing equipment is typically capital intensive and requires substantial time and effort to commission. In addition, the availability of capital is often tied to the crude oil price, which can be volatile. Being able to keep an asset in service longer, while minimizing downtime and service time, is a highly attractive and very cost-effective goal.

Abbasi et al. (2019) created a Long Short Term Memory neural network, a special kind of network architecture that can process sequences of data. It uses 14 features to predict a process variable for a motor in an air separation unit of an oil and gas facility. Using historical data as a training set, they were able to build a prediction model to generate prediction alerts if the condition of the motor suggests potential faults. The operator is given the opportunity to take action to prevent damage to the expensive assets.

- **Connected:** The neural network is trained off-line with historical data and no real-time connections are needed
- **Intelligent:** A prediction model uses historical data to determine how a motor failure can be anticipated in advance
- **Automated:** The neural network does not make changes to the operations technology autonomously but provides the operator with suggestions that can be acted on

## Case 7: Rail Accident Prevention

Rail accidents are typically highly complex and require long forensic studies to establish the multiple causes that lead to them. Systematic failures are usually the most devastating, and they are also more difficult to predict. Engineering, operation, and management can all be sources of risks to safety when systematic hazards are in play. Design defects, poor maintenance, inadequate training, carelessness or tiredness, inclement weather, and financial constraints can all contribute to unsafe conditions.

Parkinson and Bamford (2016) examined three major railway accidents to see what kinds of data may have been useful for preventing the incidents. They decided that train design information, information about close calls and safety audits, maintenance records from asset management systems, and remote monitoring were all essential data requirements. Although software systems and appropriate algorithms are not yet available to acquire and process the large amounts of data that they identified, through this analysis safety professionals were able to take the first step toward determining how big data can be used to improve rail safety. If the data had been available in a central location, operators may have been able to prevent at least one of the high fatality incidents that they studied.

- **Connected:** Wireless communications transmit remote monitoring data, asset records, audit and near-miss information, and up-to-date design to a central location
- **Intelligent:** Intelligent agents at the central location process the records and provide an alert if any leading indicators suggest that conditions are not nominal
- **Automated:** The system potentially prevents transit on identified railways, by identified cars, and/or during hazardous conditions to ensure that accidents cannot occur

## Case 8: Olive Oil Supply Chain

In this case study by Ojo et al. (2018), an olive oil manufacturer in Spain invested in Industry 4.0 technologies for the purpose of building a more sustainable food supply chain. Implementing a data platform improved the company's ability to perform monitoring, especially regarding traceability and quality control of agricultural products. Processes for communicating and coordinating with suppliers were improved to more effectively manage just-in-time (JIT) production. The production lines at two facilities were retrofit with industrial robotics and cyber-physical systems for automation and monitoring, but no intelligent algorithms were applied. This automation eliminated the need for some human employees. Additional monitoring made it easier to control energy usage, and improved coordination reduced the need for transportation, further reducing greenhouse gas emissions.

- **Connected:** Enhanced data platform and monitoring capabilities made it possible to more easily communicate with suppliers, enhancing efficiency of JIT production
- **Intelligent:** No intelligent algorithms were implemented
- **Automated:** Robotics and cyber-physical systems were employed to more fully automate two production lines

## Case 9: Medical Process Control with Blockchain

Although blockchain has been proposed as a solution for more closely guarding and protecting the privacy of electronic health records, Shifrin et al. (2019) provide another option. They explain that blockchain has the potential to help organizations manage widely distributed business processes, especially when the parties do not completely trust one another. They specify four medical processes that could be more effectively managed with blockchain: queue flows in clinical studies, organ donation, distributed treatment processes, and integrating information across organizational boundaries. They argue that the potential to self-organize the healthcare ecosystem could yield improved outcomes, including shortening the time to clinical approval of a drug and increasing participation in donor programs.

- **Connected:** Blockchain provides a mechanism to connect parties that may not necessarily trust one another; in this case, the increased connectedness directly leads to achieving desired outcomes
- **Intelligent:** No intelligent algorithms are leveraged in this case
- **Automated:** No automation is implemented in this case

## Case 10: Pharma 4.0

Because pharmaceutical design, testing, and production are data intensive, this industry is particularly well suited to reaping benefits from Industry 4.0. In fact, simple solutions may have the greatest impact. Data integrity issues are cited in the majority of warning letters issued by the U.S. Food and Drug Administration (FDA). Manzano and Langer (2018) outline some of the ways that the new and emerging technologies may help solve the pharmaceutical industry's lingering issues with data quality and integrity. They suggest that much of the industry's challenge with data is because the data is not highly structured, and that machine learning will fix this problem. Unfortunately, this is not very realistic in practice.

Other applications of new technologies that they mention do seem compelling. For example, they recommend investigating new methods for cloud-based system qualification (that is, making sure the software systems that support your business meet rigorous standards for quality management and configuration management). System and design qualification is a labor-intensive activity, even when the process is automated. Further automation on just this one process could generate substantial cost savings.

- **Connected:** Distributed innovation environments, connecting researchers at partner institutions with shared data sources, are expected to accelerate the pace of change
- **Intelligent:** Predicting trends in processes and operations, not just using the data for compliance, will find new ways to demonstrate system qualification
- **Automated:** Streamlining the compliance-driven processes of system qualification and design qualification will make them less labor-intensive

## THE BOTTOM LINE

Quality 4.0 uses traditional methods and digital technologies to accomplish quality objectives (e.g., reduce waste, reduce costs) and improve performance by:

- Enhancing connectedness (of people, machines, and/or data)
- Augmenting, increasing, or improving on human intelligence
- Increasing automation to some degree

There are multiple levels in each category (Table 5.5).

Even the most fundamental quality improvement methods will be enhanced and accelerated using connected, intelligent, and/or automated digital technologies. An original analysis of 430 abstracts uncovered the most frequent and viable topics:

**TABLE 5.5. Levels of connectedness, intelligence, and automation.**

| Connectedness | Intelligence | Automation |
|---|---|---|
| Manual process | Human intelligence | Machine executes |
| Digital process | Augmented intelligence | Machine gives options |
| Connected process | Collective intelligence | Machine suggests |
| Vertical integration | Augmented collective | Machine selects |
| Horizontal integration | intelligence | Human approves |
| Connected work systems | Machine intelligence | Human aborts |
| | | Machine reports |
| | | Machine reports (if asked) |
| | | Machine partially reports (if asked) |
| | | Complete autonomy |

- **Smart Manufacturing and Internet of Things (IoT):** smart factory, smart production
- **Environment, Health, Safety, and Security:** connected worker, connected safety
- **Lean and Supply Chain:** horizontal integration, sustainability, energy efficiency, smart supply chain
- **Digitization and Design:** vertical integration, digital design, process exploration with digital twins
- **Data-Driven Maintenance:** predictive maintenance, smart asset management
- **Quality Management Systems:** smart planning, smart auditing, optimizing human performance and human-machine interactions
- **Innovation and New Business Models:** Data as a Service (DaaS), smart cities

Product and process design will become more robust as engineers gain the ability to critically evaluate more design alternatives in advance, using simulation and digital twins. Digitization can be used to provide effective real-time feedback to adjust processes and understanding, and machines will learn to self-regulate and provide useful information to people, if and when they need it.

## ACKNOWLEDGMENTS

The contents of this chapter draw from preliminary research by Graham Freeman and Nicole Radziwill, which we expect to publish in 2020.

## REFERENCES

Abbasi, Tayaba, King Hann Lim, and Ke San Yam. (2019). "Predictive Maintenance of Oil and Gas Equipment using Recurrent Neural Network." In *IOP Conference Series: Materials Science and Engineering* 495 (1): 012067.

Barreto, L., A. Amaral, and T. Pereira. (2017). "Industry 4.0 implications in logistics: An overview." *Procedia Manufacturing* 13: 1245–1252.

Butler, T. R. (2016). "Exoskeleton technology: Making workers safer and more productive." *Professional Safety* 61 (09): 32–36.

Dombrowski, U., T. Richter, and P. Krenkel. (2017). "Interdependencies of Industrie 4.0 & lean production systems: A use cases analysis." *Procedia Manufacturing* 11: 1061–1068.

Fernández-Caramés, T. M., and P. Fraga-Lamas. (2018). "A review on human-centered IoT-connected smart labels for the industry 4.0." *IEEE Access* 6: 25939–25957.

Ibarra, Dorleta, Jaione Ganzarain, and Juan Ignacio Igartua. (2018) "Business model innovation through Industry 4.0: A review." *Procedia Manufacturing* 22 (2018): 4–10.

Manzano, T., and G. Langer. (2018, December). "Getting Ready for Pharma 4.0." *Pharmaceutical Engineering*: 72–79.

Martinelli, A., A. Mina, and M. Moggi. (2019). *The enabling technologies of Industry 4.0: Examining the seeds of the fourth industrial revolution* (No. 2019/09). Laboratory of Economics and Management (LEM). Pisa, Italy: Sant'Anna School of Advanced Studies.

Monostori, L. (2014). "Cyber-physical production systems: Roots, expectations and R&D challenges." *Procedia CIRP* 17: 9–13. http://www.sciencedirect.com/science/article/pii/S2212827114003497

Ojo, O. O., S. Shah, A. Coutroubis, M. T. Jiménez, and Y. M. Ocana. (2018, November). "Potential impact of Industry 4.0 in sustainable food supply chain environment." In *2018 IEEE International Conference on Technology Management, Operations and Decisions (ICTMOD)*, 172–177.

Parkinson, H. J., and G. J. Bamford. (2016, October). "Big data and the virtuous circle of railway digitization." In *INNS Conference on Big Data*, 314–322. Springer, Cham.

Pickett, L. (2019, August). "The state of sensors in the industrial IoT." *Quality Magazine* 58 (8): 12–13.

Rittberger, Sven, and Markus Schneider. (2018, June). "Continuous Improvement of Lean Processes with industry 4.0 Technologies." 11th International Doctoral Students Workshop on Logistics, Magdeburg, Germany.

Romero, D., S. Mattsson, Å. Fast-Berglund, T. Wuest, D. Gorecky, and J. Stahre. (2018, August). "Digitalizing occupational health, safety and productivity for the operator 4.0." In *IFIP International Conference on Advances in Production Management Systems*, 473–481. Cham, Switzerland: Springer.

Sheridan, T. B., and W. L. Verplank. (1978). *Human and computer control of undersea teleoperators*. Cambridge: Massachusetts Institute of Technology, Man-Machine Systems Lab.

Shifrin, M., A. Khavtorin, V. Stepurin, and B. Zingerman. (2019). "Blockchain as a process control tool for healthcare." *Studies in Health Technology and Informatics* 262: 172–175.

Watson, G. H. (2019, March). "The ascent of Quality 4.0." *ASQ Quality Progress*, 25–30.

———. (2020). "The infrastructure of Quality 4.0." *ASQ Quality Progress*, forthcoming.

Weking, J., Maria Stöcker, Marek Kowalkiewicz, Markus Böhm, and Helmut Krcmar. (2018, August 16). *Archetypes for Industry 4.0 business model innovations*. In 24th Americas Conference on Information Systems (AMCIS 2018), New Orleans, LA.

Zimmermann, C. M., and R. S. Bridger. (2000). "Effects of dialogue design on automatic teller machine (ATM) usability: Transaction times and card loss." *Behaviour & Information Technology* 19 (6): 441–449.

## CHAPTER SIX

# FROM ALGORITHMS
# TO ADVANCED ANALYTICS

At a conference on the social study of algorithms in 2013, a senior scholar stepped up to the audience microphone: "With all this talk about algorithms," he said, "I haven't heard anybody talk about an actual algorithm. Bubble sort, anyone?"

—N. SEAVER

N ext generation quality is enabled by connectedness, intelligence, and automation. Together, these characteristics can enhance the performance of people and the organizations they serve, making it possible to augment human intelligence in new and compelling ways. Unfortunately, marketing hype abounds, and it can be hard to tell the difference between amazing new capabilities and smoke and mirrors. The purpose of this chapter is to help you distinguish between truth and hype by clarifying the language around algorithms, analytics, business intelligence, artificial intelligence (AI), and machine learning.

Advanced analytical and statistical methods were once the domain of highly trained programmers and engineers, because coding was required to realize their value. Business decision making, driven by simulations and modeling, has traditionally been in the domain of operations research and management science for the same reason. As a result of technological innovations in software reusability, cloud computing, and algorithms that perform well even on large and streaming datasets, these techniques are now becoming more accessible and democratized. Business users are now just as likely to use "advanced methods" as engineers and programmers.

With increased accessibility comes more excitement about the topic—expressed in blog posts, marketing materials, and books. This excitement can also lead to less precise use of terminology than what was intended by the original stewards of the fields from which these techniques emerged. This is the case with *algorithm*, a word whose meaning has become less distinct in recent years, as you'll see later in this chapter.

What is the proper modern use of the word *algorithm* in the context of analytics, AI, data science, machine learning, and other methods for generating business insights? Does proper use even matter? What are the practical ramifications of these distinctions, and do they impact the practice of software quality with respect to advanced analytics? To find out, I informally interviewed data scientists with an average of ten years' experience in that field. No distinction was made between those who build models using algorithms and those who use the results of the models to make business decisions, even though the conclusions were geared toward informing quality assurance practice. With that in mind, this process revealed four considerations for quality assurance practitioners in the development of analytics and models.

Using more concise language helps reduce the risks of improperly managing expectations, which can be important when digital strategy is translated into action plans to develop analytics and machine learning models. By examining how key terms related to Industry 4.0–era analytics are used, and framing them within the current context of what it means to be (and not to be) an algorithm, practical guidance for quality assurance in analytics is uncovered.

## ANALYTICS AND BUSINESS INTELLIGENCE

In management, the core concept of analytics has been around for thousands of years. Calculating the appropriate amount of taxes to collect, for example, was noted as an issue in Egypt in 2390 BC, China in 594 BC, and England in the fourteenth century (Burg, 2004). Actuaries in seventeenth-century England needed to understand factors impacting mortality to determine appropriate prices for life insurance (Bhaduri & Fogarty, 2016).

Attention to analytics mushroomed in response to "Competing on Analytics," a 2006 article in *Harvard Business Review* by Thomas H. Davenport and Jeanne G. Harris. It was followed in 2007 by a book with the same name. Their work described numerous companies that were linking data to action plans with analytics, making sure that they had visibility into strategic goals and initiatives. By using analytics strategically, they were able to demonstrate that the most successful companies were investing in rigorous approaches to analytics that were measuring everything from customer satisfaction to supply chain efficiency.

Davenport and Harris (2007, p. 46) define analytics as "the extensive use of data, statistical and quantitative analysis, explanatory and predictive models, and fact-based management to drive decisions and actions." Evans (2012) says that business analytics is "the use of data, information technology, statistical analysis, quantitative methods, and mathematical or computer-based models to help managers gain improved insight about their business operations and make better, fact-based decisions." These descriptions describe analytics as both the activity of data-driven collaboration and the product of that activity.

Evans (2015) explains how analytics, as a discipline, came to be:

> Modern analytics can be viewed as an integration of the three fundamental disciplines: business intelligence/information systems (BI/IS), statistics, and quantitative methods/operations research (see Figure 2). These disciplines have been around for more than half a century. However, their integration, supported by various tools such as spreadsheets, statistical software, and more complex business intelligence suites that integrate data with analytical software, have led to new and more powerful ways to view, understand, and use data and information intelligently. For example, data mining can be characterized as the integration of BI/IS and statistics. Spreadsheets and formal models allow one to manipulate data to perform what-if analysis—how specific combinations of inputs that reflect key assumptions will affect model outputs. What-if analysis results from integrating concepts of BI/IS with operations research.

Analytics *is* data-driven decision making, on all scales and with all data volumes.

## From Analytics to Advanced Analytics

*Advanced* analytics is a much more recent label. In the 1980s and 1990s, Bhaduri and Fogarty (2016) explain that building and populating databases, and developing methods for easier querying and reporting, were the primary business drivers. Not until the introduction of Customer Relationship Management (CRM) in the late 1990s were executives motivated to invest in analytics. CRM systems provided a basis to characterize the unique profitability and promise of every customer relationship, and use this information to deploy resources to the areas of the business that would generate the most revenue. Even more significantly, capturing information about customer interactions at every touch point (web, phone, e-mail, and so on) meant that decisions could be made based on what each customer was *actually* doing rather than what the business *guessed* they were doing from general market research, customer surveys, and limited direct contact (Khirallah, 2001).

Although briefly mentioned with respect to geospatial analysis by Musick et al. (1997), the concept of advanced analytics did not emerge until CRM systems were beyond the fad stage. By 2002, CRM systems were so firmly entrenched that Gartner released its Hype Cycle for Advanced Analytics soon after (Linden & Fenn, 2003). Gartner describes analytics as "a very large and fragmented space . . . with roots in statistics, operations research, pattern recognition, optimization and decision theory . . . [and] mathematics." They go on to describe the specifics of what they considered "advanced":

> Analytics also encompasses many interdisciplinary schools, such as data mining, simulation, artificial intelligence, information retrieval and computational linguistics. Many of the more advanced analytics technologies, such as genetic algorithms, Bayesian approaches and fuzzy logic, while often hyped by the press, have only taken hold in niche markets. Technologies such as neural nets, data mining and mathematical programming have already matured in some areas, but in others they still lack traction and therefore it has been difficult to place these technologies precisely on the Hype Cycle. . . . This Hype Cycle shows the more advanced analytical technologies, [and] their impact on business and anticipated adoption.

The Hype Cycle for 2003 (Figure 6.1) reveals some interesting context about how business leaders perceived the value and utility of what they considered "advanced analytics" nearly 20 years ago. Viable, proven techniques that they felt would become mainstream within a couple of years included neural nets (which, in their description, could be any means of predicting or classifying), data mining, linear programming and related techniques, and "automated text categorization" (determining the documents most relevant to a particular category or concern). They saw personalization, intelligent agents, and autonomous systems on the horizon but overestimated the analytical power of video mining, swarm intelligence, and genetic algorithms. Still, their forecasts anticipated the emerging power of technologies to catalyze collaboration and engagement.

More recently, Bose (2009) says that "data integration and data mining are the basis for advanced analytics . . . [the] more information that is gathered and integrated allows for more pattern recognition and relationship identification." Barton and Court (2012) and Franks (2012) supported this orientation and reinforced the message that these insights should be tied to driving business value. There is a general consensus now that advanced analytics are (or can be)

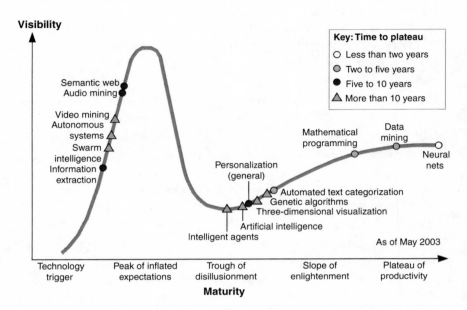

FIGURE 6.1. The Gartner Hype Cycle for advanced analytics (Linden & Fenn, 2003).

run on big data, which the National Science Foundation (2012) describes as "large, complex, and longitudinal data sets generated from instruments, sensors, internet transactions, and/or other digital sources." Duarte (2017) further explains that advanced analytics should derive value from streaming data, which he calls "data in motion."

Advanced analytics, thus, can be considered the canon of emerging methods for (and the professional practice of) generating value from large, complex, historical, and/or streaming datasets, coupled with the results produced by those methods. These datasets may be aggregated from various real-time or archived sources. As adoption increases and businesses seek new ways to achieve competitive advantage, specific methods, techniques, and algorithms will shift and evolve over time—and so too will the understanding of what "advanced analytics" involves.

## Descriptive, Diagnostic, Predictive, Prescriptive

One of the most frequently used categorizations spanning ordinary and advanced analytics is the "three perspectives" that businesses view analytics from: descriptive, predictive, and prescriptive (Evans, 2012). The additional category of diagnostic analytics also appears in many references, distinguished from descriptive analytics by context of use (Delen & Ram, 2018). These categories are illustrated in Figure 6.2. Both descriptive and diagnostic categories examine data from the past and present; predictive and prescriptive analytics use it to create new information about the future:

**Descriptive Analytics:** Virtually all organizations use descriptive analytics in some form, commonly encountered on business intelligence dashboards. These dashboards include metrics that describe things that have already happened or are in the process of happening. Both levels and trends can be captured. Descriptive analytics can reveal trends or anomalies, and by presenting information in the form of charts and graphs, provide decision makers with the basis to adjust as necessary.

**Diagnostic Analytics:** This area of emphasis uses data to identify causal relationships, determine the most significant variables or features, and uncover root causes. Diagnostic analytics can reveal why anomalies or patterns have emerged. This typically involves using descriptive analytics for a particular forensic purpose, and as a result, relies on historical data. Many practitioners lump these into the predictive category since understanding past situations can improve the accuracy of predictions (Puget, 2016).

**Predictive Analytics:** This class of analytics projects future states. It can include forecasting, model building, and incorporating diagnostic results to assess new cases. In environmental, health, and safety (EHS) operations, predictive analytics can identify conditions that have been observed leading up to past safety and health incidents, and then continuously monitor incoming data to determine whether similar conditions start to emerge. If this happens, operators or managers are alerted so they can take action.

**Prescriptive Analytics:** Techniques in this category are used to identify a recommended course of action and, specifically, "use optimization to identify the best alternatives to minimize or maximize some objective" (Evans, 2012). The action can be recommended to a human or automated and executed without intervention. These practices have been used for decades to optimize production, improve the efficiency and effectiveness of scheduling, and manage inventories and supply chains.

Examples of how researchers have grouped specific techniques into these categories are presented in Table 6.1. In practice, the techniques do not map so cleanly into the four groups (Puget, 2016). For example, machine learning (typically positioned as a means of making predictions) often involves an optimization

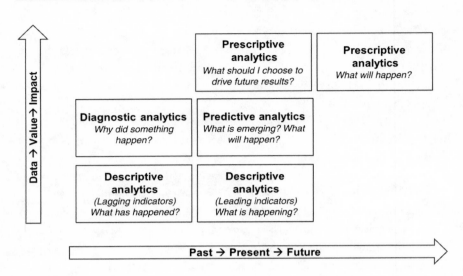

FIGURE 6.2. Categories of analytics organized by business value and future orientation.

step (which is considered prescriptive). The random forest technique creates many different decision trees and then iteratively decides which branches to keep and which to prune. The end result is a "best of class" model that can be used to predict or classify new instances. Many popular AIs (including IBM's Watson and Google AlphaGo) go beyond even the prescriptive step, not only choosing recommended actions but developing and adapting a model to update the decision process over time.

Although these four categories are sometimes presented as stages or phases of analytics, with prescriptive analytics the ultimate goal, there is no indication that the complexity or value of the business knowledge increases as you move down the rows in Table 6.1 (although computational complexity does increase as you move down). Strong return on investment can be realized by implementing analytics in any of the categories. Success will be determined more by factors like how well model results are communicated and to what degree the organization learns from those results, rather than by the complexity of the algorithms themselves.

## Business Intelligence

The term *business intelligence* (BI) was first used by Devens (1868), who described a banker whose systematic, regular practice of collecting and analyzing data gave him substantial competitive advantage. Luhn (1958), an IBM

**TABLE 6.1. Examples of methods to generate descriptive, diagnostic/ explanatory, predictive, and prescriptive analytics.**

| Domain | Category | Examples |
|---|---|---|
| Business Intelligence + Business Analytics | Descriptive | • Scorecards<br>• Descriptive statistics (e.g., mean, variance, range, confidence intervals)<br>• Histograms, scatterplots, Pareto charts, time series, boxplots, pie/waffle charts<br>• Correlation and association analysis<br>• Signal to noise analysis |
| | Diagnostic or explanatory | • Design of experiments<br>• Root cause analysis<br>• Clustering and community detection<br>• Text mining<br>• Network analysis<br>• Data reduction (e.g., PCA, SVD, factor/ discriminant analysis) |
| Data Science + Operations Research | Predictive | • Regression<br>• Classification<br>• Neural networks<br>• Time series forecasting<br>• Statistical process control (SPC)<br>• Scoring systems<br>• Risk assessment<br>• Ensemble methods (e.g., random forest) |
| | Prescriptive | • Simulations<br>• Optimization techniques<br>• Decision trees<br>• Discrete event simulation (DES)<br>• Simulation-based scenario analysis<br>• Recommender systems<br>• A/B testing (and multiarmed bandit)<br>• Multicriteria decision making<br>• Reinforcement learning<br>• System dynamics |

researcher, used BI to describe the practices associated with collecting, analyzing, managing, and reporting data. BI, as it was considered throughout the early 2000s, is a support function that adds to analytics by providing a translation layer, making them understandable and useful for decision making.

In practice, perhaps because the use of information systems is now ubiquitous, there seems to be little distinction between analytics work and BI work.

Anderson (2019) describes BI as the "combination of data, appropriate metrics, and the relevant skills, tools, and processes to make sense of what is happening in a business, and to make recommendations as to what should change or happen next. Most organizations attempt to leverage analytics to drive decision making. However, few of them are able to access the full value of what business intelligence has to offer."

## WHAT IS AN ALGORITHM?

Until the mid-twentieth century, the concept of an *algorithm* was straightforward: a sequence of goal-directed instructions that could be carried out by a general purpose computation machine. Algorithms are now ordinary and ubiquitous. Today's "computation machines" can now span multiple computers in the cloud, include computations done on edge devices in the IoT, and leverage reusable software libraries to accomplish tasks. The algorithmic sequence of instructions may instead resemble a deeply interconnected fabric of operations.

An algorithm has been defined as a sequence of instructions that can be carried out by a machine that is Turing complete—that is, a machine that is capable of performing general computations (Minsky, 1967; Savage, 1972; Gurevich, 2000). Most programming languages at present are Turing complete, but markup languages like HTML are not, because they cannot change the state of the underlying system. It is easier to frame the concept of Turing completeness by considering that modern computers were not available at the time that the idea was developed. A machine is not Turing complete if you can demonstrate that there are calculations it cannot do, which is easy for examples from analog computing.

### Analog Computing and Algorithms

Analog computing, which was dominant at the time that algorithms were first being envisioned, required building new hardware (sequences of gears, cams, and other components) whenever new types of computation were required. For example, aiming large guns and determining the time required to set fuses so that explosives detonated at the right time and in the right place used to require complex calculations. To perform these calculations, custom hardware would be built to leverage the mathematical relationships between differential gears, so that a series of input positions pointed to an output position that produced the recommended action (Clymer, 1993). Figure 6.3 shows one example

RESTRICTED

MARCH, 1945   BIF 6-1-1

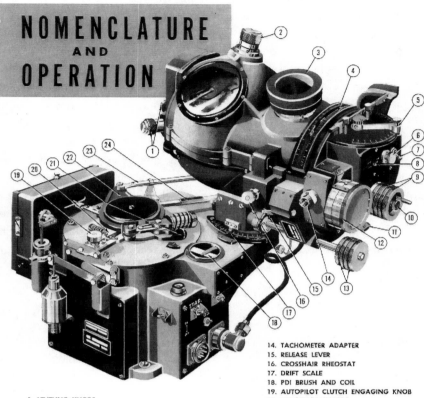

# NOMENCLATURE AND OPERATION

1. LEVELING KNOBS
2. CAGING KNOB
3. EYEPIECE
4. INDEX WINDOW
5. TRAIL ARM AND TRAIL PLATE
6. EXTENDED VISION KNOB
7. RATE MOTOR SWITCH
8. DISC SPEED GEAR SHIFT
9. RATE AND DISPLACEMENT KNOBS
10. MIRROR DRIVE CLUTCH
11. SEARCH KNOB
12. DISC SPEED DRUM
13. TURN AND DRIFT KNOBS

14. TACHOMETER ADAPTER
15. RELEASE LEVER
16. CROSSHAIR RHEOSTAT
17. DRIFT SCALE
18. PDI BRUSH AND COIL
19. AUTOPILOT CLUTCH ENGAGING KNOB
20. AUTOPILOT CLUTCH
21. BOMBSIGHT CLUTCH ENGAGING LEVER
22. BOMBSIGHT CLUTCH
23. BOMBSIGHT CONNECTING ROD
24. AUTOPILOT CONNECTING ROD

The bombsight has 2 main parts, sighthead and stabilizer. The sighthead pivots on the stabilizer and is locked to it by the dovetail locking pin. The sight-head is connected to the directional gyro in the sta-bilizer through the bombsight connecting rod and the bombsight clutch.

RESTRICTED

FIGURE 6.3. An analog fire control computer (U.S. Army Air Forces, Office of Flying Safety, 1945).

of a fire control analog computer used in World War II to position equipment for accurate bomb targeting. (This example will become important again in Chapter 14, which covers cybersecurity and risk.)

Algorithms are thus concepts that describe computable recipes. The steps to calculate travel times on multiple paths home from work, and then selecting an optimal choice, would be accomplished by an algorithm. Although there are clear inputs and outputs, the steps you use to cook dinner would be a *procedure* (and not an algorithm) because those steps are not essentially computable. The design of an analog computing machine would *not* be an algorithm, although the steps used to transform inputs to outputs within that machine would be.

Although there is a widespread tendency to equate algorithms with the implementation or automation of those algorithms, algorithms have no power until they are *enacted*—that is, combined with data and embodied in programs. Dourish (2016) explains that "an algorithm may express the core of what a program is meant to do, but that core is surrounded by a vast penumbra of ancillary operations that are also a program's responsibility and also manifest themselves in the program's code." He calls on Wirth's 1975 description of "algorithm + data = program" to express this relationship.

According to Seaver (2017, emphasis mine), "The proper definition of algorithms serves to distinguish them from typical critical concerns: algorithms are *not automation* (thus excluding questions of labor), they are *not code* (thus excluding questions of texts), they are *not architecture* (thus excluding questions of infrastructure), and they are *not their materializations* (or outputs)." He goes on to explain that they are, however, "cultural . . . because they are composed of collective human practices." Because algorithms are enacted by humans in social and cultural contexts, the technical and nontechnical aspects are fused together, making algorithms "rather unstable objects, culturally enacted by the practices people use to engage with them."

## Examples of Algorithms and Non-Algorithms

If we accept Seaver's definition of algorithms as unstable cultural objects, then sorting algorithms, optimization algorithms, combinatorial algorithms, root finding algorithms, or anything else on Wikipedia's List of Algorithms (https://en.wikipedia.org/wiki/List_of_algorithms) would count as algorithms. In addition, these more complex systems of algorithms can be considered algorithms because they are computable recipes that generate outputs based on inputs:

- Facebook's algorithm for personalizing content in the feed
- Twitter's algorithm for customizing the order and display of tweets
- Google's algorithm for selecting the most relevant documents for a search

These algorithms are only as good as the data used to build them; they may be subject to bias if the datasets they are trained on are biased. The following examples are not algorithms:

- Descriptive statistics, because they are the outputs from algorithms
- Scorecards, because they are display mechanisms for the outputs from algorithms
- Root cause analysis, because the problem-solving process is often not translatable into computable recipes
- Design of experiments (DoE), because it is a broad class of methods or techniques for establishing causal links between variables that uses computational recipes
- A/B testing, because it is a method for determining significant preference between two options

Both of the above lists should be considered illustrative rather than exhaustive. Analytics thus cannot be equated with the concept of algorithms, even though sometimes algorithms will be applied in the conduct of analytics practice. Analytics is an overarching practice that seeks to apply algorithms, evaluate models, and use the generated insights to deliver business value.

## INTRODUCING DATA SCIENCE

*Data science*, in contrast, is concerned with building, validating, and continuously improving models that adapt to (and learn from) new data. This adaptation and learning is what distinguishes data science from analytics, business intelligence, and even data mining—and for adaptation to be possible, access to newly observed and streaming data ("data in motion") must be accessible (Duarte, 2017).

Data science usually requires the use of algorithms, but not necessarily their development. Although initially envisioned by Cleveland (2001), data science did not emerge as a professional practice with a unique identity until popularized by people including DJ Patil, Drew Conway, and Hadley Wickham between 2008 and 2012.

Because the role of data science is to enable data-driven decision making across all categories of analytics (Puget, 2016), the scope of analytics practice had to expand to incorporate data capture and preparation, storage, security, and governance (Aasheim et al., 2015). More recently, the National Academies of Sciences, Engineering, and Medicine (2017) also recognized that data science encompasses data management and processing in addition to "the analytical methods and theories for descriptive and predictive analysis and for prescriptive analysis and optimization."

Data science is concerned with building, validating, and continuously improving models that integrate and incorporate computational recipes. Although many tasks faced by data scientists are predictive in nature, those predictions may require descriptive or diagnostic analytics, and may also require optimization steps. The fundamental elements of data science practice are thus:

- **Models**—Models are mathematical descriptions of the relationships between inputs and outputs, designed based on beliefs about those relationships. Models can be adapted and adjusted as new data is obtained and those beliefs change. Creating a model may require employing one or more analytical methods. Models are fit using algorithms, but are not themselves algorithms.
- **Methods or Techniques**—These procedures describe analytical techniques for problem solving and may incorporate data acquisition, assimilation, and quality assessment. Methods and techniques can be used to create models (e.g., AI techniques, computer vision techniques) or analytics (e.g., Monte Carlo methods, Bayesian methods).
- **Machine Learning Algorithms**—These are core computational recipes used to carry out machine learning methods that are used to build models. For example, the backpropagation algorithm can be used to create a neural network model for predictions and classifications based on training data.
- **Machine Learning Methods**—These techniques leverage machine learning algorithms to create models that can adapt to new data and revise themselves to increase prediction accuracy. Machine learning methods use one or more machine learning algorithms.

As expressed by P. Mohanty (personal communication, July 2019), "models + methods + data = predictions." The methods bring together the algorithms required to create the models that describe relationships within the data.

## Quality Considerations

With these distinctions in mind, how should quality assurance be conducted by practitioners of analytics and data science? First and foremost, a systematic and disciplined approach to data quality assurance and identifying suitable analytical techniques is essential (Jugulum, 2018, 2019). Because machine learning algorithms are used to discover new models based on shifting and evolving data, the quality of the input data is paramount. The model will only be as good as the data used to develop, train, and update it over time. This is not a new insight; Feigenbaum (1979) made the same assertion in his first law of knowledge engineering for developing expert systems.

Next, the analytical approach(es) selected must be aligned with the need for accuracy and speed in business decision making, as well as the risk appetite of the organization implementing the analytics. Niranjan Krishnan, head of data science at Tiger Analytics, articulated some of these in an interview with Ismail (2018). Although the "AI vs. Algorithms" dichotomy is overly simplified, he provides a distinction between the choices that should be considered (Table 6.2) when choosing among straightforward algorithms (such as those

**TABLE 6.2. Algorithms versus AI.**

|  | Algorithms: More control and transparency | AI: "Black box" or not easily explainable |
|---|---|---|
| Decision criteria | • Cost of errors is high<br>• Slower decisions are acceptable<br>• Data structures or volumes are manageable using traditional methods<br>• Explainability is more important than accuracy<br>• Environment is highly regulated with requirements for auditing or traceability | • Cost of errors is low<br>• Rapid decisions are needed<br>• Data structures or volumes do not lend themselves to traditional analysis<br>• Accuracy is more important than explainability<br>• Environment is not as highly regulated with limited requirements for auditing or traceability |
| Appropriate use cases | • Credit risk assessment<br>• Insurance underwriting<br>• Claims processing | • Marketing campaigns<br>• Targeted advertising<br>• Predictive maintenance<br>• Product recommendations<br>• Fraud detection |

Source: Adapted from Krishnan's comments to Ismail (2018).

that might be found in descriptive, and possibly diagnostic analytics), predictive model building, and prescriptive action determination.

As suggested in Table 6.2, the degree of explainability is critical. For a deep learning implementation that recognizes whether a chocolate muffin or blueberry muffin is on the conveyor belt of a production facility, being able to report the percentage of time the muffin was correctly identified may be sufficient. However, it is difficult (and usually impossible) to conceptualize the "thinking process" used by that deep learning model to make the determination. In contrast, in the case of a credit scoring model, there may be regulatory requirements that govern the degree to which the organization must be able to explain how it made its decision.

A deep learning model would not be auditable in the same way that a scoring model based on simpler methods like decision trees would be, owing to the low degree of explainability (Figure 6.4). Data scientists can tell you how well the deep learning model works but cannot articulate, in a detailed step-by-step way, how it draws its conclusions.

Finally, establish whether the analytics support learning—not just by the models but also by the people using and developing the models—and whether the organization has the communication channels in place to facilitate it. Aydiner et al. (2019) found no direct link between analytics programs and firm performance, but there was a link between business process reengineering efforts and performance. Specifically, the activities that enhanced communication, knowledge sharing, and sharing lessons learned led to improvements. This underscores the findings of Kovach and Fredendall (2013), who identified learning as the driver for the effectiveness of continuous improvement programs.

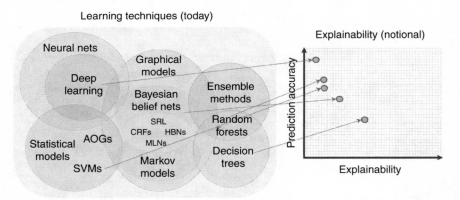

FIGURE 6.4. Degree of accuracy versus degree of model explainability (Bornstein, 2016).

## Lessons for Quality Assurance

Programs that use or combine machine learning algorithms blend descriptive with predictive techniques, and complete the modeling task using iterative optimization to prescribe actions. Those prescriptions can be automated and executed without the intervention of a human, or they can be delivered to human decision makers to augment their own intelligence. In investigating the foundations for analytics, algorithms, and AI (in particular, data science practice that focuses on model building using machine learning), the following four considerations for quality assurance emerged:

1. A systematic and disciplined approach to (a) data quality assurance and (b) identifying suitable analytical techniques must be established
2. The analytical approach(es) selected must be aligned with the need for accuracy and speed in business decision making, as well as the risk appetite of the organization
3. The degree of explainability must be considered at the time the methods are selected
4. The degree to which learning is supported, and communication channels are established to derive the most value from that learning, must also be evaluated

Classifying analytics as descriptive, diagnostic, predictive, and prescriptive does not seem to be as useful as understanding the process and models required to produce the analytics. Organizations seeking to increase their level of analytics maturity should first make sure that a solid foundation is in place for data quality assurance and systematic selection and validation of analytical methods. Next, ensure that the methods are aligned with the organization's requirements and capabilities, and build those capabilities as required for future growth.

## THE BOTTOM LINE

- **Analytics** is the practice of data-driven decision making, on all scales and with all data volumes. In research papers, there are four categories of analytics identified:

  o **Descriptive**—Levels and trends that describe things that have already happened or are in the process of happening

o **Diagnostic**—Data to identify causal relationships, determine the most significant variables or features, and uncover root causes
o **Prescriptive**—Data to anticipate future states, which can include forecasting or model building
o **Predictive**—Analytics to identify a recommended course of action

- **Advanced analytics** refers to the body of emerging methods for generating value from large, complex, historical, and/or streaming datasets.
- **Business intelligence** practitioners add value to analytics by making them more engaging, understandable, and useful for decision making.
- **Data science** is concerned with building, validating, and continuously improving models that adapt to (and learn from) new data. Those models often use algorithms.

o An **algorithm** is a sequence of instructions or activities (carried out with a goal in mind) using a machine that performs computing. Slide rules do not use algorithms, but sorting functions do.

- When data scientists build models that learn, they should consider quality at the start, and

o use a disciplined approach to development,
o identify proper analytical techniques to match the problem,
o choose an appropriate degree of explainability for the task at hand, and
o figure out how human learning will be supported by their work.

## ACKNOWLEDGMENTS

Several elements of this chapter emerged from discussions with data scientists and statisticians, including Pete Mohanty, Demetri Pananos, J.D. Long, Brandon Collingsworth, William Chase, and TJ Mahr. They shared their impressions and perceptions of the contemporary use of terminology and how they feel it is being properly and improperly addressed in practice.

## REFERENCES

Aasheim, C. L., S. Williams, P. Rutner, and A. Gardiner. (2015). "Data analytics vs. data science: A study of similarities and differences in undergraduate programs based on course descriptions." *Journal of Information Systems Education* 26 (2): 103–115.

Anderson, C. (2019). "Business intelligence." In *Data science in practice*, 97–118. Springer, Cham.

Aydiner, A. S., E. Tatoglu, E. Bayraktar, S. Zaim, and D. Delen. (2019). "Business analytics and firm performance: The mediating role of business process performance." *Journal of Business Research* 96: 228–237.

Barton, D., D. Court. (2012). "Making advanced analytics work for you." *Harvard Business Review* 90 (10): 78–83.

Bhaduri, S. N., and D. Fogarty. (2016). *Advanced business analytics*. Springer.

Bornstein, A. M. (2016, September 1). *Is artificial intelligence permanently inscrutable?* Nautilus. http://nautil.us/issue/40/learning/is-artificial-intelligence-permanently-inscrutable

Bose, R. (2009). "Advanced analytics: Opportunities and challenges." *Industrial Management & Data Systems* 109 (2): 155–172.

Burg, D. F. (2004). *A world history of tax rebellions: An encyclopedia of tax rebels, revolts, and riots from antiquity to the present*. Routledge.

Cleveland, W. S. (2001). "Data science: An action plan for expanding the technical areas of the field of statistics." *ISI Review* 69: 21–26.

Clymer, A. B. (1993). "The mechanical analog computers of Hannibal Ford and William Newell." *IEEE Annals of the History of Computing* 15 (2): 19–34.

Davenport, T. H., and J. G. Harris. (2007). *Competing on analytics*. Harvard Business School Press.

Delen, D., and S. Ram. (2018). "Research challenges and opportunities in business analytics." *Journal of Business Analytics* 1 (1): 2–12.

Devens, R. M. (1868). *Cyclopaedia of commercial and business anecdotes: Comprising interesting reminiscences and facts, remarkable traits and humors . . . of merchants, traders, bankers . . . etc. in all ages and countries . . .* Appleton.

Dourish, P. (2016). "Algorithms and their others: Algorithmic culture in context." *Big Data & Society* 3 (2).

Duarte, J. (2017). "Data disruption." *ASQ Quality Progress* 50 (9): 20–24.

Evans, J. R. (2012). "Business analytics: The next frontier for decision sciences." *Decision Line* 43 (2): 4–6.

———. (2015). "Modern analytics and the future of quality and performance excellence." *Quality Management Journal* 22 (4): 6–17.

Feigenbaum, E. A. (1979). "Themes and case studies in knowledge engineering." In D. Michie (Ed.), *Expert systems in the microelectronic age*. Edinburgh University Press.

Franks, B. (2012). *Taming the big data tidal wave: Finding opportunities in huge data streams with advanced analytics* (Vol. 49). John Wiley & Sons.

Gurevich, Yuri. (2000). "Sequential abstract-state machines capture sequential algorithms." *ACM Transactions on Computational Logic (TOCL)* 1 (1): 77–111.

Ismail, K. (2018, October 26). *AI vs. algorithms: What's the difference?* CMSWire. https://www.cmswire.com/information-management/ai-vs-algorithms-whats-the-difference/

Jugulum, R. (2018). *Robust quality: Powerful integration of data science and process engineering*. CRC Press.

———. (2019, June). "A structured approach for analytics execution and measuring analytics quality." *Software Quality Professional*.

Khirallah, K. (2001). *CRM case study: The analytics that power CRM at Royal Bank of Canada* [research note]. Tower-Group.

Kovach, J. V., and L. D. Fredendall. (2013). "The influence of continuous improvement practices on learning: An empirical study." *Quality Management Journal* 20 (4): 6–20.

Linden, A., and J. Fenn. (2003). *Hype cycle for advanced analytics, 2003.* Gartner Strategic Analysis Report. https://www.bus.umich.edu/KresgePublic/Journals/Gartner/research/114900/114961/114961.pdf

Luhn, H. P. (1958, October). "A business intelligence system." *IBM Journal.*

Minsky, M. (1967). *Computation: Finite and Infinite Machines.* Englewood Cliffs, NJ: Prentice Hall.

Musick, R., R. Fidelis, and T. Slezak. (1997). *Large-scale data mining pilot project in human genome* (No. UCRL-JC-127338; CONF-9705227-). Lawrence Livermore National Lab.

National Academies of Sciences, Engineering, and Medicine. (2017). *Strengthening data science methods for Department of Defense personnel and readiness missions.* National Academies Press. https://www.nap.edu/read/23670/chapter/1

National Science Foundation. (2012). *Core techniques and technologies for advancing big data science & engineering (BIGDATA).* http://www.nsf.gov/pubs/2012/nsf12499/nsf12499.htm

Puget, J. F. (2016, June 14). "Machine learning as prescriptive analytics." *IBM Community.* https://www.ibm.com/developerworks/community/blogs/jfp/entry/Machine_Learning_As_Prescriptive_Analytics?lang=en

Savage, L. (1972). "The Foundations of Statistics." Dover: New York.

Seaver, N. (2017). "Algorithms as culture: Some tactics for the ethnography of algorithmic systems." *Big Data & Society* 4 (2): 2053951717738104.

U.S. Army Air Forces, Office of Flying Safety. (1945, March). *Bombardiers' information file* (BIF 1-6-3). https://archive.org/details/BombardiersInformationFile

## CHAPTER SEVEN

# DELIVERING VALUE AND IMPACT THROUGH DATA SCIENCE

We dance round in a ring and suppose,
But the Secret sits in the middle and knows.
—ROBERT FROST

## THE CASCADIA STORY

Deep under the surface of the Pacific Ocean, just off the coast of Washington and Oregon, the Cascadia Subduction Zone lies in wait. It runs a thousand miles from north to south, the quiet interface between the North American plate and the tiny Juan de Fuca plate, which itself is wedged between the large continental plate and the Pacific plate.

Until recently, geologists recognized it as one of the least active regions in the seismically intense Ring of Fire, an arc tracing the coasts of the Pacific Ocean. But it's often been the subject of intense speculation because elsewhere in the world, subduction zones have hosted some of the world's most intense megathrust earthquakes: the 1755 earthquake that destroyed Lisbon, the great Chilean earthquake of 1960, the "Good Friday" earthquake in Alaska in 1964, and more recently, the 2011 Tohoku earthquake in Japan. In a subduction zone, one plate is pushed under another slowly and steadily. Tremendous pressure builds up until the moment when it can no longer be suppressed, and then, in one violent release of energy, the balance between the plates is restored in a maelstrom of death and destruction (Bryant, 2014).

Everyone thought Cascadia was a little different—at least until data science became a part of understanding the region. In recorded history, there had been no observations of a major, devastating earthquake in the Pacific Northwest. But thanks to mathematical modeling (and apocalyptic imaginations), it wasn't difficult for scientists to project what might happen if "the big one" were to hit the Vancouver, Seattle, Olympia, and Portland corridor, where 10 million live.

Historically, the area hasn't appeared prone to large earthquakes. As a result, structures built before 1994 were not constructed to withstand major tremors or a fast-moving tsunami. In a severe earthquake, over 75% of them are expected to collapse, the electrical grid will fail, and all communications will be interrupted. In a July 2015 article in the *New Yorker*, Kathryn Schulz (2015) quotes FEMA representative Kenneth Murphy, who explains that in a worst-case scenario, the "operating assumption is that everything west of Interstate 5 will be toast." That's millions dead, hundreds of thousands of square miles of devastation, and an economy that will be crippled for decades.

In a Reddit discussion in 2015, John Vidal, a seismologist for Washington state, presented a little more optimism: "Communications may black out, transportation may grind to a halt, stores conceivably could run out of goods for a while, but that doesn't constitute 'toast' in one's mind" (*Seattle Times*, 2016). Regardless of the magnitude of devastation, there's general agreement among seismologists that the big one is coming, the region is not prepared, and a period of intense chaos after the event (followed by an equally intense, and long, recovery) is likely.

## Investigating the Ghost Forest

So how do scientists know "the big one" is coming? How did they figure out that the Pacific Northwest is due for a major earthquake? Discovering the region's seismic history required a little bit of data analysis magic, and a lot of time for the pieces of the mystery to come together. Keep in mind that in the 1970s, the key question was *whether* Cascadia had ever produced a large earthquake, not *when* the last earthquake might have occurred.

First, there were environmental clues. In the 1980s, geologist Brian Atwater and his student David Yamaguchi (who studied patterns in tree rings) teamed up to investigate the "ghost forest" near Copalis Beach, Washington. Along this two-mile stretch, hundreds of lifeless cedars sit submerged alongside a river (Figure 7.1). Atwater wanted to know what killed all the ancient trees. So, together, he and Yamaguchi compared the story told by the tree rings in the ghost forest with the story told by the subsurface layers of soil, sand, and ancient mud. They concluded that sometime in 1699 or 1700, the earth abruptly sunk in the area. A saltwater tsunami, triggered by a huge quake, inundated the region around what would soon become the ghost forest.

FIGURE 7.1. Copalis ghost forest (photo from Atwater et al., 2005, p. 26).

## Insights from Historical Records

Although Atwater and Yamaguchi had solved their own problem, other scientists continued to explore problems related to the origin of the ghost forest and the mystery quake that produced it. In the 1990s, Japanese seismologist Kenji Satake made another connection based on an entirely different data source: historical records from Japan in the early 1700s described a widespread seawater flood that devastated villages, leading to fires and "massive fright." Although not coupled with an earthquake, the Japanese had experienced events like this on occasion throughout their thousand years of recorded history. It was odd, but not completely abnormal. The Japanese called these "orphan tsunamis"—not recognizing that an earthquake on the *other* side of the ocean could produce a tsunami with dire local consequences.

When Satake and his colleagues explored original historical records from each of six Japanese villages, they discovered that each place reported the exact times of arrival of large waves not associated with an earthquake, and subsequent damage (Figure 7.2). Coupling those historical records with earthquake and tsunami simulations, they determined that a magnitude 9.0 earthquake occurred a little after 9 p.m. just off the coast of Washington or Oregon on January 26, 1700.

## Lessons from Cascadia

Working with multiple disparate data sources to generate insights about what lies between them is nothing new. Scientists have uncovered new knowledge

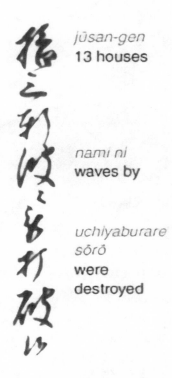

*jūsan-gen*
13 houses

*nami ni*
waves by

*uchiyaburare*
*sōrō*
were
destroyed

FIGURE 7.2. Damage report from historical records in eighteenth-century Japan (Atwater, 2005, p. 48).

this way for hundreds of years. But it took almost three decades, four different disciplines, and several people working on related problems independently to reconstruct the full story of what happened during and after the Cascadia earthquake in 1700. What if you had the ability to get that answer much more quickly—maybe even fast enough to use the information to make better urban planning decisions in coastal Oregon *today*?

This is the main premise of data science. The tools and technologies to quickly gather, store, integrate, and analyze data sources, then create adaptive models that adjust to new data and a changing external environment, are now broadly available—*and* reliable. Although domain expertise and caution are both required, profound, near-real-time insights can be generated by leveraging open source software packages that implement complex algorithms.

While the Cascadia story illustrates the slow process of scientific discovery, data science promises to speed the process up so that our organizations can discover new information and new patterns in data, and quickly act to generate business value. Data science is about aggregating data and building models to solve mysteries and make actionable predictions.

## WHAT IS DATA SCIENCE?

*Data science* is an interdisciplinary activity that encompasses all the activities required to support data-driven decision making: from collecting or obtaining data, to building and optimizing models, to generating insights, to delivering value and broader impacts. The key is building models that adapt to new data and learn. The practice of data science is not new, but the formation of a community of researchers and practitioners around a hybrid field that blends statistics, programming, data management, and domain experience is. That community only started coalescing around 2010.

### The Death of the Scientific Method

One of the catalysts for the consolidation of the community—and the expansion of data science practice—was a 2008 article in *Wired* by editor Chris Anderson. This magazine takes an interdisciplinary look at emerging technologies and their impact on the way we live, work, and play. Although Anderson's article accurately describes the new vantage point that huge amounts of data were starting to provide across many fields at the time, it was a little too enthusiastic in its willingness to throw out the lessons science had provided over hundreds of years (the article was titled *The End of Theory*):

> The scientific method is built around testable hypotheses. These models, for the most part, are systems visualized in the minds of scientists. The models are then tested, and experiments confirm or falsify theoretical models of how the world works. This is the way science has worked for hundreds of years.
>
> Scientists are trained to recognize that correlation is not causation, that no conclusions should be drawn simply on the basis of correlation between X and Y (it could just be a coincidence). Instead, you must understand the underlying mechanisms that connect the two. Once you have a model, you can connect the data sets with confidence. Data without a model is just noise.

But faced with massive data, this approach to science—hypothesize, model, test—is becoming obsolete. Consider physics: Newtonian models were crude approximations of the truth (wrong at the atomic level, but still useful). A hundred years ago, statistically based quantum mechanics offered a better picture—but quantum mechanics is yet another model, and as such it, too, is flawed, no doubt a caricature of a more complex underlying reality.

There is now a better way. Petabytes allow us to say: "Correlation is enough." We can stop looking for models. We can analyze the data without hypotheses about what it might show. We can throw the numbers into the biggest computing clusters the world has ever seen and let statistical algorithms find patterns where science cannot.

The article caused a stir among readers, with intense debate ensuing both in online and in published articles for several years thereafter. Mazzocchi (2015) provided one of the more comprehensive critiques. He reminded everyone that the value of deductive, Aristotelian problem solving had already been compared to inductive methods in the 1600s by Francis Bacon, Tycho Brahe, and Johannes Kepler. Both research driven by hypotheses and exploratory research driven by data had a critical role to play in knowledge generation and learning:

The data-driven approach constitutes a novel tool for scientific research. Yet this does not imply that it will supersede cognitive and methodological procedures, which have been refined during centuries of philosophical and scientific thought. There is no "end of theory" but only new opportunities.

Despite Anderson's provocative article, the scientific method is not dead, nor has it become less important. Instead, mass amounts of data with the computing power to learn from it quickly means that we have *more* opportunities to identify patterns and drive value.

## The Evolution of Data-Driven Decision Making

The value of data-driven decision making has been recognized for decades. When Frederick Taylor published *The Principles of Scientific Management* in 1911, he advised that all management and process improvement activities be grounded in measurable data. This approach led to more rigorous processes for quality control and quality assurance shortly thereafter, and then the birth of operations research during World War II—a discipline devoted to analytically examining work processes and making better decisions about how to design and exploit them.

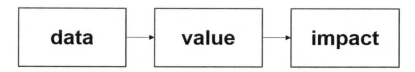

By the 1980s, many managers recognized that without knowing the full context within which a decision would be made, the data was less valuable. In addition, they realized that it was difficult to make adjustments in an organization's processes without complete engagement and buy-in, which would ensure the "stickiness" of the improvements. Total quality management (TQM), in particular, placed the responsibility for achieving quality goals with *everyone* in the organization.

In the 1990s, data-driven decision making was emphasized in many improvement-oriented business activities, including business process reengineering, change management, and knowledge management. Peter Senge's (1990) concept of the learning organization became popular during this decade, encouraging people and companies to use that data to capture best practices and evaluate opportunities for growth. During this decade and into the 2000s, lean manufacturing (for reducing waste) and Six Sigma (to reduce variation or defects), both data-driven approaches, also became more popular.

Because data science has deep roots in other fields, a definition is hard to pin down. Data science techniques are grounded in data mining and predictive analytics, well-established areas of study that focus on extracting knowledge from data. Model building has always been part of statistics, even as the data volumes and types have expanded.

> By "Data Science" we mean almost everything that has something to do with data: collecting, analyzing, modeling . . . yet the most important part is its applications—all sorts of applications. This journal is devoted to applications of statistical methods at large.
> —FROM THE JANUARY 2003 LAUNCH OF *DATA SCIENCE JOURNAL*

Perhaps the simplest and most elegant definition comes from Duarte and Dame (2019), who define data science as "the process of asking questions and getting answers from data." They also note that, using this definition, quality professionals *are* data scientists even though additional technology skills beyond those found in modern quality training would be required to meet the industry's requirements for data science practitioners.

Of course, asking questions and getting answers from data is precisely what statisticians have been doing for decades. How is data science any different? Some supporters of data science claim that since techniques are now available for inspecting and digesting all the data, there's no longer a need to take samples and infer characteristics of populations—one of the primary roles of statistical inference. But "sound statistical practices, such as ensuring high-quality data, incorporating sound domain (subject matter) knowledge, and developing an overall strategy or plan of attack for large modeling problems, are even more important for big data problems than small data problems" (Hoerl et al., 2014).

## Why Is Data Science So Popular?

If data-driven decision making has been around for over a hundred years, why is there so much buzz around data science today? The answer ties together many elements:

- **Success stories.** Early adopters have produced very promising results. Shell Oil, for example, reduced the latency of its inventory analysis from 48 hours to less than 45 minutes in a project that only took months, reducing inventory costs by millions of dollars each year (Boulton, 2018).
- **More data.** We are gathering and storing data at a more rapid pace than ever, due in large part to sensor data from the IoT and social media.
- **Connectivity and easier communications.** We have streamlined access to more data than we could have ever imagined 20 years ago, via the internet. No longer are sensors obscured by complex or specialized communications protocols.
- **Free, open source software, utilities, and services.** Many of these tools are available to help us analyze that data with sophisticated techniques and algorithms. We can use those libraries rather than programming the algorithms ourselves, enhancing quality and accelerating time-to-value.
- **A skilled workforce.** Workers equipped with interdisciplinary data science are now available to bridge the gaps between data management, data engineering, and model building.

- **Competitive pressures.** If you don't mine available data for insights and capitalize on them quickly, your competitors will.

The rising popularity of data science, and the hefty salaries that often accompany even entry-level positions, means that there is quite a bit of variation in résumés, skills, and experiences. Although there has been a renaissance in online data science training, which has done wonders for democratizing the field and reducing the barriers to entry, the extreme popularity of data science as a career means that caution is advised in hiring.

## What Does a Data Scientist Do?

Even among data scientists who are deeply trained, their backgrounds, skills, and experiences are diverse. I've worked with epidemiologists, psychologists, ecologists, physicists, astronomers, operations researchers, mathematicians, and cybersecurity specialists on data science projects, in addition to the programming-oriented statisticians that populate the field. Despite these differences in backgrounds and training, there are several characteristics that most data scientists share:

- They solve problems from the perspective of data-analytic thinking, constantly asking *How do I know that I know this?* This is similar to the data-driven decision making of Taylor's scientific management and Six Sigma.
- They are skilled at collecting, selecting, and acquiring appropriate data, integrating it with data in other formats or data that has been collected in different ways, and cleaning it prior to analysis.
- They are able to apply many different techniques and models to a problem and carefully select which approach (or combination of approaches) will yield the most accurate insights based on experimentation, optimization, and ground truthing.
- They use one or more programming languages and utilities—for example, R, Python, Hadoop, MapReduce, Spark, Pandas, NumPy, SciPy, and NoSQL databases.
- They communicate their results in compelling ways, often using visualizations, but always with a keen sensitivity to the business they are working with, and the environment within which that business is embedded.
- They love exploring data, often more than they love the discipline or domain where they originally received their training.

By its nature, data science is interdisciplinary and requires working with programmers, engineers, business leaders, and subject matter experts:

- **Data management** is a strategic role that answers questions about how to collect, document, organize, manage, and preserve data throughout its entire lifetime. The goal of a data manager is to strategically ensure that the value of data is preserved over time, regardless of who is using that data. A data management plan describes how the data will be stored and made accessible throughout its lifetime, and can be useful for companies of all sizes to plan and strategize. The DataONE project, funded by the National Science Foundation, has pioneered the development of data management plans and offers online training materials to guide effective data management at the institutional level (Allard, 2012). Its materials are freely available.
- **Data engineering** focuses on system administration, data warehouse design and administration, storage, archival, and acquisition. A data engineer may also work on data modeling (that is, designing and implementing logical structures for data storage, or even mathematical modeling and analysis). The data engineering function handles infrastructure requirements like streaming, storage, and building production-class application programming interfaces (APIs).
- **Data science** focuses on generating insights by developing and analyzing models, and translating them to business value using visualization and storytelling. Data science requires the foundations and infrastructure that are typically provided by data engineers and data managers. It provides the basis for adaptive, intelligent models that deliver demonstrable business value.
- **Data analytics and business intelligence (BI)** roles typically focus on extraction, transformation, and loading of data, writing SQL queries to one or more databases (depending on the business question being asked), and building dashboards using tools like Tableau and Microsoft's PowerBI. Analytics may incorporate forecasts and projections, but BI specialists typically do not build or deploy machine learning models.

Because the boundaries between each of these domains can be fluid, it is not uncommon to find a data scientist who is also responsible for the engineering or management aspects, especially in small companies. To map this to the quality profession, Duarte and Dame (2019) describe four additional distinctions within the data scientist role: DS 1 (strategist), who engages subject matter experts and is responsible for the overall viability of the project; DS 2 (boundary spanner), who focuses less on theory and more on the mechanics of working with multiple team participants; DS 3 (applications ninja), who understands data science within the problem domain; and DS 4 (communicator and storyteller), who can explain the findings and prepare the final reports. One person may occupy more than one role. The relationships between these distinctions are shown in Figure 7.3.

In addition, companies sometimes advertise for data scientists when all they really want is an analyst or BI specialist. Knowing the distinctions can help you staff and manage projects more effectively.

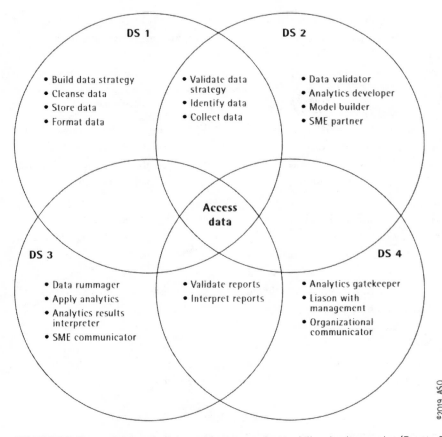

FIGURE 7.3. Types of data scientists working on a project to deliver business value (Duarte & Dame, 2019).

## DATA SCIENCE AND BIG DATA

It's no secret that people and organizations are producing and storing data faster than ever. As early as 2010, *The Economist* reported that, collectively, people and companies in the United States were generating 150 billion gigabytes of data annually. According to Kleiner Perkins Caufield & Byers, the rate of production had expanded to over 1000 billion gigabytes in 2015 (Schlein, 2015). A $125 billion industry had also emerged around big data to help companies respond to the changing technological environment.

### What Is the Data Deluge?

The exact magnitude of the collective data production isn't important, even though the progression from gigabytes to petabytes and beyond can feel overwhelming. The exciting (and potentially revolutionary) part of the story is that the focus is now shifting to automation: rapidly ingesting and combing through data to make better decisions. A competitive game of speed, smarts, and agility is brought on by this data deluge.

Some industries have been working with big data for years. Research instruments in science have been producing at the terabyte-per-day level since the early 2000s, for example, for studying pulsars in observational astronomy. As another example, the use of business analytics and BI has been steadily increasing since the 1990s, especially in finance and marketing. Larger companies were positioned early to take advantage of the emerging practice of data science, which has revealed some very tangible and practical results, including these discoveries reported by *The Economist* ("The data deluge," 2010):

- Stolen credit cards are more likely to be used for fraudulent purchases of hard liquor rather than beer
- Fraudulent insurance claims are more likely to be made on Mondays than Tuesdays
- Telecommunications companies regularly analyze subscribers' call patterns to identify up-selling opportunities and prevent churn

More recently, you've seen and experienced the fruits of data science every time you've purchased a recommended item from Amazon or watched a movie that Netflix suggested for you. You interact with recommendation systems (like those used by Amazon and Netflix), clustering systems (to determine shared interests among consumers or identify communities), classification systems (for customer segmentation), and the results of sentiment analysis (to determine how people collectively feel about products, services, and events)—usually expressed on social media on a daily basis—whether or not you realize it. "Growth hacking" to get social media followers is a thing. The innovation is not over. In fact, we are experiencing the Wild West of data science right now, as even more sophisticated mathematical techniques that will benefit businesses and organizations are in the research stage.

## Do All Data Scientists Work with Big Data?

No, not all data scientists are big data specialists. Most, however, have had education or training in handling large, distributed datasets, or alternatively, can work effectively with data engineers to solve problems using big data.

My favorite definition of big data was also the most surprising to me. In the spring of 2015, I was at a meeting at the National Science Foundation, surrounded by industry and academic leaders in high-performance computing and large-scale data processing. Sitting around the conference table eating lunch, I asked them, "Since I'm surrounded by so many experts, I want to know . . . what do you think the best definition is for 'big data'?"

First, they all laughed. They know as well as everyone else that there's lots of confusion in industry around what constitutes big data. After some discussion, one of the heads of a major supercomputing center spoke up, and soon all heads were nodding, "Big data is any volume of data that you don't have the tools or the expertise to explore right now."

That means there's no terabyte or petabyte threshold for what constitutes big data—it's relative to what *you* are accustomed to working with and what you are able to handle given current resources. Are you working with big data when your data warehouse is bigger than 100 terabytes? Maybe. Are you working with big data when you're generating more than a terabyte a day? Maybe. Are you working with big data whenever your machine can't fit the dataset into memory? Maybe. Whenever you're at the limits of what you can appreciably handle with your current computing resources, you're using big data.

Despite this relativity, it's common to think of big data as any amount of data that's too big to store in one place. There's so much of it that it can't fit on one machine or in one file. In fact, sometimes a single file is so large that it can't fit on one machine. Sometimes, one file or data collection has to reside on many machines—and in the internet age, thanks to cloud computing, those machines might be in vastly different geographical areas. Many machine learning algorithms can run on small datasets, but typically the larger the volume of data that can be processed (and the quicker the models can be updated based on newly arriving data), the more accurate the predictions or classifications will be.

Even though not all data scientists work with big data, every data scientist should have the skills to scale up—to be able to work with larger data volumes and more distributed data sources as the needs of the business expand.

## SOLID STATISTICAL FOUNDATIONS ARE ESSENTIAL

Interest and enthusiasm in data science can be providing a boost for budget allocations right now, but when building a team of data scientists, traditional statisticians can be the key to success. Without solid foundations in statistics, conclusions drawn from any volume of data—but particularly big data—can be inaccurate, misleading, or plain wrong. This can have potentially disastrous consequences.

### Survivorship Bias

Take, for example, the decidedly "small data" case of Abraham Wald, a member of the Statistical Research Group at Columbia University. During World War II, he was tasked with figuring out where to fortify B-29 bombers to prevent losses and deaths. Increasing the armor on airplanes, at the time, was labor intensive, expensive, and changed the aerodynamics of the planes—so, a somewhat risky proposition (Mangel & Samaniego, 1984). The Naval Research Group examined several planes that had returned home after battle, and discovered that the bullet holes were focused on the end of the tail, the tips of the wings, and the fuselage, as shown in Figure 7.4.

Wald, however, drew a different conclusion. He realized that the dataset of planes that had returned were the *survivors*—there were many more aircraft that were also shot at but never made it back in one piece. As a result, the areas that required fortification were the engines and other parts that had *no* bullet holes. Ellenberg (2015) shares another example of survivorship bias: "If you go to the recovery room at the hospital, you'll see a lot more people with bullet holes in their legs than people with bullet holes in their chests. But that's not

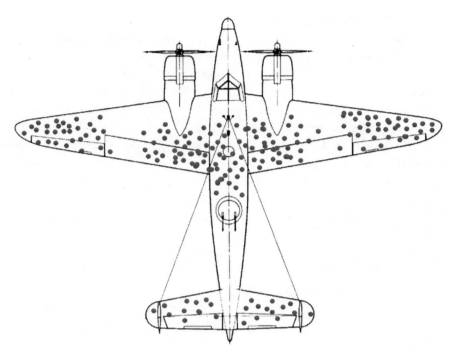

FIGURE 7.4. Bullet pattern on damaged planes that returned to base (McGeddon, 2016).

because people don't get shot in the chest; it's because the people who get shot in the chest don't recover."

Although a data scientist may be able to process hundreds of thousands of bullet hole locations over many planes or people, only one with statistical foundations will understand the assumptions that point to the real answer. Sometimes this is not an issue, and other times it's the difference between life and death. It can be difficult to make the distinction in advance.

## The Importance of Sampling

Here's another anecdote from a story I recently overheard. First, some background. It's easy to get attached to a particular programming language, especially if you spend more hours a day with it than you do with your spouse or significant other. And since the genesis of computer programming, there have been "language wars"—discussions and debates over the pros and cons of favorite languages, many of which are imbued with religious fervor.

A highly skilled data scientist with decades of experience but limited statistical background was attempting to make the following argument to his colleagues about why one language should clearly be chosen over another for their project. "I organized a conference where 80% of the accepted speakers were using Python, and almost none of them were using R," he said. "So Python is clearly the predominant language people are using for machine learning." A statistician might look a little more deeply into this assertion and discover that the announcements for the conference were sent to mailing lists and social media groups populated predominantly by Python users. With a conference committee of Python users evaluating submissions from mostly Python users, it's not surprising at all that Python appeared to be the "winner."

Which programming language is the *real* winner? Unfortunately, that's not a problem data science can solve. As with most problems quality professionals encounter, many factors have to be considered when choosing the right tool for the job, like the organization's strategy and goals, the task to be performed, and the skills and needs of the workforce.

## Lack of Theory Can Be Expensive

With companies feeling the pressure from the data science gold rush, many are pouring large amounts of money into launching data science projects or even creating departments dedicated to data science and machine learning. Although it's wonderful to see so many organizations interested in taking intelligent risks, sometimes, they're just risks. This $50 million horror story, similar to stories playing out right now in other companies, was posted on Reddit in April 2019 by user AlexSnakeKing:

At Company A, Team X does advanced analytics using on-prem ERP tools and older programming languages. Their tools work very well and are designed based on very deep business and domain expertise. Team Y is a new and ambitious Data Science team that thinks they can replace Team X's tools with a bunch of R scripts and a custom built ML platform. Their models are simplistic, but more "fashionable" compared to the econometric models used by Team X, and team Y benefits from the ML/DS (machine learning/data science) moniker so leadership is allowing Team Y to start a large scale overhaul of the analytics platform in question. Team Y doesn't have the experience for such a larger scale transformation, and is refusing to collaborate with team X.

The post goes on to say that Team X has deep domain experience and decades of experience with data, but because "they got most of their chops using proprietary enterprise tools [prior to 2010] instead of the open source tools popular nowadays," they don't completely identify with the "data scientist" label. Young, enthusiastic Team Y, which reported directly to the CEO and had no requirement to work with (or even communicate with) Team X, pitched development of a Naive Bayes classifier to their director, which was immediately approved. When Team X found out about this new project indirectly, they were initially enthusiastic, recognizing that their statistical and econometric skills (and domain experience with the models that the classifier was intended to ultimately replace) could contribute to the overall success of the project. But Team Y, according to the story, refused to collaborate and moved forward as if Team X didn't exist:

> Team X got resentful: . . . Team Y's model . . . had little chance of scaling or being sustainable in production, and they knew exactly how to help with that. Deploying the model to production would have taken them a few days. . . . And despite how old school their own tech was, Team X were crafty enough to be able to plug it in to their existing architecture. Moreover, the output of the model was such that it didn't take into account how the business will consume it or how it was going to be fed to downstream systems, and the product owners could have gone a long way in making the model more amenable to adoption by the business stakeholders. But Team Y wouldn't listen, and their leads brushed off any attempts at communication, let alone collaboration. The vibe that Team Y was giving off was "We are the cutting edge ML team, you guys are the legacy server grunts. We don't need your opinion" and they seemed to have a complete disregard for domain knowledge, or worse, they thought that all that domain knowledge consisted of was being able to grasp the definitions of a few business metrics.
>
> Team X got frustrated and tried to express their concerns to leadership . . . [but] it was impossible for them to get their voices heard. . . . Unbeknownst to Team X, the original Bayesian risk analysis project has now grown into a multimillion-dollar major overhaul initiative, which included the eventual replacement of all of the tools and functions supported by team X along with the necessary migration to the cloud. The CIO and a couple of business VPs are [now on] board, and tech leadership is treating it as a done deal.

A refusal to collaborate is sometimes (but not always) nefarious. Oftentimes, people are just busy or following the orders from their direct supervisor. Other times, they truly believe that there is no useful information or guidance to be mined from older projects. Whatever the reason, it is the job of leadership to make sure that new initiatives like this are fully informed by institutional knowledge—even if it requires additional time or expense up front. When a paradigm shift is in progress, giving team members sufficient time to progress through the forming, storming, and norming stages of team building can significantly mitigate business risks downstream.

Although the story is likely being told by a disgruntled member of Team X, many lessons can be taken from this example:

- Data science is a strategic business function, not an IT function
- A disregard for domain knowledge can be severely detrimental to success
- Data science teams should not operate independently or autonomously
- Replacing legacy teams can introduce risks to new projects, while negatively impacting employee satisfaction (and ultimately retention) for those who feel as if their voices cannot be heard

This story also provides us with direct advice about hiring data scientists:

[Data science] interviews nowadays all seem to be: Can you tell me what a p-value is? What is elastic net regression? Show me how to fit a model in [the Python package] sklearn? How do you impute NAs in an R dataframe? Any smart person can look those up on Stackoverflow or Cross-Validated [help sites for programmers]. . . . Instead teams should be asking stuff like: why does portfolio optimization use QP (quadratic programming) not LP (linear programming)? How does a forecast influence a customer service level? When should a recommendation engine be content based and when should it use collaborative filtering? (AlexSnakeKing, 2019)

Instead of focusing on the mechanics of data manipulation and cleaning, or the syntax of building models in a particular programming language, find out if candidates understand *when* and *why* models are used. This will reduce the risk of applying improper models to good data, as well as attempting to use proper models on bad data. Alternatively, make sure that data science practitioners are paired with mentors or leaders with strong domain and theoretical knowledge.

## STATISTICAL ENGINEERING FACILITATES DATA SCIENCE IMPACT

Incorporating domain expertise, statistical thinking, and business considerations to the process of building and deploying models seems to be the solution indicated to remedy these kinds of failures. But in fact, those needs were anticipated by statisticians Roger Hoerl and Ron Snee as early as 2010, when they began describing *statistical engineering*:

> The statistical engineering discipline would be the study of how to use the principles and techniques of statistical science for the benefit of humankind. From an operational perspective, statistical engineering is defined as the study of how to best use statistical concepts, methods and tools, and integrate them with IT and other relevant sciences to generate improved results.
>
> In other words, engineers—statistical or otherwise—do not focus on advancement of the fundamental laws of science, but rather how these laws might be best used for practical benefit. This is not to say engineers do not research or develop theory. Rather, it suggests engineers' theoretical developments tend to be oriented toward the question of how to best use known science to benefit society. (Hoerl & Snee, 2010)

Although the envisioned role of statistical engineering was to ensure that multiple statistical tools would be used appropriately in the strategic context of a problem, this approach can also inform data science. Hoerl and Snee (2017) describe the online competitions at kaggle.com that thousands of data scientists participate in, and ask, "Is the real problem to develop an optimal model . . . or to develop subject matter knowledge of the phenomenon of interest, or to guide intervention in the system to achieve enhanced results in the future?"

Anyone who has been a project manager or senior leader will know the answer: the model doesn't matter if it doesn't help you advance your goals or your business. As a result, the goal of statistical engineering is to build a body of knowledge and best practices that will guide teams toward practical, actionable results:

> Applying the principles of statistical thinking, particularly taking the time to understand the pedigree of data, and utilizing sequential strategies for addressing large, complex unstructured problems, such as those based on the discipline of statistical engineering, will bring the realities of big data analytics much closer to the hype. (Hoerl et al., 2014)

Without a strong foundation and strategic orientation, data science can only deliver the value that organizations dream of accidentally. A conscious approach to building data science teams and approaching large, unstructured, and often complex problems with statistical thinking as the foundation will make this deliberate.

## THE BOTTOM LINE

- Data science is concerned with building, validating, and continuously improving models that adapt to (and learn from) new data.

  o Those models often use algorithms, and the data can be stationary or in motion (streaming) in real time from sensors and other devices.
  o Data science can help us identify new data sources, discover the most important predictors, uncover patterns, detect anomalies, and build models to describe the behavior of systems.

- In a large organization, data scientists work with data managers (stewards of data over its entire life cycle), data engineers (who access, format, and clean it), and analytics or BI specialists. These specializations have arisen because the tasks have become complex.

- Although there is a misconception that data scientists don't need statistics because they have enough data to observe patterns directly, this is wrong. The scientific method, informed by a solid foundation in statistics, can help prevent data science disasters.

- Whatever roles are represented on your data science team, make sure all the bases are covered by using the PDCA cycle:

  o Strategy and a connection to business goals and drivers
  o A solid statistics foundation
  o Building and evaluating models that learn
  o Deploying those models
  o Monitoring performance and taking action to maintain the gains

- (You'll notice that this follows the PDCA cycle exactly.)

# REFERENCES

AlexSnakeKing. (2019, May). "When ML and Data Science are the death of a good company: A cautionary tale." https://www.reddit.com/r/MachineLearning/comments/beoxx8/discussion_when_ml_and_data_science_are_the_death/

Allard, S. (2012). "DataONE: Facilitating eScience through collaboration." *Journal of eScience Librarianship* 1 (1): 3.

Anderson, C. (2008). "The end of theory: The data deluge makes the scientific method obsolete." *Wired* 16 (7): 16–07. https://www.wired.com/2008/06/pb-theory/

Atwater, B., M. R. Satoko, S. Kenji, T. Yoshinobu, U. Kazue, and D. K. Yamaguchi. (2005). *The orphan tsunami of 1700: Japanese clues to a parent earthquake in North America* (Professional Paper 1707). U.S. Geological Survey. http://pubs.usgs.gov/pp/pp1707/pp1707.pdf

Boulton, C. (2018, November 14). "10 data analytics success stories: An inside look." *CIO Magazine.* https://www.cio.com/article/3221621/6-data-analytics-success-stories-an-inside-look.html

Bryant, E. (2014). "Great earthquake-generated events." In *Tsunami*, 103–129. Springer, Cham.

"The data deluge." (2010, February 25). *The Economist.* http://www.economist.com/node/15579717

Duarte, J., J. Dame. (2019). "Data science and the quality professional." *Software Quality Professional* 21 (3): 13–19.

Ellenberg, J. (2015). *How not to be wrong: The power of mathematical thinking.* Penguin.

Hoerl, R. W., and R. D. Snee. (2010). "Closing the gap." *Quality Progress* 43 (5): 52–53.

———. (2017). "Statistical engineering: An idea whose time has come?" *The American Statistician* 71 (3): 209–219.

Hoerl, R. W., R. D. Snee, and R. D. De Veaux. (2014). "Applying statistical thinking to 'big data' problems." *Wiley Interdisciplinary Reviews: Computational Statistics* 6 (4): 222–232.

Mangel, M., and F. J. Samaniego. (1984). "Abraham Wald's work on aircraft survivability." *Journal of the American Statistical Association* 79 (386): 259–267.

Mazzocchi, F. (2015). "Could big data be the end of theory in science? A few remarks on the epistemology of data-driven science." *EMBO Reports* 16 (10): 1250–1255.

McGeddon. (2016, November 11). *Survivorship bias.* Wikipedia. https://en.wikipedia.org/wiki/Survivorship_bias#/media/File:Survivorship-bias.png

Schlein, T. (2015, March 25). "A breakthrough approach to making data useful." Kleiner Perkins. http://www.kleinerperkins.com/perspectives/a-breakthrough-approach-to-making-data-useful/

Schulz, K. (2015, July 13). "The really big one: An earthquake will destroy a sizable portion of the coastal Northwest. The question is when." *The New Yorker.* https://www.newyorker.com/magazine/2015/07/20/the-really-big-one

*Seattle Times.* (2016, November 1). "Fake earthquake showed Pacific Northwest isn't ready for the big one." https://newsela.com/read/earthquake-prep-washington/id/23484

Senge, P. M. ([1990] 2006). *The fifth discipline: The art and practice of the learning organization.* Broadway Business.

## CHAPTER EIGHT

# DATA QUALITY AND DATA MANAGEMENT

In God we trust. All others must bring data.

—W. EDWARDS DEMING

O n April 26, 1986, my mother and I were away from home, visiting my aunt in Pennsylvania. The three of us, my new month-old cousin, and two of the in-laws sat in my aunt's living room, watching Peter Jennings on the nightly news. We were near the Poconos, and the air was crisp for that time of year. The fireplace was burning. It was already nearly dark.

In the stoic, direct manner of 1980s news, Jennings let us know that there had been a terrible accident on the other side of the world—at the Chernobyl nuclear plant. I'd just had a birthday the day before, and based on the dire projections on the screen, the radioactive cloud would soon be over my head.

Petrified, I interpreted the news as a death sentence: *I would surely not live to see my next birthday.* (I did; but many adults, children, and animals near the site of the incident would not.)

The event was clearly catastrophic, but the world would not comprehend its magnitude for months. Only the nuclear incident at Fukushima in Japan (after the 9.0 magnitude Tohoku earthquake in March 2011) has been comparable to Chernobyl in terms of death, damage, and global impact. In 2019, film director and screenwriter Craig Mazin captured the stories of the plant workers and residents of Pripyat, the town where the nuclear plant was located, in HBO's *Chernobyl* (Mazin & Renck, 2019). The entire five-part miniseries is fascinating, but what really caught my attention was one of the early scenes. Shortly after the meltdown, while the operators were still in the control room trying to figure out if there was a legitimate problem, one of them measured the radiation in the environment using a dosimeter. (Figure 8.1 shows a later version of a standard issue model that was available in Russia in the mid-1980s.)

When the operator turns the dosimeter on, it reads 3.6 roentgens per hour. "That's not great, but not terrible," remarks a supervisor. Minutes later, the operators find a more sensitive instrument, but as soon as they turn it on it shorts out. They consider retrieving the more sensitive dosimeter from a safe, but it's locked, and no one recalls where to find the key.

Shortly thereafter, government officials come to the facility. From a protected bunker somewhere underneath the site, they discuss the looming catastrophe and brainstorm a response plan. One official confronts the operator who took the measurements with the dosimeter. "You reported that the radiation situation was within normal limits." The operator responds yes and says that that's what the instrument showed.

At this point, nuclear chemist Valery Legasov looks visibly startled. "What value did you say the instrument reported?" he asks. The operator responds that it was 3.6 roentgens per hour—again a little high, but nothing to be terribly concerned about as long as safety protocols are followed.

*Roentgens* are the unit of measurement for how much radiation is being emitted in the environment—in German, the word itself means "x-ray." The roentgen is not a perfect measure for radiation compared with, for example, the *sievert*, which indicates the impact of radiation on the human body. But the number 3.6 is special, and when Legasov hears it, he realizes just how badly everyone has underestimated the magnitude of the disaster.

Because of his expertise and experience, Legasov knows exactly what is being measured. He also knows the context within which the measurement was taken. He is aware that the second dosimeter has shorted out, and he has observed a few cases of radiation sickness only hours after the meltdown occurred. Most critically, he knows something very important about that standard issue dosimeter.

The World Health Organization (WHO) has established limits on exactly how much radiation people can endure. For most of us, the WHO limit is half a roentgen per year—not per hour like the dosimeter measures, but per year. For nuclear engineers and operators, the limit is 5 roentgens per year. That initial dosimeter reading indicated that the radiation limit for nuclear power plant workers was exceeded in just an hour and a half. An accumulation of around 300 roentgens results in radiation sickness—and workers were already sick. By the time 500 roentgens accumulate, radiation exposure can be fatal.

FIGURE 8.1. A Russian dosimeter (Geiger Counter Virtual Museum, 2016).

The shocking insight Legasov had was that 3.6 roentgens per hour was the *maximum* value that could be read by that particular model of dosimeter. The instrument wasn't designed to go any higher. In reality, the ambient radiation produced near the reactor was between 800 and 1,500 roentgens per hour; and at the site of the meltdown, as much as 20,000 roentgens were emitted each hour. By underestimating the danger, officials failed to evacuate immediately, which led to many deaths and serious health issues.

The situation itself was deadly, but lack of understanding of the data—about what it really meant in the context of the immediate situation—cost many people their lives. The 3.6 roentgens per hour measurement was only one data point, but its impact was profound. Imagine what the situation might have been like if the operators had to monitor hundreds, thousands, or millions of monitor points.

There is no need to imagine though. Many of today's operations environments are being revolutionized by a multitude of connected sensors and actuators, collectively referred to as the industrial internet of things (IIoT). The Internet and Television Association (NCTA, 2019) estimates that in 2020 there will be 50.1 billion connected devices across consumer and industrial segments. Each of these devices will be producing a handful—or maybe hundreds—of observations every minute or hour. That's a lot of data to process and understand, especially when lives and assets are at risk.

## DRIVING VALUE FROM DATA ASSETS IN INDUSTRY 4.0

Since organizations started storing and managing data, even before the advent of electronic systems to capture and manipulate it, data has been a critical asset. While good data is necessary, bad data is expensive. As one example, Hazen et al. (2014) found that as a percentage of revenue, the cost of poor data ranged from 8% to 12% for "typical" organizations and up to 40% to 60% in service organizations.

Data is the raw material, while information is generated by critically evaluating data through the lens of meaning and purpose; knowledge examines that information in a particular context and environment and puts it into action. Some organizations have even created the new position of chief data officer to provide a dedicated focus on driving value from data, information, and knowledge.

In the Industry 4.0 era, the value of data as an asset is increasing tremendously due to several drivers. These include the following:

- The introduction of cyber-physical systems, many of which produce big data (some on a nearly continuous basis)
- Network infrastructure being more widely available, and emerging telecommunications infrastructure like 5G making it possible to transmit large data volumes and different data types, resulting in shareable data for real-time decision support

- Software libraries for advanced analytics being accessible, comprehensive, and reliable
- People, machines, and data being connected, in many cases in near real time

Cyber-physical systems, like the IoT and the IIoT, and other systems (e.g., social media), individually and collectively produce big data. Building on the 5V model from Demchenko et al. (2013) shown in Figure 8.2, big data can be described by several characteristics:

- There is lots of it (volume)
- It's coming at you fast, and may even be streaming in real time (velocity)
- It comes in different formats and sampling frequencies (variety)
- There may be huge variations in data quality, or it may fluctuate (veracity)
- There will be differences in how useful the data is (value)
- Different people or organizations produce it or own it (governance)
- It could easily change or disappear, impacting your operations (control)
- There may be restrictions on how you use it (policy)

Despite these challenges, organizations still have to respond and adapt. But how do you know if you're working with big data? You may recall the anecdote in an earlier chapter about big data being anything bigger or more complex than what your organization is currently prepared to handle. What it means is that all organizations, to some extent, are working with big data. For example, any organization that mines social media data to identify VoC is dealing with big data, even if it doesn't feel that big. And for organizations that rely on mostly manual processes, the competitive pressure of knowing that other organizations are trying to leverage big data to gain advantages may be enough to compel additional learning and investigation.

Effective data management and governance can provide strategic advantages, while lack of attention to issues regarding data can present strategic challenges. Poor data management leads to waste (when accurate, up-to-date information is difficult to find) and can slow or stall decision making. Organizations without effective data processes are less agile, spend more to deliver products and services, and are more likely to have customer satisfaction issues. Finally, the lack of high-quality information for decision making at the executive and senior leader levels can lead to costly missteps. The insights in this chapter can help you avoid some of them.

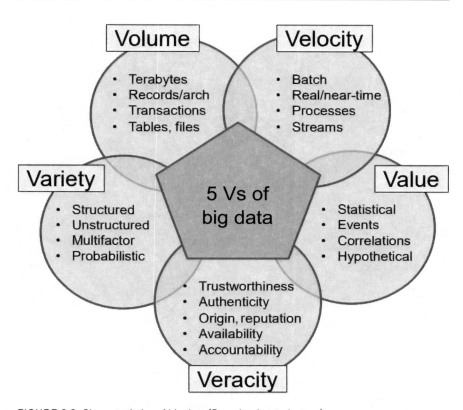

FIGURE 8.2. Characteristics of big data (Demchenko et al., 2013).

## ORGANIZING DATA MANAGEMENT AND GOVERNANCE

Particularly in the Industry 4.0 era, more data means additional care must be taken to ensure data quality, promote data management processes, and establish appropriate structures for data governance. Effective information and knowledge management practices can greatly improve cross-functional communication and overall performance.

The Data Management Association (DAMA International) provides resources and a Body of Knowledge (BoK) to organize these efforts. The framework consists of 11 data management knowledge areas, illustrated in Figure 8.3.

These 11 knowledge areas can be organized into three general categories: data quality, data management, and data governance. The tasks and responsibilities associated with each of them are as follows:

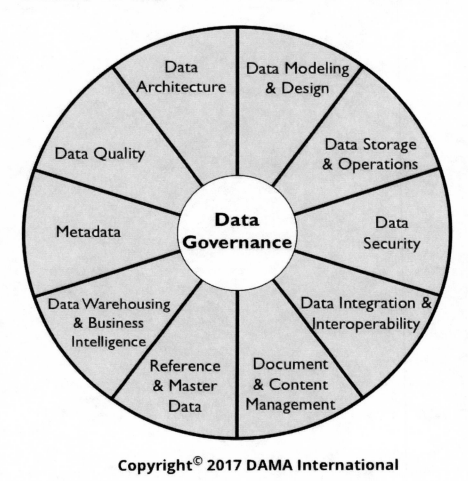

**Copyright© 2017 DAMA International**

FIGURE 8.3. The 11 data management knowledge areas, used to maximize value from data assets (Image licensed under CC BY-NC-ND; used with permission).

- **Data Quality**—defining quality and identifying important data quality dimensions for an organization; installing processes that maintain data quality over all phases of the data life cycle
- **Data Management**—overseeing practices to ensure that data meets requirements and specifications, is fit for use and purpose, and has all the characteristics needed to support enterprise decision making

o **Data Architecture**—maintaining the enterprise road map for how and where data is stored; conducting risk assessments and addressing risks; guaranteeing clarity and consistency of data strategy

o **Data Modeling and Design**—ensuring that the way the data is stored in each repository or platform is appropriate and managed; maintaining structure and architecture of data warehouse; logical modeling and conversion of logical models to physical storage models

o **Data Storage and Operations**—managing regular backups, emergency preparedness, and recovery processes; handling life-cycle maintenance, including purging old or unused data; managing types of data storage systems (e.g., RDBMS vs. NoSQL, cloud-based vs. on-premise, graph databases, streaming services, and IoT hubs)

o **Data Security**—managing privacy, confidentiality, and access controls; maintaining awareness of emerging security issues, including threats and vulnerabilities; making sure the workforce responds to these issues quickly, adequately, and responsibly

o **Data Integration and Interoperability**—providing infrastructure and processes for data acquisition, movement, transformation, and migration; managing ETL (extract, transform, and load) capabilities; managing for redundancy, provenance, and incorporating third-party data; maintaining compliance with government regulations that pertain to data

o **Document and Content Management**—implementing configuration management for physical and virtual documents; designing, updating, and using taxonomies to make information retrieval easier for the workforce; protecting confidentiality of documents and access to sensitive data within documents

o **Reference and Master Data**—managing and maintaining systems of record; maintaining reference data (e.g., topographical and land use models, zip codes, certain CAD/CAM models)

o **Data Warehousing and Business Intelligence**—managing and maintaining on-premise and cloud-based data platforms; providing services for querying, reporting, and data visualization; maintaining processes for updating the data warehouse

o **Metadata**—maintaining data definitions and dictionaries; managing information for data asset discovery; providing support for internationalization and multilingual environments

- **Data Governance**—executing oversight, policy development, and policy deployment for all data quality and data management issues; ensuring appropriate stewardship and ownership of all data; handling ethical issues and communication about data quality and data management; and overseeing data management maturity

While data management is often the responsibility of the chief information officer and IT staff, data governance is championed by the executive team and senior leaders, and data quality requires input and participation from across the organization. Figure 8.4 provides an alternative way to look at these tasks and responsibilities. While data governance, which includes establishing policies, identifying owners and stewards, and driving a quality-oriented data culture, sets the tone, managing the data life cycle depends on a solid foundation that includes security, metadata management, and data quality management.

## DATA QUALITY

*Data quality* is "fitness for use or purpose for a given context or specific task at hand" and must be quantified to be useful for management purposes (Mahanti, 2018). Ensuring data quality means protecting against negative attributes (e.g., incorrect, inconsistent, or unreliable) while promoting positive attributes (e.g., unambiguous, contextual, understandable).

The quality of every piece of data is important. To assess data quality, you need to know the following:

- The sensor, instrument, person, or machine that provided the data (who)
- The type of data that is being measured or assessed (what)
- The time that the data was collected (when)
- The location and context of the measurement or data collection (where)
- The reason why the measurement was taken or the data was produced (why)
- The process used to obtain the data, and the provenance of the data—that is, what steps were taken to clean it, transform it, produce it, or format it (how)

Data quality is even more essential in connected, intelligent, automated environments. For example, one of the key use cases for Industry 4.0 is predictive

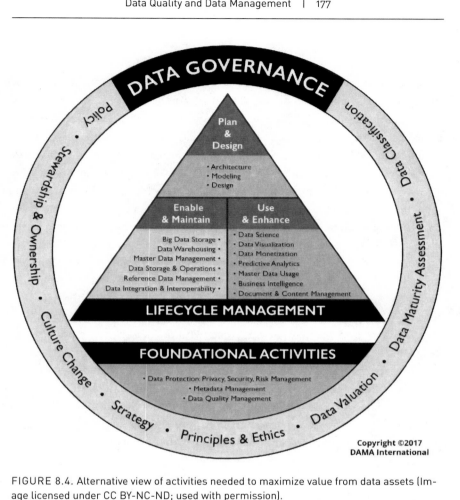

FIGURE 8.4. Alternative view of activities needed to maximize value from data assets (Image licensed under CC BY-NC-ND; used with permission).

maintenance (using advanced analytics to anticipate breakdowns and engaging in interventions to avoid downtime or damage to assets). Despite the projected benefits of predictive maintenance, including increasing the availability and reliability of machines and extending the useful life of equipment, making it happen in practice is not easy. Requirements for data quality and data access, and the need to preserve quality when data is aggregated from many sources, may ultimately impact prediction accuracy. Without data quality, predictive models may be useless (Li et al., 2016).

## Causes and Consequences of Bad Data Quality

Data quality can be impacted at any stage of its life cycle. Both humans and machines can generate or obtain data at its origin. Humans may do this by using physical senses (sight, smell, touch) or by providing data that they have access to (birthdate, address, mother's maiden name). Machines like sensors and instruments can measure things, while other machines may process those measurements, obtain other data from other sources, or aggregate or analyze information. If the human is an unreliable observer, the sensor is invalid, or the instrument is uncalibrated, this will also lead to issues with data quality. If the data is being collected by an IoT device, onboard processing or machine learning language may manipulate it before it is transmitted to a hub.

Once the data is created, it may be synthesized, processed, or combined with other data. Cleaning or transformation may be attempted to convert the data into useful features. At some point, the data may be stored in a database, repository, or other data platform. It may be archived for long-term access, and people or systems may share the data with one another. Data quality can be adversely impacted at any step along the way.

There are many reasons for data quality problems. Manual data entry, intrusions by hackers, inadequate validation, insufficient data cleaning, and issues with data migration processes can corrupt data even when it has been gathered properly. Lack of ownership, lack of understanding, and lack of processes for continuity when people leave a company can negatively impact data quality. Organizational issues like mergers and acquisitions can be damaging if data quality is not specifically managed during the transition process. Data can become less applicable or valuable over time (for example, any marketer who is trying to contact me at my address or phone number from the late 1990s will be disappointed). Finally, unintended data loss and incomplete disaster recovery can negatively impact an organization's data quality.

> Bad data, like cancer, can weaken and kill an organization.
> —R. MAHANTI

When data does not meet quality requirements, there can be consequences for individuals, the business, and even the environment or society. Decision making is always impacted by bad data quality, and this is compounded when the decision maker is unaware of it. Data quality problems can reduce trust in people, between people, and in the organization itself. Making decisions based on bad data can lead to reduced productivity and higher costs. Ultimately, bad data can also lead to compliance risk, including exposure to legal penalties and material loss if industry laws and regulations are not followed (Djali et al., 2010; Mahanti, 2018).

## Identifying and Measuring Data Quality Dimensions

What characteristics define data quality? First, an organization must examine its processes and priorities in terms of data quality dimensions. Not all data quality dimensions will be important to all organizations at all times. The Baldrige Excellence Framework (BEF), for example, calls out only accuracy, validity, integrity, reliability, and currency (NIST, 2019) even though other data quality dimensions may be critical for a particular organization. Mahanti (2018) provides 29 quality dimensions (Figure 8.5).

While some dimensions can be assessed only subjectively (accessibility, believability, interpretability, ease of manipulation), others can be evaluated objectively (accuracy, currency, volatility, precision). Data quality should be evaluated using objective measures wherever possible. For example, completeness can be measured based on the characteristics of the database (e.g., schema completeness), and accuracy can be assessed by comparing stored values with known references. Cai & Zhu (2015) provide additional indicators that can be used to guide data quality assessment, focused on evaluating the quality of big data.

Some of the dimensions are closely related to one another. For example, being able to trace the provenance of data can have an impact on its believability and trustworthiness. Data may be more interpretable if it is concise and has adequate coverage. Availability may be supported by redundancy, since if the data cannot be obtained from one source, it may be sourced from another.

Mahanti (2018) also provides recommendations for managing the assessment process:

- Master data and reference data should be given priority for data quality assessment, since problems with these values can impact hundreds or thousands of transactions.
- Owners and stewards of data should be asked to identify which data is critical for operations, decision making, or compliance.
- Industry-specific data quality issues should be examined. For example, the quality of data that describes assets is critical for water and wastewater management,

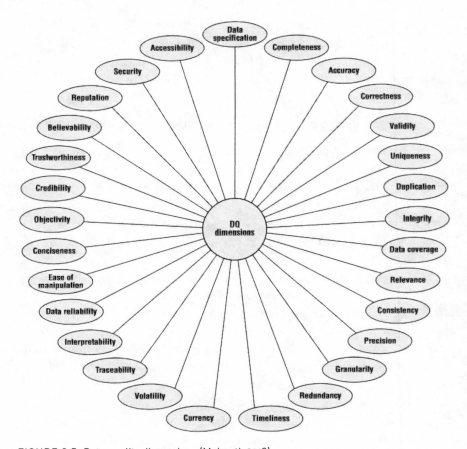

FIGURE 8.5. Data quality dimensions (Mahanti, 2018).

while the quality of reference data can significantly impact overall costs in the finance industry.

In addition to industry-specific data quality management issues, there may be additional guidance based on the specific technologies that you implement. Large amounts of data impose significant requirements on data quality.

## Data Quality in IoT Environments

Although a traditional treatment of data quality is important, it is also critical to look at how requirements for data quality management will change in con-

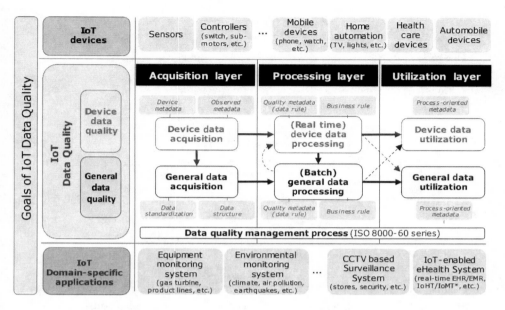

FIGURE 8.6. Data quality dimensions for IoT and smart, connected products (Perez-Castillo et al., 2018)

nected, intelligent, automated environments. Perez-Castillo et al. (2018) did just this, motivated by the threats that bad data quality could pose to the key value propositions for Industry 4.0:

> The vast amount of data in the IoT environments, gathered from a global-scale deployment of smart-things, is the basis for making intelligent decisions and providing better services. In other words, data represents the bridge that connects cyber and physical worlds. Despite its tremendous relevance, if data are of poor quality, decisions are likely to be unsound.

By examining case studies in the context of ISO/IEC 25012 and ISO/IEC 8000, these researchers developed a framework for data quality management in IoT environments (Figure 8.6). The purpose was to anticipate data quality impacts for smart, connected products. They recommend thinking about IoT data quality in terms of acquisition, processing, and utilization steps, and considering data quality issues on the device separately.

They also uncovered specific patterns in sensor data that manufacturers of smart, connected products should be aware of, as these could indicate data quality problems:

- **Constant or offset error:** Observations deviate from expected value by a constant
- **Continuous varying or drifting error:** Deviation between observations and expected value is continuously changing
- **Crash or jammed error:** Sensor stops providing any readings or gets stuck on an incorrect value
- **Trimming error:** Data is correct in some interval but incorrect outside the interval
- **Outliers:** Observations occasionally deviate from expected value
- **Noise error:** Observations randomly deviate from expected value

Finally, they proposed 23 best practices for data quality management in this environment. These include collecting sensor data sequentially, addressing security at all stages of the product life cycle, performing data quality checks on incoming sensor data, automating procedures for assessing sensor data quality, and creating policies to monitor and compare sensor data across the installed base for the product.

## DATA MANAGEMENT

Data management practices ensure that data meets requirements and specifications, is fit for use and purpose, and has all the characteristics needed to support enterprise decision making. DAMA International defines *data management* as the "development, execution, and supervision of plans, policies, programs, and practices that control, protect, deliver, and enhance the value of data and information assets" (Mosley, 2008). Consequently, data management is quality management for data.

Data management planning can have a limited scope, or a large scope that spans multiple sites and facilities, or even organizational boundaries. The DataONE project of the U.S. National Science Foundation (NSF), which helps scientists at national laboratories manage their data more effectively, recommends building a data management plan that describes the following:

- The data and its format(s)
- Metadata content and format
- Policies for access, sharing, and reuse
- Plans for storage and management over the life cycle of the data
- Budget for storage, providing access, and supporting management activities

Additional details are provided in Strasser et al. (2012). Although this describes a microcosm of the requirements that may be needed to support a large enterprise, the concepts are similar to what is needed for master data management and a supportive organizational context.

## Master Data Management

Effective decision making depends on being able to obtain data that meets your required data quality dimensions when you need it. In addition, successful business and process outcomes depend on building a culture of quality around data, which includes creating and following effective management processes. Master data management (MDM) can play this role.

One strategy for improving process quality through data is to identify master data (Figure 8.7) and decide which system of record will hold each type of master data. Master data is some of the most important data your organization has—and represents the key objects that your business is built on. In contrast with reference data, which may be the same for different organizations, master data is unique to your organization and can provide a source of competitive advantage. In general, it is nontransactional and defines the common objects and entities that are used by many business processes. Master data is not historical data, which represents old transactions. In industrial systems, transactional data is often captured by event logs, while historical data is stored on the historian or in data archives.

By strategically managing master data, you can establish a single source of truth for your most important information. According to Cleven and Wortmann (2010), master data typically includes authoritative information about the following:

- People or parties (customers, suppliers, employees)
- Things (products, services, assets)
- Locations (facilities, sites, offices)

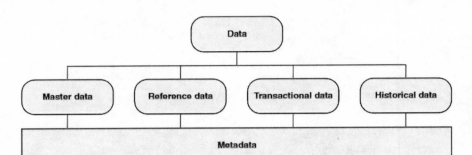

FIGURE 8.7. Relationship of master data to other data types (Mahanti, 2018).

Managing master data is similar to identifying the most trusted people in your company, being able to find them when you need them, and being able to rely on them to provide you with accurate, up-to-date information. For example, your environment, health, safety, and quality (EHSQ) software system may be your system of record containing master data for emissions, incidents, exposure data, and quality events, providing authoritative information about those processes. Master Data Management provides structure:

> Master Data Management (MDM) is the framework of processes and technologies aimed at creating and maintaining an authoritative, reliable, sustainable, accurate and secure data environment that represents a single version of truth, an accepted system of record used . . . across a diverse set of application systems, lines of business, and user communities. (Berson et al., 2011)

The absence of master data (or processes to find, use, and maintain its integrity) can result in wasted time, confused workers, faults and downtime in operations, and long recovery or repair times. Knowing where to find reliable, authoritative data will shorten the time required for anyone in your organization to make decisions, and ensure that their decisions are fact based.

## System of Record

A system of record (SoR) provides the authoritative source for each type of master data, and can be a data repository, software application, or both. This master data will be each party, thing, or location your organization identifies as important, and can be records or documents that contain data. Each of these important data types must be associated with the SoR where it will be stored and maintained. Because both people and digital systems will look to the SoRs for truth, maintaining the accuracy and timeliness of the information is important:

I met this week with the director of a large health department and her information technology (IT) manager. The team described the following challenges in how the department currently functions:

- **Multiple overlapping computer systems:** Through circumstances too frustrating to unravel, her staff were compelled to use multiple systems, each with some margin of value.
- **No system of record:** Lacking a single trustworthy system, any query required cross-checking and reconciliation with the next best source.
- **Abandoned environments:** At least one system [was not being maintained] and data were not being backed up at all! Yet, the system remained online because none of the auxiliary systems could perform.
- **Manual report reconciliation:** [Inspection reports] were so unbelievable that leadership asked the staff to return to their paper records and hand tally the true numbers.

Staff struggled to do their jobs. Managers scrambled to show progress. Leadership was . . . frustrated. Taking control of one's system implementation can make all the difference. (Booth, 2018, p. 30)

Once your organization has identified master records for its critical objects, management processes should be built to ensure that it meets all appropriate data quality dimensions. This may include cleaning data, committing new records, tracking changes to records, or auditing data management processes. How do you know if a particular type of data should be in an SoR? Hurst (2018) recommends that data be included in an SoR if several of the following characteristics are met:

- The data is proprietary
- The data supports one or more mission-critical business processes
- Many employees use or interact with it daily or weekly

- It is needed for important operational or strategic business decisions
- It captures knowledge that the business needs even if the employee leaves
- It is enhanced and improved over time

Linstedt (2006) says that an SoR has some defining characteristics itself:

1. It is a data origination point
2. It feeds other systems' authoritative data
3. It may be auditable, traceable, or cleaned, "but in all cases it provides business value in different formats and assists the business in doing business on a daily basis"

Identifying master data and the SoRs they live in can improve the efficiency of work processes and drive continuous improvement. Instead of wasting time hunting in multiple locations, people will immediately know where to go and what systems to trust. They will no longer be making decisions based on bad (or old) information.

## ISO/TS 8000

In 2011, ISO (2016) introduced a master data quality management framework, a multipart international standard for data quality and enterprise master data. The 100 series of this standard explains how to ensure quality of master data, while the 60 series outlines internal processes to gather, store, and provide quality data to stakeholders. The standard has been written with total quality in mind, and assumes that data quality is driven continuously at all levels of the organization (Benson, 2019). The utility of ISO 8000 is questioned by information scientists, who claim that "its value is limited due to its naive model of data and information" (Clarke, 2016). Regardless, it has been used by researchers to successfully navigate data quality management issues in Industry 4.0 (Cai & Zhu, 2015; Perez-Castillo et al., 2018).

## DATA GOVERNANCE

DAMA defines *governance* as "the exercise of authority, control, and shared decision making (planning, monitoring, and enforcement) over the management of data assets" (Khatri & Brown, 2010). Establishing a data governance framework is one of the most important aspects of enterprise data quality management. The primary purpose of data governance is to help the company maximize the value of its data assets for business purposes over the lifetime of the data. As a result, this effort should be driven by business leaders and from outside the corporate IT function.

Data governance provides the policies, guidelines, and approaches that the organization will use to drive value from data. IT organizations, as partners, implement them in the most efficient and effective ways. Effective data governance results in greater transparency and visibility of data, more efficient operations, and higher-quality decisions.

## Elements of a Data Governance Framework

According to Mahanti (2018), a data governance framework has six components:

- **Rules and rules of engagement:** a definition of what the program aims to achieve and the rationale for setting that goal; rules for all functions and departments to follow as they manage data; standards and guidance that will be supported (e.g., ISO 8000, ISO 27001)
- **Data policies:** documented guidelines, principles, and rules for managing data at all stages of the life cycle
- **Processes:** an articulation of decision-making processes and decision rights; how data will be defined and how data quality requirements will be established; how issues will be resolved
- **People:** defined roles and responsibilities with respect to the organization at the executive, strategic, tactical, operational, and support levels; information about how to organize and coordinate to achieve shared goals
- **Roles and responsibilities:** defined roles and responsibilities with respect to the data (e.g., data producer, consumer, owner, steward)
- **Technology:** selection of appropriate tools to support data quality and data management processes

For core principles, she recommends recognizing data as a strategic asset, establishing clearly defined accountability for data and documents, defining a statement that the rules and policies should be followed, and a call to consistently manage data according to defined policies over the entire life cycle of the data.

## Governance for Cyber-Physical Systems

Yebenes and Zorrilla (2019) extend the governance concept to specifically address the "third generation platforms" that will support Industry 4.0. These platforms provide intense computing power in any environment, the ability to ingest and process large volumes of data in real time without storing it, scaling compute power according to workload, the ability to interact with intelligent systems and intelligent environments, and AI and machine learning services.

They articulated five key processes, as shown in Figure 8.8:

- **Planning:** selecting appropriate principles, rules, and standards for the organization to follow (e.g., ISO 11179 for metadata management, ISO 8000 and ISO 25012 for data quality, and ISO 27001 for security)
- **Organization:** choosing an executive sponsor, business owners for data who are not in the IT organization, and a cross-functional Data Governance Council made up of executives from all departments
- **Operation:** enacting processes staffed by people "capable of evolving with the new technologies and dealing with the changes in the new way of operating with data and developing the automation of industrial processes" (Yebenes & Zorrilla, 2019)

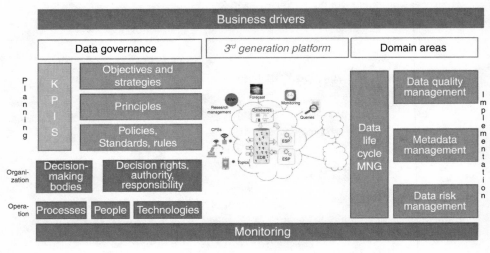

FIGURE 8.8. A data governance framework for "third generation" Industry 4.0 platforms (Yebenes & Zorrilla, 2019).

- **Implementation:** focusing on data security and risk management, data and metadata quality, and data life-cycle management, to better deal with unstructured, low-latency data
- **Monitoring:** establishing a shared decision-making process, based on mutually determined KPIs, that incorporates all stakeholders on a data platform (e.g., cloud service providers)

They note that the operating environment of Industry 4.0 and digital transformation initiatives "makes [it] more difficult to understand the context, importance, and associations of data. Moreover, applications developed on [third-generation platforms] require hiring of processing and storage services in the cloud to satisfy availability, latency and throughput requirements (among others) and all of this under demanding security and regulatory conditions" (Yebenes & Zorrilla, 2019). For this, better (and continuous) communication across cross-functional boundaries is suggested. Data governance in Industry 4.0 is a team sport.

## THE BOTTOM LINE

Data quality, management, and governance activities provide the foundation for quality in every other business process, from operations to strategy execution. Although the focus in Industry 4.0 is on ensuring the quality of big data, "small data" remains just as important. For example, Six Sigma projects will not generate reliable improvements unless they are using valid measurements; similarly, value stream maps will be less beneficial if employees are not using the same data sources to describe the value-added steps in their processes.

In addition to separating data acquisition and data analytics from the main systems where data and security are managed, there are three conceptual elements of an organization's data quality management program:

- **Data quality:** Establish critical data quality dimensions for all phases of the data life cycle, focusing on the data with the highest impact or financial value
- **Data management:** Establish master data and SoRs, and use standards like ISO 8000, ISO/IEC 25012, and ISO 27001 as necessary
- **Data governance:** Define the rules, policies, processes, and roles to guide the effort

Together, these processes reduce costs and risks, eliminate waste, and empower decision making, while proactively helping an organization address security and regulatory requirements. Information is useful only when it is generated from high-quality data and you are connected to it. Intelligence helps you understand and respond to current conditions, especially when there's lots of data to sift through. Automation can bring data and information to people and machines when it is needed, freeing up time and effort and shortening feedback loops.

Industry 4.0 and digital transformation present unique use cases that make data quality even more critical than before. For example, without data quality, predictive models may be useless. But once solid foundations for data quality management are put in place, software systems can help organizations more effectively collect, retrieve, and protect their data. The next chapter describes those logical containers that can be used to manage data.

# REFERENCES

Benson, P. (2019). *The guide to ISO 8000*. ECCMA. https://eccma.org/private/download _library.php?mm_id=22

Berson, A., L. Dubov, B. K. Plagman, P. Raskas. (2011). *Master data management and data governance*. McGraw-Hill.

Booth, D. (2018). "Building capacity by hacking your system implementation." *Journal of Environmental Health* 80 (8): 30.

Cai, L., and Y. Zhu. (2015). "The challenges of data quality and data quality assessment in the big data era." *Data Science Journal*, 14.

Clarke, R. (2016). "Big data, big risks." *Information Systems Journal* 26 (1): 77–90.

Cleven, A., and F. Wortmann. (2010, January). "Uncovering four strategies to approach master data management." In *2010 43rd Hawaii International Conference on System Sciences*, 1–10. IEEE.

Demchenko, Y., P. Grosso, C. de Laat, and P. Membrey. (2013). "Addressing big data issues in scientific data infrastructure." In *International Conference on Collaboration Technologies and Systems (CTS)*: 48–55.

Djali, S., S. Janssens, S. Van Yper, and J. Van Parijs. (2010). "How a data-driven quality management system can manage compliance risk in clinical trials." *Drug Information Journal* 44 (4): 359–373.

Geiger Counter Virtual Museum. (2016, April). "USSR Russian Bella Geiger counter—Radiation detector—dosimeter." Available from https://www.youtube.com /watch?v=3CrCTuC69kI

Hazen, B. T., C. A. Boone, J. D. Ezell, and L. A. Jones-Farmer. (2014). "Data quality for data science, predictive analytics, and big data in supply chain management: An introduction to the problem and suggestions for research and applications." *International Journal of Production Economics* 154: 72–80.

Hurst, H. (2018, November 29). "5 systems of record every modern enterprise needs." *Workfront.* https://www.workfront.com/blog/systems-of-record

International Organization for Standardization. (2016). *ISO 8000-120:2016 data quality—Part 120: Master data: Exchange of characteristic data: Provenance.* https://www.iso.org/standard/62393.html

The Internet and Television Association. (2019). *Positions.* https://www.ncta.com/positions/internet-of-things

Khatri, V., and C. V. Brown. (2010). "Designing data governance." *Communications of the ACM* 53 (1): 148–152.

Li, Z., K. Wang, and Y. He. (2016). "Industry 4.0 potentials for predictive maintenance. Advances in Economics, Business and Management Research." *International Workshop of Advanced Manufacturing and Automation* (IWAMA 2016), 42–46.

Linstedt, D. E. (2006, April). "Demystifying SoR (system of record) and MDM." http://www.b-eye-network.com/blogs/linstedt/archives/2006/04/demystifying_so.php

Mahanti, R. (2018). *Data quality: Dimensions, measurement, strategy, management, and governance.* Milwaukee, WI: Quality Press.

Mazin, C. (Writer), and J. Renck (Director). (2019). *Chernobyl.* [Television series episode]. Home Box Office (US); Sky Atlantic (UK).

Mosley, Mark. (2008). *The DAMA dictionary of data management.* Technics Publications, LLC.

National Institute of Standards and Technology. (2019). *Baldrige Excellence Framework (Business/Nonprofit): Proven leadership and management practices for high performance.* https://www.nist.gov/baldrige/publications/baldrige-excellence-framework/businessnonprofit

Perez-Castillo, Ricardo, Ana G. Carretero, Moises Rodriguez, Ismael Caballero, Mario Piattini, Alejandro Mate, Sunho Kim, and Dongwoo Lee. (2018, September). "Data quality best practices in IoT environments." In *2018 11th International Conference on the Quality of Information and Communications Technology* (QUATIC), 272–275. IEEE.

Strasser, C., R. Cook, W. Michener, and A. Budden. (2012). *Primer on data management: What you always wanted to know.* DataONE. https://escholarship.org/content/qt7tf5q7n3/qt7tf5q7n3.pdf

Yebenes, J., and M. Zorrilla. (2019). "Towards a data governance framework for third generation platforms." *Procedia Computer Science* 151: 614–621.

## CHAPTER NINE

# SOFTWARE APPLICATIONS
# AND DATA PLATFORMS

> What is it about the container that is so important? Surely not the
> thing itself . . . the value of this utilitarian object lies not in what
> it is, but in how it is used. The container is at the core of a highly
> automated system for moving goods from anywhere, to anywhere,
> with a minimum of cost and complication on the way.
>
> —MARC LEVINSON IN *THE BOX: HOW THE SHIPPING CONTAINER MADE
> THE WORLD SMALLER AND THE WORLD ECONOMY BIGGER*

Until the 1950s, the transportation industry had busy ports and bustling docks as its heart, employing millions of people worldwide and keeping the nascent global economy humming. Levinson (2016) paints the chaotic picture: warehouses and manufacturing plants operating close by, sweaty workers carrying loads on their backs and up gangplanks, cargo packed into sacks and bales and wooden boxes. There were no standards for packaging or organizing, so he describes a daily scene in which safety risks were everywhere:

> The dock would be covered with a jumble of paperboard cartons and wooden
> crates and casks. There might be steel drums of cleaning compound and
> beef tallow alongside 440-pound bales of cotton and animal skins. Borax in
> sacks so heavy it took two men to lift them, loose pieces of lumber, baskets
> of freshly picked oranges, barrels of olives, and coils of steel wire might all
> be part of the same load of "mixed cargo," waiting on the dock amid a tangle
> of ropes and cables, as lift trucks and hand carts darted back and forth.

Loading and unloading required mostly manual labor, aided by hooks and winches, and was literally backbreaking. Although the work was treacherous, demand was highly variable because it depended on the arrival times of the ships and the type of cargo—perishable goods had to be unloaded much more quickly, requiring more workers. Competition to receive an hourly wage was intense, and corruption was the norm, so getting paid was never guaranteed. Gangs took over some of the docks, and violent strikes erupted at others, eventually leading to labor laws and an imposed system for full-time employment. Still, the conditions created a tight-knit culture, group identity, and commitment to the "global fraternity" of longshoremen.

By the 1950s, these freight terminals had been acknowledged as bottlenecks, and the first cargo containers were proposed as a way to simplify packing, carrying, and unpacking. Although safety improved, there were drawbacks. The lack of weight limits made lifting more difficult, unused container space was a significant economic penalty, and it was much more expensive to ship in containers—up to 75% more. Many times, there was also an additional fee to ship the container itself back to the point of origin. Because everyone in the industry wanted to reduce the total cost of shipping, the hunt for a solution continued. And as Levinson explains, "The solution came from an outsider who had no experience with ships."

In 1934, twenty-one-year-old Malcolm McLean started driving trucks for a local transport company. Just one year later, he owned three of his own trucks and employed nine other drivers. Skilled at managing cash flow, negotiating, and finding creative ways to build business while cutting costs, McLean obtained corporate discounts from gas stations along his routes. His company was the first to invest in training programs for new drivers, and he recognized that employee involvement was the key to safety and quality: he empowered his drivers to make choices driven by safety, and awarded bonuses for serving as mentors.

In 1953, McLean noticed that the roads were becoming congested, and started looking for lower-cost ways to move his cargo. Since domestic shipping had stalled since the 1930s, he decided to try driving truck trailers onto ships that would move them up and down the U.S. East Coast. Because of drastically different regulations between land and sea transport at the time, it was a move that could make him the clear price leader in his industry—even though it required completely restructuring his company to circumvent regulations that governed ownership of a shipping company by a trucking line. The Port Authority of New York was so enthusiastic about his innovative idea that it funded his trucking port with bonds that were issued directly.

By 1955, bothered by the loss of cargo space due to the truck's wheels and engine, McLean envisioned an even more revolutionary idea. He would buy old tankers from World War II and customize them to transport only the trailer part of the truck, stacked optimally to fill the space and make loading and unloading smoother. Using a beer company as the guinea pig, McLean prototyped the system. He demonstrated that the full value stream around moving twenty tons of beer from Newark to Miami—with all costs included—was 94% less expensive than the traditional approach. This garnered plenty of interest, so he partnered with engineer Keith Tantlinger to build out the full system and demonstrate to regulators that the fully loaded ship would be seaworthy.

The real win came in 1956, when he demonstrated to the market that the cost of transporting loose cargo the old way, at $5.83 per ton, could be reduced to 15.8 *cents* per ton using standard shipping containers:

> McLean's fundamental insight, commonplace today but quite radical in the 1950s, was that the shipping industry's business was moving cargo, not sailing ships. That insight led to a concept of containerization quite different from anything that had come before. McLean understood that reducing the cost of shipping goods required not just a metal box but an entire new way of handling freight. Every part of the system—ports, ships, cranes, storage facilities, trucks, trains, and the operations of the shippers themselves—would have to change. (Levinson, 2016)

The shipping container revolutionized the transportation industry because it reoriented everyone to realize that the value came from *moving cargo*. An analogous revolution has been happening in computing, with the mass shift to cloud-based offerings using Software as a Service (SaaS), Infrastructure as a Service (IaaS), and Platform as a Service (PaaS) models (Novkovic & Korkut, 2017). Now, organizations can think of software in terms of its ability to *move information* and support business planning and operations processes, rather than as applications they need to buy, customize, and own. Moving data and information most effectively requires looking at every part of the system in a new way—as software in support of connectedness, intelligence, and automation.

Software applications (like CRM, ERP, and EHS systems) and data platforms (like IoT hubs) provide containers for various master data collections and the information generated by business or operations functions. The notion of software systems as containers has been used for decades as a way to understand, design, test, and maintain complex systems (Rational, 1987). Logical containers make it easier to manage data and information flows than if the data were not logically separated in some way. Nearly all organizations use software and data platforms to streamline data generation, collection, organization, and manipulation—to get it to the people who need it, when they need it.

This chapter uses the ISA-95 model to describe containers that hold data. This includes the software applications (many of which are now available via cloud computing) and data platforms that support typical organizations. Although much of the analysis in this chapter comes from manufacturing environments, many of the concepts can be applied across industries.

## CONVERGENCE

Breakthroughs like this occur when multiple, disparate ideas converge toward more similar or more mutually supportive ideas. *Convergence* means coming together—for example, replacing many different technologies with a single technology, or moving to a shared platform or infrastructure.

In the shipping industry, the value stream was enhanced when a single, standard shipping container became the platform to support multiple modes of transport. In business and industry, the value stream has been enhanced by several forms of convergence over the past two decades. Macaulay and Singer (2016) provide the following examples:

- **Internet Protocol (IP) Convergence**—Field devices and field controllers all supported Transmission Control Protocol (TCP/IP) by the early 2000s, in contrast with 20+ communications protocols in the 1970s and 1980s
- **Fixed-Mobile Convergence**—Field devices and field controllers that were formerly restricted to wired communications have started broadly adopting wireless
- **Triple Play Convergence**—Voice, data, and video services are all being delivered using IP, supplanting public switched telephone network (PSTN), cable TV, and analog TV
- **Blue-Sky Convergence**—New products automatically have the ability to send and receive messages using IP
- **Cloud Convergence**—Abstracting systems administration from software and business processes provides flexibility in deployment and fewer requirements for workforce capability

Convergence has many potential benefits for companies that adopt the technologies and practices that are converging:

- Cost reductions
- Capture of new revenue streams
- Increase in process productivity and efficiency
- Increase in labor efficiency through added remote monitoring
- Improvement of assets control
- Business continuity and disaster recovery enhancements

Although convergence is generally beneficial, it also creates security dependencies between parts of the system. For example, endpoints (field devices and field controllers) in an industrial control system are more vulnerable now that "security by obscurity" (uncommon communications protocols) can no longer serve as a deterrent. As the benefits of convergence continue to be realized, companies must simultaneously adapt and modernize their security practices to avoid making vulnerabilities worse.

## The ISA-95 Models

Concepts can converge just as technologies can. For example, ANSI/ISA-95 (also known as ISA-95), maintained by the International Society of Automation (ISA), is a group of technology-agnostic information models that describe relationships between business and production data. The result is a reference architecture that explains the relationships among the hardware, software, and data in a typical industrial organization.

ISA-95 emerged in the 1990s, drawing from the Purdue Enterprise Reference Architecture (PERA), as a model to help companies more easily integrate business logistics and manufacturing or production operations systems. PERA addressed people, processes (for control and information flow), and technologies (governing the production process and physical plant) and covers the entire production life cycle from concept development through asset disposal (Chacon et al., 2010).

The ISA-95 levels, including an additional level incorporated by LNS Research (Jacob, 2017) to tie operations to strategy, are listed in Table 9.1 along with examples of software systems implemented at each level. The acronyms will be described and expanded on later in this chapter. An illustration of the types of information exchanged between Level 3 (inside the dotted line) and Level 4 (outside the dotted line) is shown in Figure 9.1.

Until ISA-95, integrating ERP systems and process control was a costly, extensive, and high-risk venture. Integration projects could last many years, as teams struggled with organizational silos, different technical languages and jargon, unique departmental cultures, and software systems that were not designed

**TABLE 9.1. The automation pyramid according to PERA/ISA-95.**

| ISA-95 | Timescale | Description | Software and hardware |
|---|---|---|---|
| "Level 5" | Months/years | Strategy and Governance | Executive-level activities, including APM, CPM, GRC, knowledge management, ORM, SPI |
| Level 4 | Days | Business Planning and Logistics (*why* work is done; ensuring work meets specifications) | Management-oriented activities, including APQP, CRM, document control, ERP, GXP/PRP, HACCP, PLM (CAD/CAM, ETO, MTO), PPAP, project management, QMS (audits, CAPA/CAR, continuous improvement, management review), risk management (DFMEA, PFMEA, control plans), SCM (SQM/SRM), training and certification management |
| Level 3 | Hours | Operations Management (*how* work is done) | Workflows, including deviations, EAM, EHS, inspections, LIMS, MES/MOM, MSA, SIEM, SPC. A function is in Level 3 if it is critical to product quality, workplace safety, plant reliability, or plant efficiency or is critical to maintaining product or production regulatory compliance |
| Level 2 | Minutes | Monitoring and Supervising | Monitor and control-oriented activities, including DCS, historian, HMI, intelligent devices (e.g., cameras, scanners), SCADA, SIS |
| Level 1 | Seconds | Sensors, Field Devices, Field Controllers | Execution-oriented activities, including PLC, sensors, switches, actuators, beacons, tags |
| Level 0 | Milliseconds/ Microseconds | Physical Production Process | Hardware, physical devices, industrial robots |

Source: Adapted from ANSI, 2000, 2001, 2005; with "Level 5" added following Jacob, 2017.

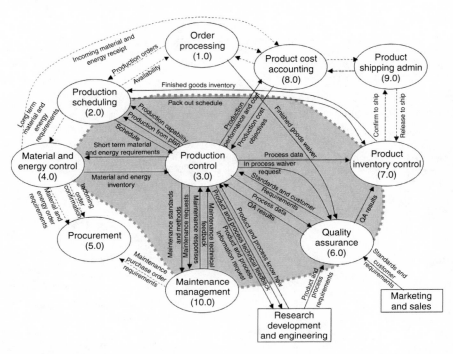

FIGURE 9.1. Purdue Reference Model for Computer Integrated Manufacturing (CIM) (Rathwell, 2006).

to be useful cross-functionally. Every time a company wanted to integrate its order management and production scheduling systems, for example, it had to go through an extensive process of requirements gathering, design, testing, and rollout. Even though all companies are different, these integrations were very similar to one another, so facilities were essentially duplicating each other's efforts. Many of these projects failed, leaving behind disparate systems and substantial sunk costs (Scholten, 2007).

Using the guidance provided by each of the ISA-95 reference models (Table 9.2), companies could benefit from the prior work and best practices identified by other organizations that had previously attempted similar integrations. The model covers the entire production life cycle and is just as useful in environments with little automation as in those with extensive automation.

ISA-95 helps organizations bridge the gap between automation providers and IT services or software vendors. In particular, mapping an organization's processes and systems to a model like ISA-95 can help people better under-

**TABLE 9.2. Descriptions of ISA-95 guidance documents.**

| Type of guidance | Model description | Key topics |
|---|---|---|
| Information exchange between business systems and manufacturing systems | ANSI/ISA-95.01-2000, Enterprise—Control System Integration—Part 1: Models and Terminology | Provides standard terminology for integrating control systems with enterprise systems, providing a basis for communication |
|  | ANSI/ISA-95.02-2001, Enterprise—Control System Integration—Part 2: Object Attributes | Describes key objects/entities and their characteristics; having a common basis for understanding these objects can reduce the risk, cost, and errors of integration projects |
|  | ANSI/ISA-95.05-2007, Enterprise—Control System Integration—Part 5: Business to Manufacturing Transactions | Describes types of information exchanged between business (Level 4) and manufacturing (Level 3) systems |
| Information exchange within and between manufacturing systems | ANSI/ISA-95.00.04-2018, Enterprise—Control System Integration—Part 4: Objects and Attributes for Manufacturing Operations Management Integration | Provides guidance for integrating production, maintenance, quality, and inventory systems within Level 3 |
|  | ANSI/ISA-95.00.06-2014, Enterprise—Control System Integration—Part 6: Messaging Service Model | Describes how message-passing facilitates information exchange between business (Level 4) and manufacturing (Level 3), and within Level 3 as well |
|  | ANSI/ISA-95.00.07-2017, Enterprise—Control System Integration—Part 7: Alias Service Model | Provides guidance for converting between global names (at sites) and local names (within areas or at facilities) |
| Activities that take place within manufacturing systems | ANSI/ISA-95.03-2005, Enterprise—Control System Integration—Part 3: Models of Manufacturing Operations | Provides a standard language for production, maintenance, quality, and inventory functions in Level 3 |

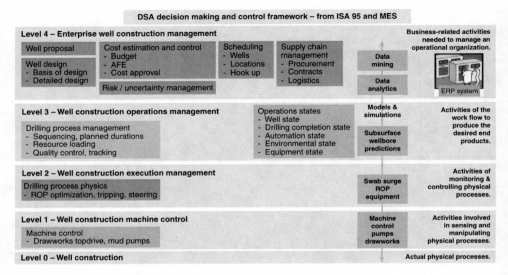

FIGURE 9.2. Well construction operations mapped to ISA-95 (De Wardt, 2016).

stand where and when to implement a system of record. A *system of record* (SoR) is a software system that stores critical proprietary data and information that the workforce interacts with on a regular (daily, weekly, or monthly) basis. It contains the single source of truth for a particular object, entity, or transaction. For example, knowing which system is the authoritative source of employee data can help managers avoid accidentally scheduling employees who have left the company or who are on vacation.

Models like ISA-95 can be used to describe the link between strategy and operations in terms of systems of record in many industries. For example, Figure 9.2 shows how well construction can be described using the different levels (De Wardt, 2016). This picture calls out the software systems needed to manage business logistics and planning (well proposal, well design, cost estimation and control, risk and uncertainty management, scheduling, supply chain management, business intelligence) and systems needed to manage operations (drilling process management, operations states monitoring).

ISA-95 continues to grow and evolve, especially in response to the changes brought on by emerging Industry 4.0 technologies. Poole (2017) shares that two additional guidance documents, Part 8 (covering Level 3 information exchange) and Part 9 (common operations events), are now in discussion. They will address new scenarios like automatically exchanging asset capability information, acknowledging work performed, expanding the definition of physical processes to include abstract and specialized cases (for example, process flows that are the same for every step except packaging), incorporating spatial and location-based information, and communicating quality targets so systems can automatically manage to them.

The remainder of this chapter walks through the different software systems from ISA-95 Level 2 (Monitoring and Supervising) to ISA-95 Level 4 (Business Planning and Logistics), and includes a "Level 5" (Strategy and Governance) as recommended by Jacob (2017). For each level, the key software systems and their functions, along with the objects, entities, or transactions they might manage as a system of record, are listed.

## ISA-95 LEVEL 2: MONITORING AND SUPERVISING

The role of Level 2 is to monitor, coordinate, and control the field devices and field controllers (Level 1) attached to physical objects or cyber-physical systems (Level 0) in industrial control systems (ICSs). An *ICS* is a "collection of personnel, hardware, and software that can affect or influence the safe, secure, and reliable operation of an industrial process" (ISA-99/IEC 62443). The ICS gathers information about the process from devices at the endpoints (Level 0 and Level 1), interprets that information in the context of production system goals, and facilitates interactions between the people who operate the system and the devices at lower levels (Table 9.3).

### Distributed Control System (DCS)

The *distributed control system* (DCS) is a special kind of active automation control system that serves as the full, state-driven "brain" of the system. Because bandwidth was historically very limited, decision making had to be done closer to the field devices at the endpoints of a process, with minimal data sent back to a central location to be monitored. The DCS performed most of the detailed work at this level and reported results back to the passive SCADA system. DCS is process-oriented, whereas SCADA mainly performs data gathering and assimilation.

**TABLE 9.3. Functions and entities in a system of record for key ISA-95 Level 2 systems.**

| ISA-95 | Software system | Function | System of record for: |
|---|---|---|---|
| 2 | Historian | • Record data about the process context like temperature, pressure, level, flow rates<br>• Record outputs, system states, or telemetry data | Process monitor data |
| 2 | Safety Instrumented System (SIS) | • Monitor control systems and abort operations if unsafe conditions are detected | Process safety requirements |

## Historian

The historian is in place to provide a record of process data and compliance data. As a replacement for chart recorders, this is an electronic system that captures telemetry data and information about the state of the system. When there is a problem, the historian can be queried for data that can be used in a root cause analysis, helping to quickly identify the source of uncommon issues.

## Human-Machine Interfaces (HMIs)

*Human–machine interfaces* (HMIs) facilitate monitoring tasks. They are the primary interfaces between human operators and the process. They display information about the state of the system, and often provide operators with the ability to adjust or manipulate controls. Because HMIs provide only a view into the data, they are a presentation layer and do not serve as a system of record for any object or entity. HMIs can be implemented as hardware (e.g., devices with physical switches and buttons), software (applications on smartphones, tablets, or screens), or a combination, depending on the environment and health and safety implications of the process.

## Intelligent Devices

Sometimes, Level 1 components contain onboard intelligence (e.g., in embedded systems) that provide local control and supervision. For example, consider access controls on a door to an authorized area. A camera is programmed to turn on every time someone or something approaches the doorway, take a picture, analyze the picture against a training set of all the authorized people (using edge computing), and trigger an alarm or incident report if they are not recognized.

## Supervisory Control and Data Acquisition (SCADA)

In contrast to a DCS, the SCADA system performs only supervisory gathering of data and reporting. In many environments, operators will refer to any HMIs as "the SCADA system." There are typically three modules that define a SCADA system (Rocha & Scholl, 2015):

- **Runtime module**—responsible for data exchange with field devices and field controllers including RTUs and PLCs; stores and analyzes data for alarm conditions; loads SCADA configurations to devices
- **Design module**—a passive component that defines the configuration for the system, including communication parameters, variables, alarms, events, and scripts
- **Client module**—the portion of the system that the user interacts with, including physical buttons and widgets on screens/devices

SCADA systems are evolving as a result of the industrial internet of things (IIoT) and cloud computing. Sajid et al. (2016) describe a web-based SCADA where multiple remote sites, made up of field devices and field controllers, are coordinated in the cloud with monitoring HMIs available in the browser. These could be used in any industry where SCADA is found today, including agriculture, food and beverage, chemical processing, nuclear power, civil administration, water and wastewater, healthcare, energy, financial systems, transportation, and aerospace.

## Safety Instrumented System (SIS)

*Safety instrumented systems* (SISs) are robust, hardened, high-reliability ICSs that have one and only one purpose: stopping or shutting down processes if unsafe conditions occur, which protects workers and assets. SISs act against random, unintended hardware faults. They often use the same technology and platforms as other parts of the ICS, so vulnerabilities are common. Because

SISs must be connected to the rest of the control system to function, there is always an attack path and the presence of an SIS does not guarantee that the system is secure.

## ISA-95 LEVEL 3: OPERATIONS MANAGEMENT

Level 3 systems manage *how* work is done, following the policies, schedules, and plans set at Level 4. Many of these systems sit within the process rather than beyond it, or are associated with checking the process to make sure it stays in control. As a result, many of these systems require configuring and managing workflows, which may include multiple handoffs or approvals, as shown in Table 9.4.

### Deviations

Particularly in high-risk or safety-critical industries like pharmaceutical manufacturing and food and beverage, any deviation from a standard operating procedure may impact safety, quality, or performance outcomes. The FDA defines this as any "deviation from current good manufacturing practice (CGMP), applicable regulations, applicable standards, or established specifications, or an unexpected or unforeseeable event (such a deviation or unexpected, un-

**TABLE 9.4. Functions and entities in a system of record for key ISA-95 Level 3 systems.**

| ISA-95 | Software system | Function | System of record for: |
|--------|-----------------|----------|------------------------|
| 3 | Enterprise Asset Management (EAM) | • Manage scheduled and predicted maintenance <br> • Manage workflows from maintenance events | Equipment (unless in an APM system) |
| 3 | Environment, Health, and Safety (EHS) | • Track emissions and environmental impacts to meet compliance obligations <br> • Manage workflows to resolve the cause of safety issues or incidents | Emissions incidents; EHS audits/ inspections |
| 3 | Statistical Process Control (SPC) | • Monitor processes and trigger nonconformances if special cause variation is detected | N/A |

foreseen event is referred to hereafter as an 'event') that may affect the safety, purity, or potency [of a product]" (Biological Products, 2014). Many organizations track deviations using specialized software that can keep records, alert operations, or provide an audit trail if issues are found later.

## Enterprise Asset Management (EAM)

EAM systems issue work orders based on predicted, planned, or triggered maintenance events, and helps workers make sure they are following standard operating procedures and keeping the right records for compliance purposes. EAM is different from Asset Performance Management (APM) systems. APM, in contrast, monitors real-time equipment data to obtain optimal performance over the life cycle of the asset.

## Environment, Health, and Safety (EHS)

Companies need to make sure they are protecting their people and the planet. Tracking environmental impacts, like greenhouse gas (GHG) emissions, can help them ensure that they are complying with legal obligations. Protecting workers is also important, especially when they operate in hazardous or high-risk environments. EHS systems can track exposure to contaminants, make sure workers are not exposed to extreme heat or cold, design safe material handling systems to prevent musculoskeletal disorders, schedule safety checks and inspections, track results, and make sure action items that result from those activities are completed quickly. Techniques like Job Hazard Analysis and Process Hazard Analysis can empower organizations to design danger out of their work systems.

## Laboratory Information Management System (LIMS)

If your company is a laboratory or uses the services of a laboratory, you may need a LIMS. For example, industrial hygiene data that is gathered to determine whether workers have been exposed to contaminants is often not processed on site but is sent to a specialized lab. Similarly, water treatment facilities and food processing plants often gather data about their processes that must be sent out for analysis. Sample management, provenance (chain of custody), workflow management, and electronic data exchange with one or more laboratories are commonly handled by LIMSs.

## Manufacturing Execution System (MES)

While MES refers to all the subsystems that carry out (and report on the results of) production processes, *manufacturing operations management* (MOM) is a newer label to describe the broader context that includes quality management, planning, forecasting, materials management, and analytics. In practice, MES and MOM vendors offer products that cross these boundaries, making the distinctions less important.

## Measurement Systems Analysis (MSA)

Processes cannot be controlled without accurate data about whether those processes should be adjusted. Measurement systems analysis (MSA) examines the bias and repeatability of the gage or measurement device (MSA Type I), or alternatively, the repeatability and reproducibility of measurements by different operators (MSA Type II). Process capability analysis and analysis of variance (ANOVA) drive these methods (Roth, 2011).

## Security Incident and Event Management (SIEM)

The analog to EHS systems that monitor environmental, health, and safety related incidents and near misses for security is security incident and event management (SIEM). SIEM systems continuously monitor access and event logs and capture information about security breaches and potentially malicious events (like failed logins or malware signatures).

## Statistical Process Control (SPC)

Production processes must be monitored to detect when (if ever) a problem occurs that could negatively impact the product. SPC software examines quantitative variables (e.g., length, width, diameter) or attributes (e.g., percent of defects/nonconformities in a sample) to determine whether the observed issues are due to random variation or something more serious. The goal is to intervene only when special cause variation is observed and get the process back on track as quickly as possible to minimize the costs associated with waste and rework.

## ISA-95 LEVEL 4: BUSINESS PLANNING AND LOGISTICS

Level 4 systems establish *what* work will be done and *why* it will be done, linking strategic goals to production objectives. The processes associated with carrying out those instructions are in the domain of Level 3. While Level 3 emphasizes the workflows themselves, Level 4 addresses the processes that support those workflows (Table 9.5).

### Advanced Product Quality Planning (APQP)

Sometimes, developing a new product requires close collaboration among tens or hundreds of suppliers, each of which contributes a part to the final assembly. In the mid-1990s, U.S. Tier 1 automotive suppliers collaborated to solve the problem of how to communicate and share controlled documentation, and Advanced Product Quality Planning (APQP), a collaborative project management tool to prevent production issues before they arise, was born. This is a stage-gate process with five phases: planning, product design and development, process design and development, validation, and production. Many APQP systems focus on providing document management and support for workflows and approvals across organizational boundaries that aligns with the five-phase structure.

### Asset Performance Management (APM)

Straddling Levels 3, 4, and 5, APM systems catalog high-value systems and assets, identify asset failures before they occur, and create risk-based maintenance strategies that maximize return on assets. In contrast with EAM systems that issue and track work orders based on maintenance events, APM is used to monitor real-time equipment data, helping senior leaders define policies and strategies to drive optimal performance over the life cycle of the assets.

### Customer Relationship Management (CRM)

From the time someone first makes contact with your company as a prospect, it is important to keep a record of all the interactions anyone in your company makes with them. Providing a single image across all departments and customer touch points (e.g., phone, e-mail, web, social media) will positively impact their experience. A good experience can move them more smoothly through the sales

**TABLE 9.5. Functions and entities in a system of record for key ISA-95 Level 4 systems.**

| ISA-95 | Software system | Function | System of record for: |
|---|---|---|---|
| 4 | Asset Performance Management (APM) | • Manage equipment characteristics<br>• Manage records for equipment performance<br>• Manage records for equipment maintenance | • Equipment<br>• High-value systems |
| 4 | Customer Relationship Management (CRM) | • Manage contacts<br>• Manage sales funnel<br>• Manage multiple customer touch points | • Contacts<br>• Customers<br>• Documents |
| 4 | Document Control | • Manage access to key documents<br>• Manage approval workflows<br>• Track changes to documents | |
| 4 | Enterprise Resource Planning (ERP)—Finance | • Financial planning and analysis<br>• Manage accounts payable and receivable<br>• Keep records for financial compliance | • Financial transactions<br>• Balance sheets |
| 4 | Enterprise Resource Planning (ERP)—Human Resources (HR) | • Manage employee information<br>• Track onboarding, training, and professional development | • Employee |
| 4 | Enterprise Resource Planning (ERP)—Information Technology (IT) | • Manage IT requests and field service<br>• Manage IT processes and security<br>• Track problems, alerts, and outages | • IT processes<br>• Security controls<br>• IT service tickets |
| 4 | Enterprise Resource Planning (ERP)—Order Management | • Process orders from submission to fulfillment<br>• Request that goods are produced or services are scheduled | • Sales orders<br>• Bill of materials (BoM) |
| 4 | Quality Management System (QMS) | • Record and track product and process issues<br>• Manage the problem-solving process to resolve those issues and/or sustain improvements | • Quality controls<br>• Quality issues |
| 4 | Training and certification management | • Proactively ensure that all workers keep required training up to date<br>• Track certificates and courses | • Training schedules<br>• Certifications<br>• Course completion |

funnel, and a bad experience can send them to your competitors. A Customer Relationship Management (CRM) system serves at least one (and sometimes all) of these roles: managing contact information, managing records of interactions at each touch point and supporting documentation, supporting the sales process as the contact progresses from a lead to an opportunity to a customer, and then tracking that relationship through its life cycle. Some CRM systems help you identify and nurture the customers that contribute most to your business.

## Document Control

For the most critical documents in your business, it is important to know whether you are looking at the most recent versions. Although most document control systems just provide mechanisms for securely storing documents, managing access controls, and keeping track of changes, some have more advanced features. For example, electronic signature capability is sometimes provided to comply with regulations like the FDA's 21 CFR 11. Other document control systems make it possible to electronically collaborate on documents and manage workflows and approval processes.

## Enterprise Resource Planning (ERP)

The most complex information management system in organizations today is the ERP system. ERP was originally built to handle production planning, scheduling, and inventory management but rapidly grew into a collection of subsystems to manage all of an organization's resources. Today, there are many different ERP modules:

- Financial and management accounting
- Human resources management
- Order management and fulfillment
- Supply chain management
- Product management
- Manufacturing execution (e.g., production, quality, product life-cycle management)
- Customer management

ERP systems very commonly serve as the system of record for objects, entities, and financial transactions. For example, ERP provides employee information to other systems that need to grant access, and sales orders or bill of materials (BoM) to manufacturing execution systems (some of which may themselves be ERP modules). After production is complete, ERP receives information about product results, inventory levels, and process performance.

## Good Manufacturing Practice (GMP)/Prerequisite Programs (PRPs)

Prerequisite programs (PRPs) are part of Good Manufacturing Practice (GMP). These practices, which include sanitation and pest control, are intended to be applied at all levels of the supply network and are required by most standards governing the food and beverage and pharmaceutical industries. Regulatory agencies understand that quality assurance is not possible in some industries without establishing and ensuring clean, safe conditions first.

## Hazard Analysis and Critical Control Points (HACCP)

Hazard analysis and critical control points (HACCP) is a seven-step process for assessing and mitigating risks to food safety. It begins with conducting a hazard analysis, then using the hazards identified to uncover ways to control or prevent each of them. A monitoring system to regularly evaluate these critical control points (CCPs) is put in place, and from this, corrective actions may be identified. Performance of the system, in terms of the variables that are monitored using the CCPs and their limits, is tracked and recorded to support compliance audits.

## Product Life-Cycle Management (CAD/CAM, ETO, MTO)

Managing the design and production of complex products can require tracking hundreds (or thousands) of documents and variables—including 2D or 3D computer-aided design (CAD)/computer-aided manufacturing (CAM) models (which can be annotated), BoMs, part attributes, revisions, and engineering change requests—and collaborating across teams that may be geographically distributed. Make to Order (MTO) systems require tighter synchronization among product definitions, facility capabilities, and production orders. Engineer to Order (ETO) systems require sophisticated integrations, because change requests may be issued while production is ongoing. Some product life-cycle management (PLM) systems incorporate MES/MOM functionality, ensuring a seamless transition between product definition and production.

## Production Part Approval Process (PPAP)

Before engaging a supplier, sometimes it is important to get additional information about its product and the quality management systems used to maintain the production process. In some industries, including automotive and aerospace, this communication process is accomplished by the production part approval process (PPAP). PPAP software helps you select the right controlled documents to include in Part Submission Warrants (PSWs), and send the documents and the PSW (sometimes electronically) to the organization that is performing the design qualification on you. Alternatively, you may be performing a design qualification on an item from another organization and need to request a PPAP from the organization. PPAP packages can include many other artifacts, including FMEA, control plans, and inspection reports.

## Quality Management System (QMS)

Software for a QMS helps track requirements, identify and resolve issues that prevent requirements from being met, ensure progress toward continuous improvement, and drive a quality-focused culture. Some of the functions supported by QMS software include the following:

- **Nonconformance Reporting.** A nonconformance arises when variation exceeds the normal limits in a process, or a product is produced that does not meet a requirement. The requirement that was violated, along with its source, evidence, and the source of the evidence, is captured, triaged, and contained.
- **Customer Complaints.** These are nonconformances raised by customers.
- **Continuous Improvement/Corrective and Preventive Action (CAPA).** If there is a broader problem that an immediate countermeasure cannot resolve, a corrective action is launched. Problem-solving processes, such as root cause analysis, are applied in response to the severity and complexity of the nonconformance.
- **Audit Management.** Especially for organizations that maintain certifications to ISO standards, internal and external audits are used to ensure that each clause of the standard is being effectively addressed. Audit systems help reviewers keep track of requirements, evidence, and findings and carry out action items to respond to those findings.
- **Management Review.** A regular, cross-functional leadership review of an organization's QMS is required to obtain some certificates (including ISO 9001,

ISO 13485, and ISO 27001). Management review software gathers audit results, customer feedback, process performance results, the status of nonconformance reports and corrective actions, follow-up actions from previous reviews, and external changes that could impact the performance of the QMS.

In addition to continuously improving products and production processes, QMS processes are used to improve the QMS itself.

## Risk Management (DFMEA, PFMEA, Control Plans)

Both product design and production processes have risks, many of which can be addressed or mitigated with appropriate planning. Failure modes and effects analysis (FMEA) provides one approach for conducting this due diligence. A design FMEA (DFMEA) articulates potential failure modes for each function of a product, while a process FMEA (PFMEA) outlines potential failure modes at each step of its production. DFMEA can be used to draw out requirements for test and verification plans, while PFMEA is used to identify effective process controls that are captured in control plans.

## Supply Chain Management (SCM)

There are many different types of supply chain management (SCM) applications. These include supplier relationship management (SRM) systems that manage only supplier contacts, certifications, and supplier-issued nonconformances, and supplier quality management (SQM) applications that go deeper, populating scorecards and performing risk monitoring to make sure that suppliers are meeting their obligations. Suppliers that do not meet requirements can quickly be identified, and restorative actions taken. More complex SCM applications address supply chain planning, including managing against variability in demand and balancing production requirements across multiple sites and facilities.

## Training and Certification Management

Many organizations need to keep track of who is qualified to perform what job, whose training is up to date for high-risk work (e.g., confined space training), and who is qualified to audit others' work (particularly for internal ISO auditors). Training systems, which are sometimes integrated with employee records in ERP, can be obtained to make sure training and qualifications are up to date before a worker performs a task. This is important to maintain a safe work environment and to demonstrate compliance with certain standards.

## LEVEL 5: STRATEGY AND GOVERNANCE

Although not called out by the ISA-95 model, there are other information management systems that operate on timescales from months to years. These are the systems that monitor performance at a high level and provide information to executives and senior leaders regarding how well policies are performing and how well the strategy is being deployed. Key systems are described in Table 9.6.

### Corporate Performance Management (CPM)

CPM systems track information that is needed by the CEO and his or her closest senior leaders to keep the business running. This includes financial planning and projections, performance to budget, and performance of initiatives to the strategic plan. There is overlap between CPM systems and financial modules in many ERP systems.

### Governance, Risk Management, and Compliance (GRC)

GRC systems help organizations comply with financial regulations (e.g., SOX), data protection regulations (e.g., GDPR), labor laws, and management system standards (e.g., ISO 9001, 14001, 27001, 45001). Although GRC systems are intended to provide a view into these functions from the executive and board levels, there is a clear overlap between GRC systems and QMSs. GRC software reduces the risk of fraud and increases the ability to detect malicious attacks in other software systems (e.g., ERP).

### Knowledge Management

Knowledge management systems capture and organize information, and often create taxonomies and expose utilities like search to make information easier to find. Many software applications brand themselves as knowledge management, and in fact, any software system that manages data could fall into this category. Content management systems, wikis, document repositories, and messaging systems for team support all provide ways to manage knowledge.

**TABLE 9.6. Functions and entities in a system of record for key strategy and governance systems.**

| ISA-95 | Software system | Function | System of record for: |
|---|---|---|---|
| Not Available | Corporate Performance Management (CPM) | • Budgeting<br>• Planning<br>• Modeling<br>• Forecasting | • Strategic plan<br>• Scorecards<br>• Budget<br>• Financial reports<br>• Operations reports |
| Not Available | Governance, Risk Management, and Compliance (GRC) | • Monitor information access against privileges<br>• Maintain records for financial and legal/regulatory compliance (e.g., SOX, GDPR), including audits<br>• Maintain records for management systems (e.g., ISO 9001, 14001, 31000, 45001), security (ISO 27001), and IT compliance (ITIL, COBIT) | • Controls<br>• Processes (sometimes)<br>• Records<br>• Financial audits<br>• Legal audits |
| Not Available | Organizational Risk Management (ORM)/ Integrated Risk Management (IRM) | • Identify hazards<br>• Assess risks<br>• Address risks | • Hazards<br>• Risks<br>• Risk management plans |

## Organizational Risk Management (ORM)

Risk management often requires intense coordination across functional boundaries. Especially if an organization is using the ISO 31000 standard for guidance, addressing risks requires establishing an organizational context, deciding on risk communication plans, outlining hazards, assessing risks (which includes identifying, analyzing, and evaluating them), and addressing risks. This last step can include a decision to avoid or ignore the risk, accept or amplify the risk to capture opportunities, remove the risk by changing organizational processes or business context, or shift the burden of the risk (e.g., by getting insurance). Because the environment changes regularly, this catalog of risks and how they are being handled changes on a regular basis. ORM software, which is sometimes called Integrated Risk Management (IRM), keeps track of these elements.

## Sustainability Performance Indicators (SPIs)

Nearly 75% of the S&P 500 in the United States prepare a sustainability report each year, describing their company's performance with respect to the economy, the environment, and society. Reporting is governed by Global Reporting Initiative guidelines. Economic indicators include proportion of spending on local suppliers and wage ratios across genders. Environmental factors include GHG emissions, energy consumption, water quality discharged into the environment, and the proportion of recycled materials used as inputs. Social factors include human rights, fair labor practices, and handling of customer and employee privacy. This data is often captured as Sustainability Performance Indicators (SPIs) in a Sustainability Performance Assessment system. SPIs are sometimes included in EHS software (Paun et al., 2016).

## DATA PLATFORMS

Before Industry 4.0, when companies referred to a "data platform" they typically meant the enterprise relational database (RDBMS). This was where all mission critical data was stored. It was assumed that the data volume and velocity were not that large, and it imposed requirements on the structure and types of data that could be stored.

With the advent of broader data types (image, audio, spectrum, video, unstructured), the traditional RDBMS could no longer support all the organization's needs. Rather than force-fitting unstructured data into the RDBMS, for example (which would be better served by a NoSQL database), the concept of a wider, distributed data platform emerged—an amalgam of on-premise and cloud-based services that together could satisfy new use cases stimulated by big data:

I now see the "Data Platform" as much broader than ever before and includes many other "non-traditional" data services . . .

- Relational Database Platform (RDBMS) (i.e., SQL Server, Azure SQL DB/DW, Oracle, SAP Hana, MySQL, etc.)
- NoSQL (i.e., DocumentDB, MongoDB, Cassandra, etc.)
- Big data Solutions (i.e., Hadoop, Data Lake, etc.)
- Intelligent Data (i.e., Cognitive Services, Machine Learning, Deep Learning, etc.)

- Data Ingestion/Management (i.e., Event Hub, IoT Hub, Stream Analytics, Polybase, Data Catalog, Data Factory, etc.)

Just outside of the periphery of these are additional services such as Bots, and Workflow (i.e., Logic Apps, Flow, etc.) which are not "Data Platform" per se but worth mentioning. Some will probably successfully argue these could and should be part of the Data Platform? Time will tell. (Tesmer, 2017)

The relationship between types of technologies that form the data platform and ISA-95 levels is shown in Table 9.7. Although there may be differences between organizations and between industries, this should provide a general view of where the technologies will be used. Organizations will need to examine their data platform at all levels of the ISA-95 hierarchy, with their specific business planning and production needs in mind, to ensure that all scenarios for data storage and retrieval are covered.

Together, software applications and data platforms can be used to more easily manage the information flows between levels of the ISA-95 hierarchy (Figure 9.3). This model can also be used to make sure that the data required to establish business results flows from its source to the level where it is analyzed and acted on.

## THE BOTTOM LINE

Software applications and data platforms provide containers that hold logical subsets of an organization's knowledge and process intelligence. Every organization should have a clear picture of which software systems provide the system of record for each key object, entity, and transaction. This helps reduce the costs and risks of information management and increase the quality of deci-

**TABLE 9.7.  Relationship between data platforms and ISA-95 levels.**

| ISA-95 | Relational database | Nonrelational (NoSQL) database | Big data/ distributed processing | Intelligent data/cognitive services | Data ingestion/ streaming |
|---|---|---|---|---|---|
| Level 5 | X | X | | | |
| Level 4 | X | X | X | X | |
| Level 3 | X | X | X | X | |
| Level 2 | | | | | X |

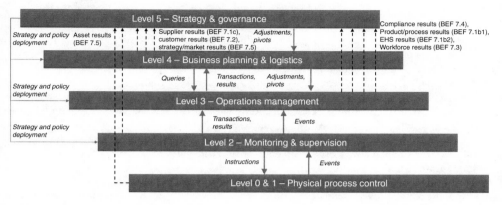

FIGURE 9.3. Relationships between ISA-95 levels, information exchange, and business results in terms of elements from Criterion 7 in the Baldrige Excellence Framework (BEF).

sions made using that information. The ISA-95 hierarchy can be used to describe any organization's work systems, recognizing that the IIoT is making it possible for systems and devices to easily communicate across levels if desired:

- Levels 0 and 1—Physical Process Control/Gemba (where the work is done)
- Level 2—Monitoring and Supervision
- Level 3—Operations Management
- Level 4—Business Planning and Logistics
- "Level 5" (Jacob, 2017)—Strategy and Governance

Even though this chapter was focused on understanding the links between software systems, not all these information systems must be supported by software or be resident on a data platform. In most (if not all) organizations, there remain manual processes or processes supported by rudimentary technologies (like Excel). For effective data management, the important part is to have an awareness of *where* these processes are taking place, *what* systems are supporting them, and *where* accurate and complete master data (in systems of record) can be found. This will help prevent waste and rework.

## REFERENCES

ANSI/ISA-95.00.01-2000. (2000). *Enterprise-Control System Integration. Part 1: Models and terminology*.

ANSI/ISA-95.00.02-2001. (2001). *Enterprise-Control System Integration. Part 2: Object model attributes.*

ANSI/ISA-95.00.03-2005. (2005). *Enterprise-Control System Integration. Part 3: Activity models of manufacturing operations management.*

Biological Products: General, 21 C.F.R. § 600 (2014). https://www.govinfo.gov/app/details/CFR-2014-title21-vol7/CFR-2014-title21-vol7-part600

Chacon, E., A. Eisula, A. Carnevalli, E. B. La Hechicera, J. Cardillo, and A. A. Carnevalli. (2010). "Evolution of integrated automation approach." *Advances in Computational Intelligence, Man-Machine Systems and Cybernetics,* 274–288.

De Wardt, J. (2016, February). *DSA decision making and control framework.* http://dsaroadmap.org/wp-content/uploads/2016/03/Poster-Feb-29-2016-DSA-Decision-frame-Final-Released.pdf

Jacob, D. (2017, January 4). "5 times to change your supplier quality management strategy." *LNS Research Industrial Transformation Blog.* https://blog.lnsresearch.com/5-times-to-change-your-supplier-quality-management-strategy

Levinson, M. (2016). *The box: How the shipping container made the world smaller and the world economy bigger.* Princeton, NJ: Princeton University Press.

Macaulay, T., and B. L. Singer. (2016). *Cybersecurity for industrial control systems: SCADA, DCS, PLC, HMI, and SIS.* Auerbach Publications.

Novkovic, G., and T. Korkut, T. (2017). "Software and data regulatory compliance in the cloud." *Software Quality Professional* 20 (1).

Paun, D., S. Bray, T. Yamaguchi, and S. You. (2016). "A sustainability performance assessment tool: The SPA system." *Journal of Sustainability Education* 12: 1–20.

Poole, B. (2017, January/February). "Advancements in ISA-95: Adopted worldwide, ISA-95 improvements continue adding value." *InTech Magazine.* https://www.isa.org/intech/20170205/

Rathwell, G. (2006). *ISA 95—Setting the stage for integration of MES & ICD systems.* Enterprise Consultants International Ltd. http://www.pera.net/Standards/S95_Presentations/Warrick_Univ_v6-full.pdf

Rational. (1987). "Rational environment training—large system deployment." *Rational.* https://datamuseum.dk/w/images/7/7e/R1000_Training_Large_System_Development.pdf

Rocha, C. R., and M. Scholl. (2015). "Conceptual design of an open source SCADA system." In *Proceedings of the 25th International Conference on Flexible Automation and Intelligent Manufacturing* (2): 182–189.

Roth, T. (2011). "Teaching measurement systems analysis to engineers using R." *UseR!* https://www.r-project.org/conferences/useR-2011/TalkSlides/Poster/7-Roth.pdf

Sajid, A., H. Abbas, and K. Saleem. (2016). "Cloud-assisted IoT-based SCADA systems security: A review of the state of the art and future challenges." *IEEE Access* 4: 1375–1384.

Scholten, B. (2007). *The road to integration: A guide to applying the ISA-95 standard in manufacturing.* ISA.

Tesmer, R. (2017, April 19). "What exactly is the 'data platform' nowadays?" *Mr. Fox.* https://mrfoxsql.wordpress.com/2017/04/19/what-exactly-is-the-data-platform-nowadays/

## CHAPTER TEN

# BLOCKCHAIN

The blockchain revolution is coming, but you might not see it.
—BRIAN BEHLENDORF, EXECUTIVE DIRECTOR OF HYPERLEDGER
(BLOCKCHAIN PLATFORM)

Maersk is a cargo network founded and originally headquartered in Denmark. Today, the company operates in 120 countries, is over 100 years old, and is still family owned and operated. In fact, Maersk is in its fourth generation of leaders from the same family. As of 2019, the company has grown to employ nearly 80,000 people. It transports shipping containers by sea and by land, manages logistics for those shipments, forwards freight that originates on other networks, and operate terminals at ports. Maersk is the largest global transport company and has been an all-around technology leader since the 1980s, when it operated its own communications network, including a private way to transfer data.

One Tuesday afternoon, in an office at one of the main facilities in Copenhagen, an employee glanced at her laptop and noticed the screen was black. She checked to make sure the power plugs were correctly attached—when the cable slips out, the whole battery can quickly drain. But the power cords were firmly in place, and the green power light was on. A minute or two later, she noticed a tiny little text warning in the upper left-hand corner that said "Repairing file system on the C drive." She hadn't received any e-mails about maintenance from the IT department, and it wasn't a good time for an update. There were too many e-mails to finish before closing time.

No matter how many times she hit Enter or pressed Escape, the message just stayed on the screen. The hard drive wasn't whirring at all, and there were no other clues to be found. She left her office to visit the IT desk.

When she got there, the line was out the door and down the hallway. She found out that the other people standing in line were all there because their laptops had also gone dark. The man in front of her explained that some rogue process had encrypted all of his files, and the newest error message said he wouldn't get them back unless he sent bitcoin. There were varying levels of responsiveness from the laptops people in line were holding.

All of a sudden, every computer in the building shut down. People were disoriented and confused, and some started to panic—maybe something was really wrong. An announcement went out over the building's speaker system, directing everyone to unplug or turn off all machines immediately. The quiet soon overtook every room and hallway. In a completely connected company, there was nothing to do—everyone was told to go home.

What these Maersk employees didn't realize was that this same scenario was playing out in every Maersk office in 130 countries. Not too long after the shutdown, trucks in transit started backing up at the ports, making lines along the road back from the security gates. Some of them had cargo that needed to be refrigerated, and yet only had so much fuel to maintain the chillers. Drivers were getting angry. The ports had shut down, and trucks could not even pull out of the line and head elsewhere because there was no room to navigate. Everyone was stuck.

Nearly all of Maersk's operations were at a standstill. Over a million dollars was being lost every minute. Although it hadn't traced the problem back to its source yet, someone had installed accounting software on one computer in the Maersk office in Odessa. It was software specific to Ukraine operations, and it also happened to be infected with ransomware that was programmed to lay dormant until it received its orders. And on that Tuesday, the orders were silently—and electronically—issued to the machines in wait.

This was the beginning of the NotPetya cyberattack, which, within hours, knocked out the entire Maersk infrastructure. Although the incident resulted in $200–$300 million in losses, theft of a shipping line's data could have had even more serious consequences. For example, pirates or enemy nations could use ship location data to plan and execute cargo theft, take seamen hostage, or worse.

There is a happy ending to the Maersk story. All of the company's domain controllers had been wiped out, and its recovery plan assumed that a backup could be obtained from another live site. With zero live domain controllers, this was impossible. Fortunately, a power outage at the Accra office in Ghana had knocked its domain controller off-line, and serendipitously, it had not been impacted by NotPetya. Maersk was able to recover, but only thanks to this fortunate accident (McQuade, 2018).

The reason Maersk operations came to a standstill is that its information resources were centralized in databases connected to one another by a network of domain controllers. Once the ransomware got inside the Maersk network, it was able to take the whole thing down as soon as it was activated because all the data repositories were connected. Distributed data management systems that do not rely on a single network or infrastructure, like cloud-based systems, could have protected its data from a catastrophic attack more effectively—but even secure cloud systems do not have the level of security needed to support a business the size of Maersk.

In 2015, Maersk partnered with IBM to implement a system based on permissioned blockchain (which will be described later in this chapter) to connect the containers, shippers, carriers, and the ports. A heavily paper-based process was digitized, with records captured in the blockchain. In addition to adding transparency, the improvement removed process inefficiencies and reduced losses due to errors, tampering, and fraud (Hackius & Petersen, 2017).

Learning from this proof of concept, Maersk joined the TradeLens blockchain consortium in 2019, in which members "gain a comprehensive view of their data and can collaborate as cargo moves around the world, helping create a transparent, secured, immutable record of transactions" (Maersk, 2019). Although the link between the attack and the choice to invest in blockchain was never explicitly made, the company's decision came only months after realizing the losses.

There are many more cases like this one. Prototypes and pilots based on blockchain, particularly when they use permissioned systems, are beginning to demonstrate clear and compelling business impact. In this chapter, you will learn what a blockchain-based system is and understand when it should (and should not) be considered as one of the elements of your organization's digital transformation.

## WHAT IS BLOCKCHAIN?

A *blockchain* is a shared, digital ledger that contains transaction data. Each transaction is joined to the sequence of prior transactions like a link in a chain, and the data structure containing them cannot be changed once a new record has been logged and verified. The algorithms used to create each new link in a blockchain mathematically guarantee that, once accepted, the details of the transaction in the ledger cannot be altered without applying an immense (and impractical) level of computing power.

All parties to the transaction, and neutral third parties that are members of the supply network or business ecosystem, maintain their own copy of the ledger. This ledger is a blockchain. The nature of this data structure, and the unique computational processes used to create it, means that it is impossible to fake information or cheat on a transaction.

As the name implies, the blockchain data structure is made up of blocks of information, linked together in sequence. The content of each block is determined by the nature of the transactions that need to be stored. With bitcoin, each block contains a list of recent transactions of bitcoin moving between digital wallets. For supply chain tracking, a block could contain information about an event (e.g., arrival, departure, inspection) and characteristics of the event (e.g., time, status, environmental conditions), or any transaction of information, materials, or money. A block can contain any information that can be represented digitally, including photographs, video, audio, or text.

## Permissioned Versus Permissionless

There are different ways blockchain can be implemented, depending on the degree of anonymity that is required and the degree of trust that is present in the network or ecosystem. As a result, blockchains can be public or private, and "permissioned" or "permissionless."

In public blockchains, no one controls the blockchain (maintenance is shared among participants), whereas private blockchains are more like your company's corporate network. In permissionless blockchains, anyone can join the transaction network at any time (as in the bitcoin and Ethereum cryptocurrencies, for example). Permissioned blockchains, in contrast, can be managed and manipulated only by trusted partners that have been granted access to the blockchain.

The blockchain concept emerged from the concepts of distributed systems and peer-to-peer networks. In distributed systems, participants are well known and their behavior can be somewhat controlled. Peer-to-peer systems, in contrast, let any participant join the network and provide distributed access to files or resources (e.g., Napster for music, BitTorrent for videos). Because a request can be fulfilled by multiple participants (sometimes even within a single transaction), they are robust and provide high availability, but are particularly prone to cyberattacks. Peer-to-peer networks are thus essentially "permissionless" systems—they assume that most of the participants are honest. This assumption reduces the barriers to entry but increases the risks associated with cyberattacks (Benton & Radziwill, 2017).

## Bitcoin Versus Blockchain

Although often confused with one another, the bitcoin cryptocurrency (though based on a permissionless blockchain) is not the same as blockchain itself. Bitcoin, which is one example of an application that is enabled by this emerging technology, runs on a public, permissionless blockchain. New bitcoins are "mined" by solving a prescribed, difficult class of computational problem, a task that requires extreme processing power that is inaccessible to typical computer users (called *Proof of Work*). Due to the computational nature of the bitcoin mining game, only 21 million bitcoins can ever be obtained; by mid-2019, nearly 18 million were already in existence.

Although bitcoin was published as an idea in 2008 and initially released as software in 2009, it attracted interest because it was not backed by any government and was purely digital. Trading with bitcoin began shortly thereafter, its independence from government currencies making it the method of choice for illegal purchases. The price of a bitcoin hovered under US$10 and then spiked to $100 in early 2013 and $1000 in late 2013. At the end of 2017, a frenzy ensued when the price surged to $19,783.06 on December 17. The frenzy amplified when bitcoin tumbled, losing over half its value in just two weeks. (Bitcoin prices are available from http://www.coindesk.com and http://bitcointicker.co.)

Even though bitcoin is compelling, there are drawbacks. First of all, if you lose your password or access to your bitcoin wallet, your investment is inaccessible forever—it cannot be recovered. As a public, permissionless blockchain, bitcoin is subject to the "51% attack," in which a group of affiliated bitcoin miners gain control of the ability to approve new blocks on the bitcoin blockchain and can prevent miners who are not in their group from verifying the computations used to mine new bitcoin (Boddy, 2019).

## Growing the Blockchain

As time goes on, and events and transactions occur in a supply network or business ecosystem, information about them is stored in the public or private blockchain that supports the network. Much like the adoption of fax machines in the 1970s and 1980s, the value of any particular blockchain (especially private, permissioned blockchains) will increase as the number of participants in the network increases.

Figure 10.1 shows an example of the data stored in a single block in the bitcoin blockchain. This is representative of the types of data you might see stored in any blockchain. The number of times bitcoin was bought or sold is included (Number of Transactions), as well as details about the event when the block was recorded (Timestamp, Received Time, Relayed By) and even energy usage (Weight). Details about each of the 1,121 transactions recorded in this block are at the bottom, off the screen. In the right-hand column, two key fields that define this collection as a blockchain can be found: a long string of letters and numbers called a *hash* under Hash, and another long string of letters and numbers (also a *hash*) under Previous Block.

The process of constructing the next block in the blockchain goes something like this:

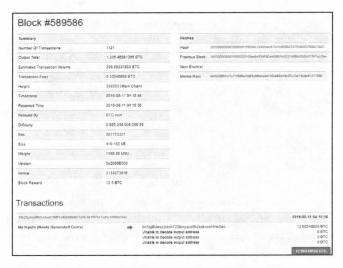

FIGURE 10.1. A block in the bitcoin blockchain (http://blockchain .com).

1. Gather information to store in fields (Number of Transactions, Output Total, etc.)
2. Copy the hash from the previous block (Previous Hash) into a field in the new block
3. Hash the entire collection of information (fields + Previous Hash) to get a new hash
4. Store the new hash in the Hash field

This is great, but what's a hash and how do you get one? That's the subject of the next section.

## HOW BLOCKCHAIN WORKS

Blockchains depend on *hashing*, a cryptographic, mathematical process that turns an object into a sequence of unique numbers and letters. (The root *crypto-* in the word *cryptocurrency* also reflects the importance of cryptography in building the blockchains that support new currencies.) In fact, you are already familiar with cryptography, even if you don't realize it.

### Hashing Algorithms

Most of the password-protected systems you use do not know your actual password. When you save a new password, the system hashes your password to create a random-looking sequence of numbers and letters. It stores that sequence, but not your actual password. When you log in, your input is hashed and compared with the hash of your real password that the system knows. If the two hashes match, your login is successful. This is why so many systems require you to reset your password if you forget it—it's not that they won't give it to you but that they don't actually know what your password is. They just know the hash, and it is impossible to un-hash a sequence to recover the original. Examples of some hashes are shown in Table 10.1.

There are numerous hashing algorithms, each producing different lengths of sequences. As you can imagine, the longer the sequence, the less chance that if you hash two different data objects, the same string of letters and numbers will be produced. As a result, the greatest collision avoidance for the hashing algorithms in Table 10.1 comes from SHA-256, SHA-1, and MD5 respectively. This is why SHA-256 (which stands for "secure hashing algorithm") is often used in production systems.

**TABLE 10.1. Example of hashes for "Hello World!" using different hashing algorithms.**

| Hashing algorithm | Hash |
|---|---|
| SHA-1 | a7cfe0581825aaeb63231804f8ef181e54305a10 |
| CRC32 | 32253911 |
| MURMUR32 | ef6abc31 |
| MD5 | 7737f5add04daf9160355b65338a5caa |
| SHA-256 | 9374f2c6f404965fb4ef8299642a8ede7b1cbe5a5e58f65a05d7a52e4e20ed91 |

## Hash My Cat

Because you can hash nearly any data type, in the following example we will hash my cat, Lexington, whose picture appears in Figure 10.2. To do the hashing, we use the digest package from the R Statistical Software. (You can get the image data to walk through this example yourself at https://github.com/NicoleRadziwill /Data-for-R-Examples/blob/master/kitty.7z on GitHub. First, download and un-zip the file to a directory on your hard drive. The four files in this zip archive are on my D: drive in the Test directory.)

The hashing process encapsulates all the information in a data object, no matter how big it is, in a string. For SHA-256, this resulting string is a 256-bit, 64-character sequence of numbers and letters. Applying a SHA-256 hash to the picture of my cat in Figure 10.2 results in a 64-character string starting with dcd239 (Figure 10.3).

Using this SHA-256 hash, we can easily solve problems that would other-wise be difficult. In Figure 10.4, there are four pictures of my cat. One of them has been manipulated and is not the same as the others. Can you tell which one? The hashes for each picture are in Figure 10.5.

Looking at the code in Figure 10.5, it is easy to see that the third picture is the deviant (the one with the hash that begins with fe5751). But when you go back to the collection of images in Figure 10.4, can you tell what the difference is? Probably not, and this is one of the limitations of hashing: you can quickly pick out which object is different, but it takes extra work to figure out exactly what the difference is. (In the third picture, a single pixel on the back left cor-ner of the cutting board to the right of the cat has been turned to white. If you zoom in or use a magnifying glass, you may see it.)

FIGURE 10.2. Lexington the cat, as a five-month-old kitten
(filename: kitty.jpg).

```
library(digest)
library(jpeg)
kitty <- readJPEG("D:/Test/kitty.jpg")


> digest(kitty,"sha256")
[1] "dcd239ba6a09080eb61b7310a5428753f63d05ae2b282bf81dc0182f7552f60d"
```

FIGURE 10.3. R code to hash the picture in Figure 10.2.

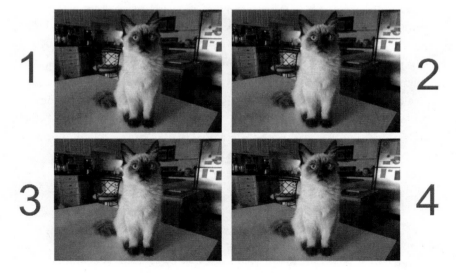

FIGURE 10.4. One of these images is not like the others. How can we tell which one?

```
library(digest)
library(jpeg)
kitty <- readJPEG("D:/Test/kitty.jpg")
kitty2 <- readJPEG("D:/Test/kitty2.jpg")
kitty3 <- readJPEG("D:/Test/kitty3.jpg")
kitty4 <- readJPEG("D:/Test/kitty4.jpg")


> digest(kitty,"sha256")
[1] "dcd239ba6a09080eb61b7310a5428753f63d05ae2b282bf81dc0182f7552f60d"
> digest(kitty2,"sha256")
[1] "dcd239ba6a09080eb61b7310a5428753f63d05ae2b282bf81dc0182f7552f60d"
> digest(kitty3,"sha256")
[1] "fe5791ee490693d7d7b25379278b2374c3afda25c76aec5f3aa17e7e8b184362"
> digest(kitty4,"sha256")
[1] "dcd239ba6a09080eb61b7310a5428753f63d05ae2b282bf81dc0182f7552f60d"
```

FIGURE 10.5. R code quickly highlights which of the four images in Figure 10.4 is different.

## The Value of Blockchain

Tiny changes in data that would be undetectable to even the most careful human will generate easily detectable changes in a hash. Because each new block of records includes the unique hash from the previously recorded block, no new data is recorded without keeping a snapshot of all the data that has ever been recorded in the blockchain, in sequence, in the form of the previous block's hash. As a result, any attempts to tamper with data in a blockchain will disrupt the sanctity of the hashes. The blockchain will immediately alert its administrators if fraud has occurred anywhere along the chain of transactions (Figure 10.6).

Blockchain is an immutable peer-to-peer transaction record, transparent to everyone in its network and instantly auditable. In fact, since rules in the form of "smart contracts" are in place in most blockchain-based systems to evaluate and validate the data before it can be committed to the blockchain, bad data shouldn't even be able to get in. Thus, blockchain has the potential to help organizations significantly improve data quality, reduce fraud and tampering, and substantially improve visibility and transparency.

## CASE STUDIES

Competing in a fast-moving global business environment means that the quicker you can get accurate, complete information, the better decisions you will make. Unfortunately, with so many different ways work processes can be performed,

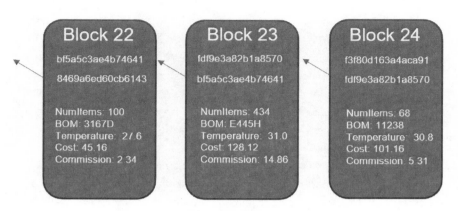

FIGURE 10.6. A sequence of blocks of data, chained together in a blockchain.

and so many different ways data can be captured, it can be a challenge to synchronize processes across organizational boundaries. This is true even when supply network or business ecosystem partners share the same spoken language.

Blockchain-based systems provide the opportunity to standardize the language of exchanging information without necessarily constraining its content or structure. This section describes four examples, from research and practice, of how blockchain prototypes are already smoothing business processes that were formerly extremely complicated.

### Walmart: Safer Food with Instant Traceability

Consumers are accustomed to hearing about occasional food safety issues in the news, including the up-front delay as investigators rush to figure out the source of the outbreak. The situation can become dire when deaths are involved and the affected food items have not yet been isolated. In the United States, every year 48 million people become ill due to foodborne illness, while 128,000 are hospitalized. From this group, 3,000 will ultimately die (Freeman, 2019).

The more quickly organizations can respond to a health crisis that results from food contamination, the fewer number of people will be affected, and the less likely people will become ill or die. Consequently, finding the source of the problem as quickly as possible is the priority. Being able to respond more quickly can save lives and corporate reputations when food safety is concerned.

One day [in December 2016], Frank Yiannas went to a Walmart store near company headquarters in Fayetteville, Ark., and picked up a package of sliced mangoes. Yiannas is Walmart's vice president of food safety, and the fruit was part of a crucial experiment. He brought the mangoes back to his office, placed the container on a conference table, and gave his team a mission. "Find out where those mangoes came from," he ordered, setting a timer. It took six days, 18 hours, and 26 minutes to get an answer. . . . In the event of an outbreak of foodborne illness—one in which a suspected pathogen is tied to mangoes somewhere—a lag that long could be painfully costly. By that point, Walmart might have had to pull every package of every mango product off its shelves, as a precaution; farmers, distributors, and Walmart itself would take the hit. (Hackett, 2017)

But it's hard to manage information within an organization, and exponentially harder to keep track of information that goes beyond organizational boundaries. In an increasingly globalized marketplace, scouring the records from what may amount to hundreds of organizations can be not only time-consuming but potentially impossible.

Traceability can improve response time by making the investigation step nearly instantaneous: Walmart's prototype solution for mangoes reduced tracking time from a week to 2.2 seconds (Kamath, 2018). These pilot programs demonstrated value to all stakeholders and motivated players across all parts of the supply network to collaborate further. Moving forward, Walmart required its suppliers of leafy greens to join the IBM Food Trust, a supply chain traceability network built using a permissioned blockchain, by September 2019 (Lin, 2019).

## IBM Food Trust: A Hyperledger Exemplar

The Food Safety Modernization Act (FSMA) was signed into law by President Obama in 2011. Organizations immediately began a multiyear process to comply with the new standards. FSMA establishes a broad vision, driven by safety, quality, and sustainability. It compels organizations to think more holistically about the interrelationships among risk management, decision making, and the environment, and is raising the bar for what can and should be done in food and beverage operations, even outside the United States. To meet these heightened requirements, some organizations are turning to blockchain-based solutions.

The IBM Food Trust, one of these efforts, is built on the Hyperledger blockchain platform pioneered by an open source effort led by the Linux Foundation. This blockchain initiative, from its inception, has been driven by actual and business needs of more than a hundred large organizations that are committed to becoming early adopters.

> I have all the information I need to prevent an outbreak and I can't see it because of all the noise. Because everything is on paper and I can't connect the dots.
> —NATALIE DYERSON, VP OF FOOD SAFETY & QUALITY, DOLE

IBM Food Trust is designed to serve small, medium, and large enterprises and is one of the most active enterprise blockchain networks in operation. Growers, suppliers, and retailers can all cooperatively share details about food origin, processing events, handoffs, shipping details, and transit conditions. The end result is a degree of transparency that serves as a cornerstone for holistic and integrated food safety and quality management systems. Not only can this ease FSMA compliance, but it is tremendously beneficial for responding to food safety incidents quickly and accurately to prevent injury and illness (Newsroom, 2018; Lin, 2019).

## Circular Economy: The Textile Industry

In addition to enhancing traceability of events and transactions, many companies are being pressured to ensure traceability of their products as well. Sometimes, this pressure comes from consumers who demand information about what they buy; other times, the pressure is from regulatory bodies that need to enforce legal requirements in their countries. Whether motivated by competitive pressure or compliance, many industries are pursuing emerging technologies like blockchain to enhance transparency and enable new kinds of decision making.

Rusinek et al. (2018) explored how blockchain might be used to enable circular and sustainable textile supply chain management. A *circular economy* is an economic system, like a supply chain, that is designed to minimize or eliminate waste. The sustainable life cycle for textiles involves reuse, recycling, and remanufacturing, all to avoid the landfill (which leaves waste) or incineration (which requires energy) at end of life. The value-add for a blockchain solution in textiles—in contrast with food production, where safety is the driver—is providing information to supply chain partners so they can more effectively choose suppliers that share their core values:

> All actors involved in the supply chain—from raw material producers to consumers—can see the full environmental and social impacts of their choices and use this information as the basis for exploring alternative procurements. (Rusinek et al., 2018)

Their solution used a requirements-based approach to envision the structure of a system to support the specific needs of the textile industry. A collection of blockchains, each supporting a particular goal of the textile, was proposed. Figure 10.7 shows potential economic, environmental, social, and other data objects that would be stored on blocks. This framework could be used to start a blockchain network like IBM Food Trust, but for textiles.

## Energy Demand Management

Blockchain is also being investigated to support new business models. These include energy markets where residential, commercial, and industrial consumers may also produce energy to contribute back into the market (ASQ, 2015) and demand management in these distributed energy systems.

Wang et al. (2019) prototyped a system to explore the latter case of demand management. They implemented smart contracts to ensure data quality surrounding programmable user behaviors, and perform checks to make sure transaction records were credible and legitimate as compared with historical data. They concluded that peer-to-peer power trading is certainly feasible but, using their specific design, may not be robust to high-frequency trading and scalability to larger service areas.

## The Future of Auditing

Audits in finance and healthcare have also been identified as candidates for breakthrough improvement using blockchain solutions. The impact of blockchain on quality audits has not yet been explored by researchers and remains an open area for investigation.

In finance, Vishnia and Peters (2019) built a prototype to explore resilience given malicious actors, efficiency for high-volume transactions, and the abil-

| Metric | Information to include on block |
|---|---|
| Economic | a. Smart contract–executed transactions (e.g., payments and deliveries) [3, 4, 7, 8]<br>b. Bank access to network [3]<br>c. Insurance information [1]<br>d. Age of material or resource<br>e. Market resources and commodities prices |
| Environmental | a. Relevant environmental certifications (e.g., EU Ecolabel, FSC certification, chemical certifications like OEKO-TEX, GOTS, Cradle to Cradle) [5]<br>b. LCA impact data<br>c. Higg MSI impact data [6]<br>d. Raw materials used [2]<br>e. Chemicals used<br>f. Amount of water used<br>g. GHG emissions<br>h. Waste, by-products, and coproducts produced<br>i. Biodegradability, compostability |
| Social | a. Relevant certifications (e.g., Fair Trade, GOTS, OEKO-TEX, SA8000)<br>b. Living wages (120 percent of minimum wage) [6]<br>c. Worker age and hour restrictions; freedom to organize [6]<br>d. Gender equality [5]<br>e. Responsible care instructions [5]<br>f. Responsible disposal instruction [5] |
| Functional | a. Intended use [2]<br>b. Capabilities (e.g., heating, cooling, data tracking, water resistant, antimicrobial, UV protection) [9]<br>c. Design for X (e.g., environment, disassembly; privacy) [10]<br>d. Warrantee information [1]<br>e. Repair information<br>f. Quality control information |

@2018, ASQ

FIGURE 10.7. Block contents for sustainable, circular textiles (Rusinek et al., 2018).

ity to support dark pools (trading executed on private networks that are not accessible to the trading public). Based on their results, they recommended a "governed blockchain," with distributed nodes at each trading venue and optionally at the regulator as well. Ortman (2018) concluded that the biggest benefit for applying blockchain to financial audits may actually be financial and come from the ability for firms to dispatch smaller audit teams to accomplish typical jobs:

> Auditing, the examination of financial transactions, has always been an essential part of the regulation of markets. Proving the order of events and verifying who the action takers, i.e., traders were in each event, is fundamental to ensure that the markets are operating successfully without interruptions and added risk to market participants . . . [regulators and] individual exchanges ensure veracity and validity of the reported trade actions. (Vishnia & Peters, 2019)

In healthcare, blockchain has also been explored for record keeping, particularly for electronic health records (EHRs) (Anderson, 2018). Angraal et al. (2017) explored using blockchain to improve authenticity and transparency for several use cases, including EHR and claims processing. Although the desire for improved auditability of health records was clear, they found that several issues may delay implementations, including concerns about scaling, cost, storing personally identifiable healthcare data in a ledger that may be publicly exposed, and the potential to "re-identify" patients by making inferences.

## IS A BLOCKCHAIN SOLUTION APPROPRIATE?

Despite the excitement and promise of blockchain applications, there are many cases in which a traditional relational database solution will be more appropriate than a blockchain solution. Wüst and Gervais (2018) provide guidance on how to tell the difference in Figure 10.8. They explain that the decision depends primarily on five factors:

- **Statefulness**—If you need to exchange information but not store it, a blockchain solution is not appropriate (in these cases, a database may not be required either)

- **Multiple writers**—If a single person, organization, or agent commits new records to the data repository, blockchain is probably not necessary
- **Trusted third party (TTP)**—If an external agent is always available to verify transactions and you do not need to rely on distributing the task of verifying data, blockchain may not be needed
- **Trust in parties that will commit data**—If you can monitor and verify all people and agents that will be committing data (for example, through logins on a corporate network), blockchain may provide a solution that is too heavy for your needs
- **Public verifiability**—Sometimes members of the general public will want or need to verify details of any transaction that is stored in the system; if all the other conditions apply, this determines whether a public or private permissioned solution is better.

The default position for organizations should be that they will not use blockchain unless a compelling business case can be made, preferably using the guidelines above. If blockchain is an option, small pilots or prototypes should be conducted, with data entry, validation through smart contracts, and access to blockchain analytics all thoroughly investigated prior to a larger commitment.

## THE BOTTOM LINE

A blockchain keeps an immutable record of events, transactions, or both. Blockchain solutions have the potential to help organizations significantly improve data quality, reduce fraud and tampering, and substantially improve visibility and transparency. As a result, they are ideal for solving large-scale monitoring problems that require trust and cooperation across organizational and global boundaries. A blockchain solution is best when multiple organizations that do not all trust one another need to interact with a system, and change states and data, but cannot agree on a TTP to validate all transactions:

- Blockchains are good for maintaining records of events and transactions.
- A cryptographic hash is a unique string of characters and numbers that can be produced from a data object. Although you can reliably convert the object into a hash, you can never reverse the process and recover the object from the hash. A common hashing algorithm is SHA-256, which generates 64-character

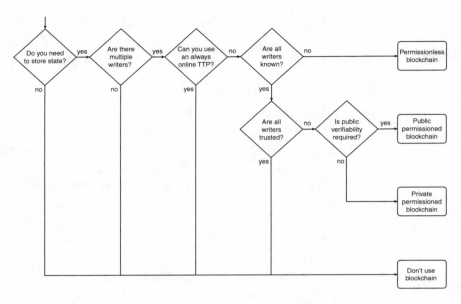

FIGURE 10.8. A flowchart for determining whether you need blockchain
(Wüst & Gervais, 2018).

hashes that are nearly always unique (that is, they have good collision avoidance).

- A blockchain is formed by creating a new data structure (a block) that contains a unique cryptographic hash of the block that came before it, as well as the hash from the new block of information that contains the previous hash.
- The ledgers from public blockchains are visible to everyone, while those from private blockchains are only visible to authorized parties.
- Permissioned blockchains accept updates only from trusted participants.

Blockchain solutions provide supply network or business ecosystem partners a single source of truth for event and transaction data at any point in time, and across the entire network of participants, while letting data producers maintain ownership and control over their own information. Blockchain hype, however, means that many organizations are investigating blockchain when a relational database would be more appropriate. To determine whether you should consider blockchain, use the guidance from Wüst and Gervais (2018). If you need to store information that will persist, collect information from multiple writers, and cannot rely on a TTP to be online all the time to validate transactions, a blockchain may be appropriate.

## REFERENCES

American Society for Quality. (2015). *ASQ future of quality study.* https://asq.org /quality-resources/research/future-of-quality

Anderson, J. (2018). *Securing, standardizing, and simplifying electronic health record audit logs through permissioned blockchain technology* (Technical Report TR2018-854). PhD thesis, Dartmouth College.

Angraal, Suveen, Harlan M. Krumholz, and Wade L. Schulz. (2017). "Blockchain technology: applications in health care." *Circulation: Cardiovascular Quality and Outcomes* 10 (9): e003800.

Benton, M., and N. Radziwill, N. (2017). "Quality and innovation with Blockchain technology." *Software Quality Professional* 20 (1).

Boddy, M. (2019, May 25). "Two miners purportedly execute 51% attack on Bitcoin cash blockchain." *CoinTelegraph.* https://cointelegraph.com/news/two-miners-purportedly -execute-51-attack-on-bitcoin-cash-blockchain

Freeman, G. (2019). "Food integrity: A practitioner's guide to navigating food quality and safety standards." *Intelex.* https://www.intelex.com/resources/insight-report/food -integrity

Hackett, R. (2017). "Why big business is racing to build blockchains." *Fortune Magazine.* https://fortune.com/2017/08/22/bitcoin-ethereum-blockchain-cryptocurrency/

Hackius, N., and M. Petersen. (2017). "Blockchain in logistics and supply chain: Trick or treat?" In *Proceedings of the Hamburg International Conference of Logistics (HICL)*, 3–18.

Kamath, R. (2018). "Food traceability on blockchain: Walmart's pork and mango pilots with IBM." *The Journal of the British Blockchain Association* 1 (1): 3712.

Lin, C. F. (2019, May). "Blockchainizing food law: Implications for food safety, traceability, and sustainability." In *Conference on Food Law and Policy: Food Safety and Technology Governance.* Taipei.

Maersk. (2019, July 2). "TradeLens blockchain-enabled digital shipping platform continues expansion with addition of major ocean carriers Hapag-Lloyd and Ocean Network Express" [Press release]. https://www.maersk.com/news/articles/2019/07 /02/hapag-lloyd-and-ocean-network-express-join-tradelens

McQuade, M. (2018, August 22). "The untold story of NotPetya, the most devastating cyberattack in history." *Wired.* https://cyber-peace.org/wp-content/uploads/2018 /10/The-Untold-Story-of-NotPetya-the-Most-Devastating-Cyberattack-in-History -_-WIRED.pdf

Newsroom. (2018, October 8). "IBM Food Trust expands blockchain network to foster a safer, more transparent and efficient global food system." *Lovely Mobile News.* https://lovelymobile.news/ibm-food-trust-expands-blockchain-network-to-foster-a -safer-more-transparent-and-efficient-global-food-system/

Ortman, C. (2018). *Blockchain and the future of the audit.* Senior thesis, Claremont McKenna College. https://scholarship.claremont.edu/cgi/viewcontent.cgi?referer =https://scholar.google.com/&httpsredir=1&article=2903&context=cmc_theses

Rusinek, M. J., H. Zhang, and N. Radziwill. (2018). "Blockchain for a traceable, circular textile supply chain: A requirements approach." *Software Quality Professional* 21 (1).

Vishnia, G., and G. Peters. (2019). "AuditChain: A trading audit platform over blockchain." https://papers.ssrn.com/sol3/papers.cfm?abstract_id=3391634

Wang, X., W. Yang, S. Noor, C. Chen, M. Guo, and K. H. van Dam. (2019). "Blockchain-based smart contract for energy demand management." *Energy Procedia* 158: 2719–2724.

Wüst, K., and A. Gervais. (2018, June). "Do you need a blockchain?" In *2018 Crypto Valley Conference on Blockchain Technology (CVCBT)*, 45–54. IEEE.

## CHAPTER ELEVEN

# PERFORMANCE EXCELLENCE

*The great enemy of communication, we find, is the* illusion *of it. We have talked enough; but we have not listened. And by not listening we have failed to concede the immense complexity of our society—and thus the great gaps between ourselves and those with whom we seek understanding.*

—WILLIAM H. WHYTE IN *FORTUNE* (1950)

To serve your community, you aren't limited to volunteering in soup kitchens or making donations to charity. In 2009, I led a five-person team of undergraduate and graduate students working on a "service learning" project—donating their new skills and expertise in quality and process improvement to help the local community. Our focus was Social Services, an agency long challenged by an extended placement time for foster children.

At the beginning of the project, the agency shared its as-is process performance for child placement. It took (on average) 157 days to place a child in out-of-home care, even when the assigned guardians were relatives. That's nearly four months. Because the majority of placements were the result of court orders or serious safety concerns, any delays could seriously impact the well-being of the child. Understandably, Social Services wanted to improve its process.

Documenting the way an organization works gets everyone—literally—on the same page. And working with Social Services, we learned this lesson the hard way.

There were five key people at the agency working on this process, located at three different buildings around town. I matched each of my students to one of these key players and gave them three weeks to go meet their partner and document the process flow from that person's perspective. Usually, this activity results in a collection of mostly similar process flows. When the group reconvenes, they use the slight differences to call out places where there is confusion or places where process participants have slightly different expectations.

You can use these gaps to identify issues that might be negatively impacting the process, and then fix them during the improvement process.

But this time was different. The five process flows looked nothing like one another, and it was clear that there would be no reconciliation. I had the students swap agency partners, and sent them back to duplicate the same assignment. Three weeks later they came back with the exact same results—five completely different processing and handoff processes.

Although we had grand plans for using discrete event simulation to investigate and improve this process, I had to abandon that goal. We needed a single process flow diagram, and we needed to confirm that each of the five key players was meeting the expectations of the others.

It took six more weeks and two in-person meetings (with everyone around the same table and collaborating at whiteboards) to come up with that single sheet of paper. By the end of the sessions, the five key people were apologizing to each other, saying that they had no idea they weren't doing exactly what the others were expecting, but that they were happy to start working from the process flow they had all created as soon as possible.

Months later, we found that the placement times had dropped to an average of 21 days. That's nearly three potentially life-saving months per child. We made a real impact by doing something simple: getting all the players on the same page:

> The truth is that having data, simply for the sake of having it, doesn't benefit quality improvement efforts. What manufacturers need is real-time process control and data interrogation. . . . Effective process control in dairy production is all about consistency, not only in taste profile and packaging, but also in how everyone works on the plant floor. Critical to achieving this consistency is making sure everyone is on the "same page." . . . This means standardization in one quality intelligence system, overall production processes and even naming conventions for parts, features and processes. The right people need to get the right information at the right time—otherwise, too many opportunities for improvement will be lost. To aid in consistency, workflows can provide prescriptive steps to walk users through timely response. That way, in the event of an issue, plant personnel always follow the same best practices. (Weisbrod, 2019)

Aligning workers around a shared understanding of the work may require "coming to terms" with one another. Especially when emerging technologies are concerned, it is important to clarify key terms to confirm that everyone is talking about the same thing, no matter what your industry:

> Take a public safety-first responder project I was involved in years ago. We stakeholders weren't speaking the same language. The municipality referred to "edge" as the data visible to dispatchers at its network operations center. The drone company defined "edge" as the computing and AI capability running in real time on the drone. We considered "edge" the transport and compression capabilities to determine response time in the cellular network and on the drone. Technically speaking, all these are edge components. (Allgood, 2019)

Lack of a common vernacular, a basis for understanding one another, or mutual trust can threaten any project or initiative. Roles and responsibilities should be clearly articulated at the individual and team levels. Inputs, outputs, and decision-making processes should be clarified and written down so that no one has to guess or make assumptions. Project details, initiatives, agreements, and updates should be captured and formalized to eliminate subjectivity as much as possible. *The greatest enemy of communication is the illusion of it.*

## WHAT IS PERFORMANCE EXCELLENCE?

A *model* is a map for understanding a complex system (Edgeman, 2019); frameworks for performance excellence are models that help build understanding and trust in organizations. As defined by the Baldrige Excellence Framework and explained in detail by Thürer et al. (2018), performance excellence is

> an integrated approach to organizational performance management that results in
>
> 1. delivery of ever-improving value to customers and stakeholders, contributing to ongoing organizational success;
> 2. improvement of your organization's overall effectiveness and capabilities; and
> 3. learning for the organization and for people in the workforce.

These frameworks go by several names, including business excellence, operational excellence, and organizational excellence. They provide guidance for describing work, aligning objectives with strategy, and monitoring progress so that organizations can more easily adjust dynamically. These frameworks share common features, including a focus on the customer, core values that include efficiency and effectiveness, and an emphasis on risk-based thinking and data-driven decision making. They provide a basis for establishing a quality culture, navigating the dynamics of control and power, and providing the discipline that leads to order and innovation (Freeman, 2019). Most significantly, using an excellence framework increases opportunities for genuine communication across functional and organizational boundaries, and focuses everyone in the organization on continuous improvement.

> Competitive advantage comes from the combined impact of improvements made over time by each and every employee. If you and your organization aren't continuously improving, you are being left behind.
> —REBECCA SIMMONS, ASSISTANT PROFESSOR OF INTEGRATED SCIENCE AND TECHNOLOGY, JAMES MADISON UNIVERSITY

This chapter examines nine frameworks from the perspective of the shifting technological landscape: ISO 9001:2015, TQM, lean management (which incorporates practices from the Toyota Production System and kaizen), Six Sigma (and Lean-Six Sigma), the Shingo model, Capability Maturity Model Integration, agile methods, the European Foundation for Quality Management model, and the Baldrige Excellence Framework.

## CORE VALUES

Organizations pursue excellence to achieve sustainable, customer-driven results, and to provide a solid foundation for innovation (Thürer et al., 2018). To be successful, a delicate balance among autonomy, coordination, and control must be achieved. Selecting a quality management philosophy or excellence framework that best resonates with a company's needs, given its stage of growth and maturity, can support this journey.

In Table 11.1, core values from the nine frameworks considered in this chapter are shown. While some models are holistic (addressing strategy, operations, and competitive context), others are not (and address only aspects of performance improvement specific to, for example, operations processes).

**TABLE 11.1.  Core values associated with quality philosophies and excellence frameworks.**

| Approach to excellence | Holistic | Nonprescriptive | Core values |
|---|---|---|---|
| ISO 9001:2015 | | X | Customer focus, leadership, engagement of people, process approach, improvement, evidence-based decision making, relationship management |
| Total Quality Management (TQM) | X | X | Customer focus, total employee involvement, process focus, integrated systems, strategic and systematic approach, continuous improvement, fact-based decision making, effective communications |
| Lean management | X | X | Long-term thinking, value, flow, pull, continuous improvement, teamwork, customer focus, respect for people, information sharing, management by facts, management commitment, honesty, responsibility |
| Six Sigma/Lean-Six Sigma | | | Reducing variation, data-driven decision making, continuous improvement, breakthrough improvement, use of statistical tools, top management commitment, stakeholder involvement, customer focus |
| Shingo model | X | X | Respect every individual; lead with humility; seek perfection; flow and pull value; ensure quality at the source; focus on process; embrace scientific thinking (data-driven decision making); think systematically; create constancy of purpose |
| Capability Maturity Model Integration (CMMI) | X | | Better process leads to a better product, discipline, standardization |
| Agile methods | | | Customer responses, minimal overhead, requirements refinement; from Agile Manifesto: individuals and interactions over processes and tools, working software over comprehensive documentation, customer collaboration over contract negotiation, responding to change over following a plan |

(continued)

**TABLE 11.1.** (*continued*)

| Approach to excellence | Holistic | Nonprescriptive | Core values |
|---|---|---|---|
| European Foundation for Quality Management (EFQM) model | X | X | Adding value for customers; creating a sustainable future; developing organizational capability; harnessing creativity and innovation; leading with vision, inspiration, and integrity; managing with agility; succeeding through the talent of people; sustaining outstanding results |
| Baldrige Excellence Framework (BEF) | X | X | Systems perspective, visionary leadership, customer-focused excellence, valuing people, organizational learning and agility, focus on success, managing for innovation, management by fact, societal contributions, ethics and transparency, delivering value and results |

Source: Adapted from Hellsten & Klefsjö, 2000; Chen et al., 2016; Van Dun et al., 217; Shingo Institute, 2017; Elshafey & Galal-Edeen, 2008; van der Wiele et al., 2000.

Many of the models are nonprescriptive, meaning that they capture principles associated with best practices and not the practices themselves.

Many of these philosophies and approaches share common elements:

- **Value and impact.** Most frameworks assume that the reason people organize (as businesses or nonprofits or even informal clubs) is to provide value to someone, somewhere. These are often customers and stakeholders, but they can also be members of a community, society in general, or nonhuman stakeholders like the environment. Impact is the degree or extent to which that value is delivered.
- **Empathy and customer focus.** Nearly all models assume that to deliver value to customers, an organization must dedicate effort to understand their needs, respond to them, and adapt to changes in those needs on a continuous basis. Empathy, "the caring, individualized attention given to customers," is an important element of this process. The SERVQUAL instrument (for capturing customer perceptions of service quality) includes empathy as one of its five key drivers of those perceptions (Parasuraman et al., 1988). In addition, Coo

and Verma (2002) found that empathy was also tightly linked to the success of strategic planning and the quality of market focus.

- **Data-driven decision making.** High-performance organizations are committed to finding (and making decisions based on) truth, even when it is uncomfortable. Kendall and Bodinson (2016) tell the story of a financial services organization that had lots of financial data but no market data, benchmarks, or in-process performance measures. By gathering and making these other data sources transparent across the organization, it was able to cultivate a culture where "data would not be used in a punitive way but as a source of information that would lead to process improvement."

- **Systems thinking.** If two people can dig a five-foot hole in two hours, how deep will four people dig in six hours? If you started calculating in your head, you're missing some of the elements that systems thinking might bring into play. For example, what are the requirements for digging the hole, the environmental conditions, or the characteristics of the workers? Are they all trained? Do they have the capability and motivation to complete the task? Do we have a permit for digging? Will this negatively impact other nearby projects? Is this digging legal, acceptable, or in a protected area? There are lots of unanswered questions that will all influence how deep the hole can get with our four people in six hours. Systems thinking requires us to step back and outline as many of them as possible before we start digging.

- **Learning.** Continuous improvement happens when we learn, collectively, about ourselves, our work, our interactions, and the competitive environment we do business within. Kovach and Fredendall (2013) found that learning is what makes a continuous improvement successful, not just the presence of the structures or practices for improvement.

Each of these core values is overlaid onto a basis for action in the performance excellence frameworks. Defining actions, examining whether those actions yield desired results (and if not, making appropriate adjustments as quickly as possible), and balancing efficiency and effectiveness will all still be essential (Thürer et al., 2018). For example, an organization that implements lean management may decide to accept some waste in processes if it leads to enhanced customer satisfaction. Although a digital transformation initiative to replace a human customer service team with chatbots and voice-response systems may have a substantial ROI and quick payback period, if customers become unhappy as a result, holistic performance has not truly improved.

## PERFORMANCE EXCELLENCE FRAMEWORKS

Every organization is a complex assortment of people, processes, and technologies. When a performance excellence framework is used as the basis for communication, coordination, and improvement across organizational boundaries, the workforce shares a common concept of what is required. In addition, leaders can more effectively identify gaps in processes or issues with organizational design that may be negatively impacting performance.

Determining which framework is best is always contextual. An organization's choice should be influenced by the size of the company, its life-cycle stage (e.g., startup, growth, expansion, maturity), the company's core values and culture, its immediate needs, and the background and capabilities of the workforce:

- Highly technical frameworks like Six Sigma can be inaccessible to workers on shop floors, in contrast with the principles of lean management, which are broadly understandable (Mika, 2006).
- Small, high-growth startups would benefit less from the rigor and complexity of CMMI but could experience substantial performance improvements by implementing some agile practices (Staples et al., 2007; Pino et al. 2008).
- Software companies may struggle to see performance benefits from ISO 9001, which can only help organizations achieve a baseline level of process maturity as described by CMMI (Ijaz et al., 2016). Software companies may benefit more from domain-specific alternatives like SPICE and ISO/IEC 29110—systems and software engineering (Sanchez-Gordon, 2017).
- Some frameworks are more common in particular industries. For example, the BEF is well known and understood in healthcare, education, and government/nonprofits.

With these things in mind, nine frameworks are briefly summarized. The role and relevance of each framework in the context of Industry 4.0 and emerging technologies are also explored.

### ISO 9001:2015

The ISO 9001:2015 quality management standard is "a set of criteria that, when satisfied by an organization, enables it to demonstrate their capability and in doing so give customers confidence that they will meet their needs and expectations" (Hoyle, 2017). There are ten clauses (criteria categories) in ISO 9001:2015. In the text of the standard, detailed guidance is provided on the following:

1. Scope
2. Normative references
3. Terms and definitions
4. Context of the organization
5. Leadership
6. Planning
7. Support
8. Operation
9. Performance evaluation
10. Improvement

Clauses 1 through 4 establish the context, while Clauses 5 through 10 identify requirements for strategy development, execution, and improvement. In the latest update, there are fewer requirements on documentation (for example, a quality manual is no longer strictly required), but processes must still be documented and records must be maintained and retained. ISO 9001:2015 was written to align with the 10-clause Annex SL standard, making it look and feel like the other standards that use Annex SL. This update makes it easier to integrate environmental management, occupational health and safety, and quality processes, improving performance across the board while streamlining audits.

Although risk management has always been part of ISO 9001, it takes a much more prominent role in ISO 9001:2015. For example, Clause 4 requires that you characterize your organization's capabilities and desired outcomes in the context of internal and external stakeholders' needs. The purpose is to help organizations recognize the need to tie structures for management and continuous improvement to business outcomes.

Each organization can determine the appropriate level of rigor for its risk management practices, and ISO 31000 can be used to supplement ISO 9001 for this purpose. Once the organizational context is identified (Clause 4), the risk assessment process has three parts (identification, analysis, and evaluation), followed by risk treatment. Treatment means making a choice: you can accept the risk (or just agree to let it happen), avoid the risk (by ignoring it or somehow externalizing it), transfer the risk (for example, by buying insurance), or implement controls to reduce the risk. At the core of this decision is identifying the potential negative consequences, their likelihood, and the impact on your organization.

> ISO 9001 is a common sense framework that can be used in any industry, environment, or organization to support, scale and evolve your operational processes. It provides a solid starting point for any digital transformation initiatives you may want to pursue.
> —NICKY JAINE, DIRECTOR OF QUALITY AND CONTINUOUS IMPROVEMENT AT INTELEX TECHNOLOGIES ULC

In the past, ISO 9001 has been criticized because companies can "do all the things right"—that is, operate according to documented processes and procedures—and still not be guaranteed to "do all the right things." Many organizations have reported that it was very easy to achieve conformance even without a viable business model using earlier versions of ISO 9001 (Priede, 2012; Aba et al., 2015). Many of the changes in ISO 9001:2015 were made to remedy this. In the context of Industry 4.0 and emerging technologies, ISO 9001 should accommodate improvements in intelligence and automation well, although it may not be as robust in its support for enhanced connectedness.

## Total Quality Management (TQM)

In the 1980s, the notion of performance excellence began to grow from the more limited roots of the quality practice in manufacturing. TQM applied to all industries, fueled by scientific management and incorporating principles from Deming's system of profound knowledge, including psychology and systems thinking. Evans and Lindsay (2005) contrasted TQM with the Six Sigma philosophy, explaining that TQM

- focused more on worker empowerment and teams,
- provided tools for quality improvement activities within functional areas or departments,
- promoted tools and concepts that were straightforward and mostly qualitative, and
- emphasized improvement for improvement's sake, rather than the impact of process and system improvement on business outcomes.

Although extremely popular at the time, TQM became a catch-all label for all quality management practices rather than an actionable collection of tools and best practice. Even today, researchers such as Bergman and Klefsjö (2010) consider Six Sigma, ISO 9000, and lean management all part of the TQM concept. This breadth and conflation may have contributed to its downfall as a management approach:

> Considerable confusion arises from the rhetoric of the quality movement. Many confuse quality as a competitive issue with the particular name we assign to the efforts taken to achieve that end at any given time. In the course of the 1980s, the term total quality management became dominant. For some, TQM means simply doing quality control in an environment of good communication and feedback. For others, it requires participative management with strong leadership from the top. Still others think of quality as doing things right the first time or conceive of it exclusively in terms of defect-free products. Many equate it with customer satisfaction. Still others see quality as a toolkit of quality methodologies. Finally, some see it as a management model. . . .
>
> With management and media dynamics so close in America, almost as soon as such a term spreads widely, it becomes a symbol for everything that goes wrong under that rubric. By the mid-1980s, predictably, articles began appearing with the inevitable title "Beyond Quality." By the early 1990s, the very name TQM was already stigmatized and increasingly out of favor. . . . [Many] had come to believe that the name was a liability. (Cole, 1999)

Although TQM fell out of favor, the principles and practices it espoused are solid and many provide the foundation for other excellence frameworks. In fact, van der Wiele et al. (2000) note that the ISO 9000 series grew from the foundation of TQM, providing organizations with an actionable route to demonstrate and validate adherence to quality principles. Today, TQM is rarely mentioned, but the predominance of "modern" frameworks like agile and lean suggest that its core principles are deeply embedded in the corporate zeitgeist.

## Lean Management

The concepts of lean production and lean management emerged from practices developed at the Toyota Motor Company in Nagoya, Japan, by Eiji Toyoda and Taiichi Ohno. After observing the production system at Ford's Rouge plant in Detroit in 1950, they felt they could improve it. In addition, they were aware that Ford's mass production environment would not suit Japanese workers, who expected to be treated well and would not put up with the substandard work environment that the American "guest workers" would tolerate.

By the late 1950s, Toyota had discovered many things. First, Ohno perfected a process for quick exchange of the stamping dies used to form steel auto components. This process, which could be performed by the workers, made it possible to produce smaller batches. Being able to produce smaller batches of stampings reduced inventory holding costs and revealed quality problems much earlier in the process. Empowering anyone to call out potential quality issues yielded far less waste and rework, even when the production lines were halted.

"To make this system work at all—Ohno needed both an extremely skilled and a highly motivated workforce" (Womack et al., 1990). The salary structure in Japan dictates that senior workers receive much higher pay. Long tenures at a company are expected, since switching companies resets the worker's position on the pay scale regardless of age. To create this skilled and motivated workforce, Toyota offered lifetime employment and began strategically cultivating workforce capabilities.

All the mechanics of the Toyota Production System (TPS) had taken form by the late 1960s. Today, companies that use lean management as their organizing framework emphasize the same principles that Toyota did many decades ago:

- **Identify value.** What does the customer want and need? All definitions of value should come from the customer and should be informed by an examination of the full experience that a customer has with a product or service. This often requires looking beyond the boundaries of your own product or service to gain an appreciation for the customer's perspective.
- **Map the value stream.** To draw out value-adding and non-value-adding steps from a process, it should be mapped from a cross-functional perspective. This process helps you identify and remove wastes.
- **Create flow.** Once wastes are removed, interruptions or delays should be removed from the process to generate the best performance from the value-adding steps. By removing distractions, too, you can create a work environment that helps people "get into flow," removing waste from the cognitive processes that support operational processes (Csikszentmihalyi, 2013).

- **Establish pull.** The waste generated by waiting can be eliminated when actors in a process can "pull" what they need for their process step at any time. Lean management seeks to create just-in-time systems where pull is enabled.
- **Seek perfection.** Lean thinking and a spirit of continuous improvement (*kaizen* – 改善) should be "baked into" your corporate culture. Where there is an opportunity to improve, it should be pursued. When everyone in the organization is tuned to reducing waste, improving flow, and capturing opportunities for improvement, progress is rapid and visible.

Reducing waste (*muda* – 無駄) is also a common theme in lean practices. This includes the waste associated with transport, inventory, motion, waiting, overproduction, overprocessing, defects, and underutilization of skills (these can be recalled using the mnemonic TIMWOODS). In addition, confusion (people not sharing a common understanding of their goals or process, or believing that they do when in fact they do not) can be a particularly pernicious waste. Reducing overwork (*muri* – 無理) and unevenness (*mura* – 斑) is also important. In fact, anytime you see people in your office performing heroic efforts to solve problems or meet deadlines, that is waste—it is to be avoided in lean by developing processes that respect effective resource loading.

> Toyota/Lean principles remind us that we should not try to automate a bad process. We should use well-tested technologies that serve our people and our processes.
>
> —MARK GRABAN, AUTHOR OF *MEASURES FOR SUCCESS AND THE EXECUTIVE GUIDE TO HEALTHCARE KAIZEN*

Some major companies, like the conglomerate Danaher (and its spin-off, Fortive, also a conglomerate), have customized their recipe for deploying TQM and lean principles throughout the workforce. The Danaher Business System (DBS) and Fortive Business System (FBS) are applied at each of the group's operating companies, providing a shared conceptual model for how to organize and improve work. Forms and templates for several lean practices have been customized to match the types of businesses they specialize in, and employees trained at one operating company can easily shift to another, maintaining a strong culture of quality:

The [Danaher] operating model creates value by emphasizing discipline and continuous improvement. This is particularly important for the Danaher businesses with high average gross margins. Unless well-managed, such margins have a habit of being self-destructive, as they tend to encourage lax management practices. Danaher also generates considerable value by applying DBS to newly acquired businesses. The company has repeatedly improved operating margins by seven percentage points or more in what were already high-margin businesses at the time of acquisition. For example, after Danaher's acquisition of Tektronix in 2007, its sales grew by 14.9% and margins increased to 15.8% in 2008. (Pidun et al., 2019)

Lean is not just about efficient production; it's also about the effective flow of information from strategy through to action. By creating more value with fewer resources, organizations benefit, but so will customers, partners, and suppliers. These factors make lean management ideal for operating Industry 4.0 and other digitally enabled business models.

## Six Sigma/Lean Six Sigma

Strictly speaking, Six Sigma is a collection of statistical tools for reducing variation (including process control, process capability, and design of experiments). But after Motorola won the Malcolm Baldrige National Quality Award (MBNQA) in 1988, Six Sigma was increasingly adopted as a management system and philosophy focused on fact-based, data-driven decision making, supported by statistical methods.

While the Six Sigma philosophy does place quality culture at the forefront, it does not specifically address leadership, governance, strategic planning, or the link between strategy and execution to the extent that other models do. Even so, many organizations with a strong commitment to Six Sigma at all levels have demonstrated performance improvements and better business results (Pyzdek, 2001).

More recently, Pepper and Spedding (2010) recommended a combined approach: to apply the business improvement philosophy and tools of lean while leveraging the statistical tools of Six Sigma to drive optimal results. They used the example that reducing inventory levels can lead to greater variability in demand satisfaction, and higher risk exposure, when lean management is applied without broader awareness of the production system's context in mind.

## Shingo Model

The Shingo model is based on four categories into which ten guiding principles are classified (Shingo Institution, 2017). The influence of lean management and Deming's principles are evident:

- **Cultural enablers:** lead with humility, respect every individual
- **Continuous improvement:** follow the principle of flow and pull, ensure quality at the source, focus on the process, embrace scientific thinking, seek perfection
- **Enterprise alignment:** create constancy of purpose, think systematically
- **Results:** create value for the customer

The model is maintained by the Shingo Institute at Utah State University, which administers the Shingo Prize for Operational Excellence to applicants worldwide. Established to honor Shingo's contributions to the TPS, the award process examines the extent to which the ideal behaviors of a quality culture are demonstrated. Joseph A. DeFeo, chairman and CEO of Juran Global, said, "I like to tell people on your way to being excellent in your manufacturing company . . . win the Shingo prize first then go on to win Baldrige. There's a pretty good chance it is a stepping stone" (quoted in Bailey, 2016).

## Capability Maturity Model Integration (CMMI)

In the mid-1980s, the Software Engineering Institute at Carnegie Mellon University partnered with MITRE Corporation to develop a framework to help organizations improve their software processes. Motivated by the need to promote process maturity for government contractors, the TQM-based Software Capability Maturity Model (SW-CMM) was initially released in 1991. It was based in part on Crosby's Quality Management Maturity Grid and its six stages of measuring quality in an organization (Table 11.2).

CMMI grew from what has become the "standard" five (plus one) level maturity model:

- **Level 0: Incomplete.** Work is ad hoc and may or may not get done.
- **Level 1: Initial or Chaotic.** Work gets done, but may be costly or incomplete. Workers are stressed and anxious.
- **Level 2: Repeatable.** Processes and projects are defined, managed, and controlled.

**TABLE 11.2. Crosby's six stages of quality measurement.**

| Measurement category | Uncertainty | Certainty |
| --- | --- | --- |
| Management understanding | Leadership has no concept of how a quality system improves outcomes | Leadership believes that quality systems are an essential part of operations and directly contribute to delivering results |
| Quality organization status | Hidden | Thought leader in quality |
| Problem handling | Problems are fought | Problems are prevented |
| Cost of quality as a % of sales | 20% | 2.5% |
| Quality improvement actions | No organized quality activities | Quality is a normal and continuous activity |
| Company quality posture | "We don't know why we have problems with quality" | "We know exactly why we do not have problems with quality" |

Source: Adapted from Paulk, 2009.

- **Level 3: Defined.** Organization is proactive rather than reactive, and demonstrates alignment across functional areas and organizational levels.
- **Level 4: Managed.** Organization is data-driven; processes are predictable, monitored, adjusted. Outcomes meet customer and stakeholder needs.
- **Level 5: Optimized.** Organization is stable and flexible, and continuous improvement is a way of life.

Although the Capability Maturity Models for Integration (CMMI) has since been retired, the CMMI emerged in its place. This is a family of models, including CMMI-DEV for software development, that consist of various practice areas (PAs). CMMI v2.0, released in March 2018, has 20 PAs (Table 11.3). In a CMMI assessment, each of the PAs is evaluated against the maturity levels.

Paulk (2009) notes that "one drawback to the use of maturity levels, however, has been the dysfunctional behavior associated with organizations more concerned with assessment results than improving against business objectives."

**TABLE 11.3. PAs in CMMI v2.0.**

- Causal analysis and resolution (CAR)
- Decision analysis and resolution (DAR)
- Risk and opportunity management (RSK)
- Organizational training (OT)
- Process management (PCM)
- Process asset development (PAD)
- Peer reviews (PR)
- Verification and validation (VV)
- Technical solution (TS)
- Product integration (PI)

- Supplier agreement management (SAM)
- Managing performance and measurement (MPM)
- Process quality assurance (PQA)
- Configuration management (CM)
- Monitor and control (MC)
- Planning (PLAN)
- Estimating (EST)
- Requirements development and management (RDM)
- Governance (GOV)
- Implementation infrastructure (II)

**TABLE 11.4. The Agile Manifesto, established in 2001.**

We are uncovering better ways of developing software by doing it, and helping others do it. Through this work we have come to value:

- Individuals and interactions over processes and tools.
- Working software over comprehensive documentation.
- Customer collaboration over contract negotiation.
- Responding to change over following a plan.

That is, while there is value on the items on the right, we value the items on the left more

Source: Beck et al., 2001.

## Agile Methods

Agile methods emerged as a way to improve product and process quality in software engineering in the late 1990s. Agile was originally envisioned as a response to the cumbersome waterfall model of development (a stage-gate approach where requirements are gathered, followed by design, implementation, test, and release). The waterfall approach was slow, error prone, and often resulted in a product that no longer met customer needs (because they had changed during the time it took to build the software). In 2001, a group of 17 software engineers met to envision a new way of designing quality into software, which they called the Agile Manifesto (Table 11.4).

Since 2001, many agile practices have emerged to make these principles actionable. These include daily stand-up meetings, short development iterations, maintaining close ties with the customer, pair programming, maintaining a product backlog by splitting the work into tiny slices, test-driven development, and release planning. Other practices, like agile documentation (drawing on whiteboards and taking pictures of flowcharts rather than drawing them in software programs), have also taken hold. Kanban boards, signaling the flow of work from planning to release, are now as common in marketing departments as they are in software development teams.

Agile was not intended to provide organizations with the ability to rationalize a lack of planning or documentation. In fact, Paulk (2002) concluded that "agile methodologies *imply* disciplined processes, even if the implementations differ in extreme ways from traditional software engineering and management practices; the extremism is intended to maximize the benefits of good practice. The SW-CMM tells *what* to do in general terms, but does not say *how* to do it; agile methodologies provide a set of best practices that contain fairly specific how-to information—an implementation model—for a particular kind of environment." Agile methods are thus excellent for rapid prototyping and daily work management, but less adept at helping organizations manage the link between strategy and execution or the special nature of Industry 4.0 initiatives.

## European Foundation for Quality Management (EFQM)

The quality model most commonly used in Europe is the European Foundation for Quality Management (EFQM) Excellence Model. It consists of nine

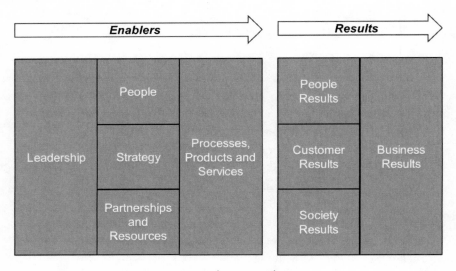

FIGURE 11.1. The EFQM Excellence Model (EFQM, 2013).

elements: five "enablers" that build the quality system and four results categories (people, customers, business, and society). It is the basis for the European Quality Award and is based on the foundations of TQM (Figure 11.1).

The self-assessment process (and award application approach) used by EFQM is flexible so that applicants can modify their approach according to their unique organizational context. Nagyova and Markulic (2016) describe the assessment approach used to evaluate a public university in the Slovak Republic. This applicant wrote five sections, one for each of the enablers. Within each section, the applicant articulated self-assessed strengths, opportunities for improvement, and evidence to support each of the claims. There are no specific aspects of EFQM that differentiate its applicability for initiatives involving emerging digital technologies.

## Baldrige Excellence Framework (BEF)

Creating a strategy that tells your organization where it needs to go is easy, but actually getting there can be hard. The Baldrige Excellence Framework (BEF) has been developed and continually improved by senior leaders from all industries over three decades, in part to support the process for identifying winners of the annual MBNQA. Like many of the other models, it draws heavily from TQM but makes the principles actionable.

Emphasizing the links among culture, strategy, execution, and results, BEF provides hundreds of self-study questions for organizations to critically examine seven interconnected areas:

1. **Leadership.** Senior leaders set the tone and model the behavior they wish to see throughout the organization. They establish the vision, mission, and values, develop protocols for governance and communication, and create the conditions for success.
2. **Strategy.** BEF provides criteria questions for developing, deploying, and implementing a strategy across an organization. This process must be supported by development of relevant core competencies, development and execution of action plans, and establishment of timetables that are matched to resource availability.
3. **Customer Focus.** Figuring out what customers want should be a data-driven, holistic, continuous process. Mapping customer needs to product and service offerings that appeal to defined customer segments, and action plans to engage customers via two-way communication, will impact the potential for success.

4. **Data and Knowledge Management.** Measuring and improving performance using data is also critical. BEF criteria questions help organizations increase transparency, avoid data silos, and promote decision making based on accurate and complete information.

5. **Workforce Management.** Because strategic objectives can't be realized without capable, engaged, motivated employees (who have manageable workloads), a solid workforce management plan will directly address capability building and retention. Effective workforce management will also help the entire organization adapt to changing needs and changing demands from the market.

6. **Operations.** Understanding what you do, and what other departments in your organization do, is the basis for communication and collaboration. Written procedures for standard work also provides a basis for growth and innovation, and can be used to align with processes in the broader supply network.

7. **Results.** Governance, work processes, and support systems are only as good as the results they can achieve. The final category in BEF links the six process categories to overall performance in the seventh category focusing on results. Outcomes are expressed in terms of product and process results, customer-focused results, workforce-focused results, leadership and governance results, and financial and market results.

Many organizations that adopt the BEF also implement ISO 9001:2015, ISO 14001:2015, and ISO 45001:2018 management systems. Even with these other systems in place, BEF encourages organizations to think holistically and helps draw out gaps that may impact strategy execution. By highlighting the connections among people, processes, data, and technologies, BEF can help eliminate silos as performance is improved. Because of its emphasis on connectedness, and recent work to incorporate critical capabilities like cybersecurity, BEF is particularly well suited for helping organizations achieve success in their digital transformation efforts.

## WHAT MAKES THEM WORK

Consistently applied, broadly deployed structures to align daily work with strategic objectives and long-term stretch goals can provide the basis for an organization to achieve and innovate. This foundation is provided by the consistent application and reinforcement of behaviors that support quality and performance in digital cultures:

> Risk aversion, weak customer focus, and siloed mind-sets have long bedeviled organizations. In a digital world, solving these cultural problems is no longer optional.
>
> Shortcomings in organizational culture are one of the main barriers to company success in the digital age. That is a central finding from McKinsey's recent survey of global executives, which highlighted three digital culture deficiencies: functional and departmental silos, a fear of taking risks, and difficulty forming and acting on a single view of the customer. (Goran et al., 2017)

According to this McKinsey survey, culture and behavioral challenges are the most significant roadblocks for digital effectiveness. Because excellence models are drivers for culture and behavior, effective application with consistent commitment from senior leadership can remove these barriers. Key elements of the value system include:

- **Values made tangible.** An organization's culture is not enacted by statements of core values but by the way those core values play out in day-to-day interactions and decisions. By compelling organizations to critically examine the way they structure and carry out their work, opportunities arise for an organization's value system to be continuously revealed through everyone's actions.
- **Patterns for discipline.** The practices associated with several of the excellence frameworks establish a discipline for daily management that supports the attainment of strategic goals. Implementing these models helps free up cognitive energy that can be dedicated to tasks that require insight, creativity, or quick responses.
- **Collective effort.** Each model seeks to unify the workforce, aligning them to achieve strategic goals by adopting a shared approach. Although grassroots change is possible, broad, collective change is more likely to be sustainable.
- **Empowerment and engagement.** Toyota was the first to recognize that quickly identifying and resolving problems would require everyone to actively participate, not just the managers. Its practice (dating back to the 1960s) of empowering workers to halt production as quality issues emerge set the standard for broad involvement. An organization with a stand-alone quality department will not be as successful as one where the commitment to quality is designed into every job.

- **Phased approach.** None of these excellence frameworks require comprehensive adoption, nor will they provide benefits overnight. It takes time for the behaviors promoted by the models to take hold and become part of the organization's DNA. For example, organizations can use Baldrige one criterion at a time. Radziwill and Mitchell (2010) used Criterion 5 to rapidly develop a workforce management plan, building on mature human resources (HR) practices to identify gaps and implement improvements.

There are, however, limitations to these frameworks. The need for trade-offs is not well represented in excellence models, and most models are generic enough to be applied to any organization. Design decisions that arise when implementing excellence frameworks must be made with the unique characteristics of the organization in mind (in ISO 9001:2015, this is the organizational context in clause 4; in Baldrige, it is the Organizational Profile that precedes the criteria questions). Best practices specific to industries or domains must be identified in other ways (Thürer et al., 2017).

**TABLE 11.5. Relevance of excellence frameworks to Industry 4.0 and digital transformation efforts.**

| Approach to excellence | Connectedness | Intelligence | Automation | Notes |
|---|---|---|---|---|
| ISO 9001:2015 | | X | X | Although the updates from the previous version of the standard focus more on interconnectedness of processes and holistic risk management, this approach may be less effective than others in organizing for digital transformation. However, it can be used in conjunction with Baldrige and EFQM, which would close this gap. |
| TQM | X | | | Emphasizes connections, processes, and simple improvement tools; not as rigorous as Six Sigma (Evans & Lindsay, 2005). Many other models (ISO 9001, Six Sigma, CMMI, EFQM, Baldrige) are based on TQM. |

(*continued*)

**TABLE 11.5.** (*continued*)

| Approach to excellence | Connectedness | Intelligence | Automation | Notes |
|---|---|---|---|---|
| Lean management | X | X | X | Value stream maps make connections visible. The five lean principles (value, value streams, flow, pull, perfection) are strongly aligned with digital transformation goals that can be realized through Industry 4.0 and emerging technologies. |
| Six Sigma/ Lean-Six Sigma | | X | | Promotes use of statistical tools and models to optimize process improvement; often embedded in a TQM-based philosophy that emphasizes reducing variation (and waste, if Lean-Six Sigma). |
| Shingo model | X | | | Considered by some to be a "light" version of Baldrige; not prescriptive and not complex. |
| CMMI | X | X | X | Although CMMI has expanded beyond its initial domain as a quality management framework for software engineering, it is detailed, costly, more prescriptive than other models, and can be overly bureaucratic. High-performance, safety-critical, or highly regulated environments may benefit from the rigor of CMMI. |
| Agile methods | X | X | | Best used in conjunction with other frameworks like CMMI or Baldrige. Provides tools to simplify work processes in support of digital initiatives. |
| EFQM Model | X | X | | Provides general nonprescriptive guidance. There are fewer examples of EFQM applied to digitally oriented initiatives than all the other models except Shingo. |
| BEF | X | X | X | Particularly well suited to supporting strategies that emphasize digital transformation, in part because it is continually improved on a regular basis to incorporate critical concepts like cybersecurity. |

## THE BOTTOM LINE

Because culture and behavioral challenges are the most significant roadblocks for digital effectiveness, digital transformation efforts benefit from a solid foundation in quality. Excellence frameworks can be used to drive culture and behavior, and coupled with consistent commitment from senior leadership, behaviors that support the success of digital transformation initiatives can be established and reinforced. Holistic, nonprescriptive models may be most effective in supporting enhancements to connectedness, intelligence, and automation that are brought by emerging technologies (Table 11.5). However, more prescriptive models like CMMI may better satisfy the needs of high-risk, safety-critical, or highly regulated environments with complex software engineering requirements.

Quality management philosophies and excellence models describe *what* organizations should be focusing on and doing, but not *how* to apply practices and tools. Ultimately, it is not the tools that matter but the clarity and transparency they bring to the organization, coupled with the discipline of groups working together to follow (and continually improve) their work.

## REFERENCES

Aba, E. K., M. A. Badar, and M. A. Hayden. (2015). "Impact of ISO 9001 certification on firms financial operating performance." *International Journal of Quality & Reliability Management.*

Allgood, K. (2019, July 12). "Why wireless will drive industrial IoT." *Ericsson Blog.* https://www.ericsson.com/en/blog/2019/7/wireless-will-drive-industrial-IoT

Bailey, D. (2016, February 11). "Which way to operational excellence?" *Blogrige: The Official Baldrige Blog.* https://www.nist.gov/blogs/blogrige/which-way-operational-excellence

Beck, Kent, Mike Beedle, Arie van Bennekum, Alistair Cockburn, Ward Cunningham, Martin Fowler, James Grenning, Jim Highsmith, Andrew Hunt, Ron Jeffries, Jon Kern, Brian Marick, Robert C. Martin, Steve Mellor, Ken Schwaber, Jeff Sutherland, and Dave Thomas. (2001). "Manifesto for Agile Software Development." https://agilemanifesto.org

Bergman, B., and B. Klefsjö. (2010). *Quality from customer needs to customer satisfaction.* Studentlitteratur AB.

Chen, C. K., K. Anchecta, Y. D. Lee, and J. J. Dahlgaard. (2016). "A stepwise ISO-based TQM implementation approach using ISO 9001:2015." *Management and Production Engineering Review* 7 (4): 65–75.

Cole, R. E. (1999). *Managing quality fads: How American business learned to play the quality game.* Oxford, UK: Oxford University Press.

Coo, Low S., and Rohit Verma. (2002). "Exploring the linkages between quality system, service quality, and performance excellence: service providers' perspectives." *Quality Management Journal* 9 (2): 44–56.

Csikszentmihalyi, M. (2013). *Flow: The psychology of happiness.* New York: Random House.

Edgeman, R. (2019). *Complex management systems and the Shingo model: Foundations of operational excellence and supporting tools.* Productivity Press.

Elshafey, L. A., and G. H. Galal-Edeen. (2008). "Combining CMMI and agile methods." In *Proceedings on the 6th International Conference on Informatics and Systems,* 27–39.

European Foundation for Quality Management. (2013). *EFQM excellence model.* https://www.efqm.org/index.php/efqm-model-2013/

Evans, J. R., and W. M. Lindsay. (2005). *The management and control of quality.* 6th edition. Mason, OH: Thomson-Southwestern.

Freeman, G. (2019). *Culture of quality: Achieving success with tools, processes, and people.* Intelex. https://www.intelex.com/resources/insight-report/culture-quality-achieving-success-tools-processes-and-people

Goran, J., L. LaBerge, and R. Srinivasan. (2017, July). "Culture for a digital age." *McKinsey Quarterly.* https://lediag.net/wp-content/uploads/2018/05/0-Culture-for-a-digital-age.pdf

Hellsten, Ulrika, and Bengt Klefsjö. (2000) "TQM as a management system consisting of values, techniques and tools." *The TQM magazine.*

Hoyle, D. (2017). *ISO 9000 quality systems handbook.* 7th edition. Updated for the ISO 9001: 2015 standard. Routledge.

Ijaz, Q., H. Asghar, and A. Ahsan. (2016, August). "Exploratory study to investigate the correlation and contrast between ISO 9001 and CMMI framework: Context of software quality management." In *2016 Sixth International Conference on Innovative Computing Technology (INTECH),* 388–391.

Kendall, K., and G. Bodinson. (2016). *Leading the Malcolm Baldrige way: How world-class leaders align their organizations to deliver exceptional results.* New York: McGraw-Hill Professional.

Kovach, J. V., and L. D. Fredendall. (2013). "The influence of continuous improvement practices on learning: An empirical study." *Quality Management Journal* 20 (4): 6–20.

Mika, G. (2006). "Six Sigma isn't lean." *Manufacturing Engineering* 137 (1).

Nagyova, A., and S. Markulic. (2016). "Practical experience of the EFQM model implementation in the conditions of public university." *QMagazin.* http://katedry.fmmi.vsb.cz/639/mj116-en.pdf

Parasuraman, A., V. A. Zeithaml, and L. L. Berry. (1988). "Servqual: A multiple-item scale for measuring consumer perc." *Journal of Retailing* 64 (1): 12.

Paulk, M. C. (2002). "Agile methodologies and process discipline." *CrossTalk, Institute for Software Research* 3: 15–18.

———. (2009). "A history of the capability maturity model for software." *Software Quality Professional* 12 (1): 5–19.

Pepper, M. P., and T. A. Spedding. (2010). "The evolution of lean Six Sigma." *International Journal of Quality & Reliability Management* 27 (2): 138–155.

Pidun, U., A. Richter, M. Schommer, and A. Karna. (2019). "A new playbook for diversified companies." *MIT Sloan Management Review* 60 (2): 56–62.

Pino, F. J., F. García, and M. Piattini. (2008). "Software process improvement in small and medium software enterprises: A systematic review." *Software Quality Journal* 16 (2): 237–261.

Priede, J. (2012). "Implementation of quality management system ISO 9001 in the world and its strategic necessity." *Procedia-Social and Behavioral Sciences* 58: 1466–1475.

Pyzdek, T. (2001). *The Six Sigma handbook.* New York: McGraw-Hill.

Radziwill, N. M., and L. Mitchell. (2010, July). "Using the Baldrige criteria for observatory strategic and operations planning." In *Observatory operations: Strategies, processes, and systems III* (Vol. 7737, p. 77370E). International Society for Optics and Photonics (SPIE).

Sanchez-Gordon, M. L. (2017). Getting the best out of people in small software companies: ISO/IEC 29110 and ISO 10018 standards. *International Journal of Information Technologies and Systems Approach (IJITSA)* 10 (1): 45–60.

Shingo Institute. (2017). *Shingo model.* http://sapartners.com/wp-content/uploads /2017/07/Shingo-Model-booklet.pdf

Staples, M., M. Niazi, R. Jeffery, A. Abrahams, P. Byatt, and R. Murphy. (2007). "An exploratory study of why organizations do not adopt CMMI." *Journal of Systems and Software* 80 (6): 883–895.

Thürer, M., I. Tomašević, M. Stevenson, L. D. Fredendall, and C. W. Protzman. (2018). "On the meaning and use of excellence in the operations literature: A systematic review." *Total Quality Management & Business Excellence,* 1–28.

Van Der Wiele, A., A. R. T. Williams, and B. G. Dale. (2000). "Total quality management: Is it a fad, fashion, or fit?" *Quality Management Journal* 7 (2): 65–79.

Van Dun, D. H., J. N. Hicks, and C. P. Wilderom. (2017). "Values and behaviors of effective lean managers: Mixed-methods exploratory research." *European Management Journal* 35 (2): 174–186.

Weisbrod, E. (2019, June 19). "Turn dairy production data into quality intelligence: Manufacturers need real-time process control and data interrogation." *Dairy Foods.* https://www.dairyfoods.com/blogs/14-dairy-foods-blog/post/93680-turn-dairy -production-data-into-quality-intelligence

Womack, J. P., D. T. Jones, and D. Roos. (1990). *The machine that changed the world.* New York: Simon and Schuster.

## CHAPTER TWELVE

# ENVIRONMENT, HEALTH, SAFETY, QUALITY (EHSQ), AND CYBERSECURITY

There is no greater threat to the future of our species than the
environmental and safety impacts of the companies we operate.
—MARK JAINE, CEO OF INTELEX TECHNOLOGIES ULC

The centrifuges at the uranium enrichment plant outside Natanz in central Iran had been failing at an alarming rate. Although it was not uncommon to replace about 10% of the facility's fragile centrifuges in any
given year, unconfirmed reports by anonymous European diplomats who had
visited the plant in 2010 put that number closer to 20%–25%. Inspectors from
the International Atomic Energy Agency, who had been assigned to monitor
the facility by the United Nations, could not confirm or deny these reports. As
part of their inspection responsibilities, they verified that each centrifuge was
not harboring rogue nuclear material as a part of decommissioning, but they
were not responsible for determining or tracking the root cause of the failures.

Occupational health and safety are critical at all gaseous diffusion plants (the
innocuous name for facilities that produce enriched uranium for generating
power and building weapons). Exposure over short and long terms to uranium
hexafluoride, neptunium, plutonium, ionizing radiation, trichloroethylene and
other chemicals, and toxic metals (arsenic, beryllium, chromium, nickel) can
lead to higher rates of adverse health outcomes. These include lung, lymphatic,
brain, and pancreatic cancers, as well as leukemia (Chan et al., 2010).

Construction workers in particular are prone to health hazards at these
plants, and the majority of them are contractors who work short periods and
move between sites. Workers routinely handle transite and asbestos thermal
insulation, are exposed to dust, radon, and welding fumes, and deal with harsh,
noisy conditions on a regular basis. Hearing loss, degenerative joint disease,
and asbestos-linked lung diseases have been observed in a large proportion of

workers studies on a retrospective basis (Wages et al., 2003). Without effective management of worker and job data over (at least) the lifetime of the worker, when health impacts are uncovered—which can be years after the construction project has ended—it can be difficult to track down the worker to inform him or her of the exposure.

Safety at a uranium plant includes implementing security and access controls, protecting workers during operations and decommissioning phases, and designing work processes to prevent safety hazards. Personal protective equipment (PPE) and work processes that put distance between the workers and the hazards are common. Since structures and buildings can degrade over time, controls can be implemented such as "installation of modular work platforms, lifelines for accessing process piping ducts, corbel/beam repairs, cross bracing for wind loads, imposition of significant floor loading restrictions, and an intensive structural inspection program" (Kopotic et al., 2013). Disposal of radioactive waste using processes similar to single-piece flow can prevent safety issues associated with giant debris piles, and on-site disposal facilities with dedicated hauling roads can limit exposure to local communities.

Uranium enrichment plants can also have adverse environmental impacts, not only when processes and equipment fail but also when the plants are running smoothly. Equipment, vehicles, diesel generators, and boiler rooms can all generate air pollution. Plant operations frequently contaminate the sites where they are constructed—for example, by polychlorinated biphenyl (PCBs), trichloroethylene (a degreaser), and technetium-99 (a fission product), which can contaminate groundwater in the vicinity. Equipment is often contaminated due to uranium exposure. When mined uranium is ground and processed with acid, a radioactive sand-like byproduct (uranium tailings) is pumped as a slurry into impoundments (Carvalho, 2011). Studies have also shown that in the vicinity of uranium mining and milling operations, radiation may exceed the recommended limits for the general population. So in addition to protecting workers, there is a responsibility to protect local and regional communities, during regular operations as well as in the unfortunate event of a nuclear accident or meltdown.

At any uranium enrichment plant, highly sensitive centrifuges have to be carefully controlled and regularly monitored. This is critically important because the tiniest issues could cause the devices to explode. In 2006, the head of the Atomic Energy Organization in Iran noted that before technicians wore gloves to assemble the centrifuges, germs on their hands could degrade the materials to the point where they would disintegrate on spin-up (Zetter, 2014).

In Natanz, the human-machine interfaces (HMIs) in the control room showed no issues during the inspection. Despite the high failure rates, the monitor data indicated regular activity patterns and production output. But behind the scenes, reality was much different. Many of the centrifuges were dutifully following rogue commands, executed by Programmable Logic Controllers (PLCs), that made them spin up to well beyond their design specifications long enough to destroy the devices at a higher than anticipated rate—and ultimately slow the progress of uranium enrichment.

What was going on? The field devices in the Iranian plant, including the centrifuges, were controlled mainly by a particular model of Siemens PLCs. When an engineer needs to change a control sequence on a machine on the factory floor (which is usually not accessible via the internet to provide security in the form of an "air gap"), this requires creating a PLC program using a language called ladder logic, storing it on a USB drive, and walking it to the machine for direct upload.

But at the Natanz plant, several of these project files carried the malicious (and now well-known) computer worm Stuxnet. "By infecting [the PLC] project files and investing Stuxnet with the power to jump the air gap as a USB stowaway, the attackers had essentially turned every engineer into a potential carrier for their weapon" (Zetter, 2014). While the Stuxnet infections resulted in major asset damage and geopolitical impacts, similar cyberattacks could just as easily affect safety-critical systems (and endanger workers) or release toxins or pollutants into the air, soil, or water in the vicinity of the plant. The potential impacts of cyberattacks like this are staggering, in particular because it is a relatively new frontier and presents substantial payoffs to rogue actors, enemy nation states, and terrorist organizations.

Environment, health, safety, and quality are all interconnected, even more so in a world where additional risks are presented by the potential for cyberattacks. In an industrial scenario, risk is the synergistic result of threats, vulnerabilities, and the possible consequences that can emerge from them (Figure 12.1). The stakes are high because these risks can impact the product, the customers, the company, countries, the environment, human civilization, or a combination of all these things (Boyce et al., 2011; Foglietta et al., 2015; Agrafiotis et al., 2018).

This chapter looks at environment, health, safety, and quality (EHSQ)—which is quickly becoming an identifiable discipline with its own Body of Knowledge (BoK)—and cybersecurity, which can make or break EHSQ outcomes as the Stuxnet case shows. Addressing these areas in a holistic manner

| Threat | Vulnerability | Consequence |
|---|---|---|
| • Cyberattacks<br>• Physical breaches<br>• Insider threats<br>• Natural disasters<br>• Mechanical failures<br>• Process failures<br>• Geopolitical instability | • Human error:<br>  o Fatigue<br>  o Attention<br>  o Confusion<br>  o Environment<br>  o Training<br>  o Situation awareness<br>  o Usability<br>  o Failure to follow<br>    procedures OT/IoT design<br>  o Unpatched systems | • Asset damage or loss<br>• Injury or death<br>• Environmental impacts<br>• Reduced profits<br>• Stalled growth<br>• Damage to reputation<br>• Damage to supplier trust<br>• Suspended certifications<br>• Fines and investigations<br>• Geopolitical instability |

FIGURE 12.1. Risks in production systems.

will be critical for digital transformation and Industry 4.0 success for many organizations. Future EHSQ systems will provide "smart assistance for compliance managers" (Thimm, 2017) and help leaders discover issues and identify appropriate containment strategies before incidents occur.

## HUMAN FACTORS THAT INFLUENCE QUALITY

Many of the risks in Figure 12.1 are driven by or impacted by human factors, and people are central to quality practice. Quality management systems align people, processes, and technology to efficiently and effectively achieve shared goals. In the Industry 4.0 era, this increasingly involves planning for how humans, machines, and intelligent agents will work together. Many FACETS (fatigue, attention, confusion, environment, training, situation awareness) of human-machine interaction that influence safety and security in complex systems should be included in this planning process, since each can impact product quality and operational performance:

- **Fatigue**—Body and mind can be tired, overworked, or otherwise weary
- **Attention**—Clear, engaging displays, proper training, and sound body and mind can ensure that appropriate signals are received
- **Confusion**—Ambiguous signals or improper training can lead to incorrect interpretation
- **Environment**—Adverse environmental conditions can compound fatigue, reduce attention, add to confusion, or interfere with situation awareness

- **Training**—Ineffective or incomplete training can adversely impact attention, lead to confusion, and complicate situation awareness
- **Situation awareness**—An inability to accurately characterize a situation and project it forward in time can negatively impact decision making

Here's another way to think about it. For you to optimally advance your organization's quality and performance goals, your body must be fit for purpose (it's hard to do work when you have the flu), your brain must be up to the task (it's difficult to do anything when you're distracted, exhausted, or jet-lagged), and you must be trained for the work and prepared to respond and adapt as needed (context). The relationships among these internal human factors that drive quality are shown in Figure 12.2.

There are also external drivers. Workplace conditions impact the people in it (including exposure to toxins and hazardous materials, heat and cold stress, musculoskeletal stress and strain, and cognitive demands). Similarly, cyberattacks can potentially impact health and safety (for example, if hacked PLCs release hazardous chemicals at times when workers are not equipped to handle it).

## ENVIRONMENT

In many countries, companies are required to comply with laws and regulations that govern how they interact with the environment. Regulatory bodies enforce these requirements, and penalties can be both civil and criminal, ranging from settlements and cleanup enforcement to fines and jail time. The laws are in place not only to protect the environment from harm but also to protect other diverse economic and commercial interests and the global economy itself. Areas governed by environmental compliance laws include the following:

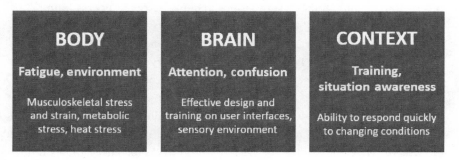

FIGURE 12.2. Internal human factors that drive quality.

- **Air:** pollution, particulate matter, ozone
- **Emergencies:** oil spills, release of hazardous chemicals
- **Materials:** asbestos, lead, mercury, PCBs
- **Waste disposal:** solid, hazardous, radioactive
- **Water:** drinking water, fracking, mining operations
- **Wildlife:** land, marine, endangered species

Demonstrating compliance can be labor intensive, and often requires keeping track of production on a very granular level to be able to accurately calculate emissions and releases for annual reporting. Many organizations have EHSQ software systems that make this process less overwhelming, which can also facilitate a strategic shift in thinking from compulsory compliance to sustainability as a strategic advantage.

## Compliance

In the United States, the Environmental Protection Agency (EPA) routinely monitors organizations to ensure that they obey environmental laws and regulations. This is done by interviewing workers, performing inspection, and training inspectors. At the same time, the EPA provides a mechanism for anyone to report potential environmental violations online, which allows the whistleblower to specify the affected entity (land, water, air, or workers) and pertinent incident (e.g., illegal dumping or releasing, a spill, or falsified documents or permits).

In the United States, the main compliance monitoring effort relates to greenhouse gas reporting in March and April. The Greenhouse Gas Reporting Program requires facilities that generate or receive over 25,000 metric tons of carbon dioxide a year to declare and justify their activities. Each July, facilities with 10 or more full-time employees in manufacturing, energy, mining, or waste management, and federal facilities are also required by the Toxics Release Inventory to declare the release of several chemicals or waste products. In Canada, the National Pollutant Release Inventory establishes that releases, disposals, and transfers of specific wastes or pollutants must be reported each June by facilities where wood preservation, fuel operations, pits or quarries, or waste or sewage incineration takes place. The European Union (EU) also requires facilities in 28 member states to declare waste and wastewater transfer, in addition to the release of pollutants and other substances, on an annual basis (Sarnowski, 2019).

## Sustainability

Many organizations take a "beyond compliance" approach, and quality management systems can support this reorientation toward sustainability. Siva et al. (2016) explains that the recent alignment of the ISO 9001 (quality management), ISO 14001 (environmental management), and ISO 45001 (health and safety) standards to the high-level Annex SL structure will help promote sustainability by making it easier to maintain integrated EHSQ management systems. In addition, these systems should make it possible to integrate sustainability considerations into daily work and process design, while better supporting stakeholder management and a focus on the customer.

In addition to a sustainability emphasis, some companies have started to pursue "green product innovation" to catalyze growth, recognizing that "environmental value and social good is fostered by market-driven product innovation and new technologies rather than regulation alone" (Dangelico & Pujari, 2010). This is one step along the continuum from compliance to sustainability to eco-design to sustainable design:

> Most of the focus [in organizations] is on eco-design (the integration of environmental considerations into product design and development) rather than sustainable design (the integration of a balanced approach to social, environmental and economic considerations into design and development), and is focused on product-related environmental compliance rather than innovation or the creation of new business models. The social component of sustainability is still largely missing from product design and development, outside of "bottom of the pyramid" and ethical product discussions. . . . Commercialisation still remains weak. . . . Tackling the softer organisational issues associated with implementing product sustainability will be a growing issue for those companies wanting to move from a compliance to an innovation mindset. (Charter, 2016)

To begin supporting this continuum from the compliance side, the recent ISO 14001 revision expanded its focus on sustainability. It includes additional management and leadership responsibilities, risk-based thinking, and a broader look at environmental impact throughout the supply network and across all phases of the product life cycle.

Sustainability is not limited to a single organization but is a process that connects all parties in the production ecosystem. While connectedness can make it easier to reduce energy requirements and optimize processes to advance sustainability goals, it introduces cybersecurity risks and the potential for human error:

> The sustainability of cybermanufacturing systems can be extended to stakeholders along the entire product life cycle including material suppliers, manufacturing systems, distributors, customers, and material recovery facilities which all benefit from digitalization of traditional information handling. Three relations are particularly improved: customers with factories, suppliers with factories, and the production control within manufacturing systems. Cyber systems enable customers to communicate directly with factories on design, processing, and other customization needs which, however, requires a flexible platform to coordinate and plan resources for production towards economic and environmental sustainability. Suppliers are able to receive information (e.g., inventory, quality feedback) from production processes in a timely manner. Within the manufacturing system digitalization and dematerialization of information handling creates opportunities for efficient energy planning and process optimization.
>
> This type of system, however, faces risk of attacks at both physical (e.g., machine, materials, power) and cyber elements (e.g., server, programs) of the factory, becoming a new challenge in future manufacturing. In the aspect of social sustainability at the shop floor level, the evolution of cybermanufacturing changes the interaction between human and machine systems, reducing physical hazard environment health impact but increasing risks of user interface misoperations." (H. Zhang, 2019)

## HEALTH AND SAFETY

Although mechanization and automation have reduced the need for manual lifting and material handling in industry, and exoskeletons promise to ease labor even more, many jobs still require physical exertion. Any strenuous or repetitive physical activity can present the potential for serious and costly injuries. Musculoskeletal disorders (such as lower back injuries, muscle strain, and carpal tunnel syndrome) and heat stress are two of the most common afflictions. In 2011, in the United States alone, these cost businesses over $50 billion in expenses associated with industrial accidents, medical expenses, and workers' compensation claims, and loss of productivity as a result of injuries.

These costs, which represent nearly 2% of the U.S. GDP, have been relatively consistent since the 1970s (Garg et al., 1982; Shi et al., 2015).

## Ergonomics in Industry 4.0

*Ergonomics* is the study of people engaged in physical and cognitive work. Ergonomists design work systems to eliminate discomfort and risk of injury and reduce the physical and cognitive harm impacts of environmental stress. They design safety and physical comfort into the workplace while keeping the costs associated with providing them low. Ineffectively designed work environments, in addition to presenting hazards, can lead to waste due to excessive motion and strain.

To increase worker safety and prevent injury, enhance productivity, and improve human well-being in work environments, physical tasks must be designed so that the physical capabilities of the workers are not exceeded. In particular, manual materials handling (MMH) tasks (e.g., lifting, carrying, holding, and placing) must be designed so that the physical requirements of a task do not exceed the physical capabilities of the workers who must perform those tasks. That is, task demands and worker capacity must be balanced to ensure quality, performance, and safety. This is usually the domain of occupational safety and health professionals, "biasing managers to link its role to safety and not to effectiveness, performance or costs" even though the discipline of ergonomics shares many goals with both lean and Six Sigma (Nunes, 2015, p. 14).

People need active, capable muscles (supported by strong bones and connective tissues) to do physical work. For muscles to function, they must be supported by healthy motor neurons, which tell the muscles when to contract and relax, and they must have energy available to consume in the form of adenosine triphosphate. The energy that those muscles need can come from myoglobin within the muscles (although only a tiny bit is available), creatine phosphate (which provides energy for another one or two seconds after muscle activation), anaerobic glycolysis (used once the body warms up, and also when oxygen is not available), and aerobic glycolysis, which requires an abundance of oxygen (and is thus the most common pathway to support sustained muscular work).

Making oxygen available to the muscle tissues throughout the body is the job of the lungs and circulatory system. The efficiency of these systems, from the perspective of physical ergonomics, depends on (1) the volume of oxygen that an individual can pull in, which can be improved through physical activity and training, and (2) the amount of that oxygen that can be effectively transported by hemoglobin in the blood to the muscles that need it to function. If core body temperature or blood acidity increases, then the hemoglobin will not be as effective

**TABLE 12.1. Ergonomic considerations and key questions.**

| Ergonomic consideration | Key questions |
|---|---|
| *Biomechanical:* Is the worker's musculoskeletal system capable of supporting this task? | Are the compressive forces and shear forces on the lower spine (especially at the L5/S1 vertebrae), which depend on the postures an individual uses within a job, less than the National Institute of Occupational Safety and Health (NIOSH) limits? |
| | What is the lifting capacity for this task, depending on the physical specifications for the motion and the frequency with which the task will be performed? |
| | What is the muscle force required for a particular task? |
| | What portion of an individual's maximum muscle capacity (called %MVC) is used by a particular task? |
| *Physiological:* Are the worker's circulatory and respiratory systems fit to support this task? | Is the metabolic energy required for a job matched to what the individual can safely provide? |
| | Is the required heart rate less than 220 beats per minute minus the person's age? |
| | Is the heart rate variability between work and rest periods less than 40 beats per minute? |
| | Are core body temperature and blood acidity stable? (Note: Blood acidity may not be possible to easily measure in a business or industrial setting.) |
| *Psychophysical:* Does the person feel comfortable and capable of doing the task? | Is the person comfortable with the physical requirements of the job? |
| | Is the person comfortable with the cognitive requirements of the job? |
| *Environmental:* Will environmental conditions increase the physical burden on the body, or is the environment neutral? | Are environmental conditions too hot or humid for physical labor? |
| | Are environmental conditions too cold for physical labor? |
| | Is the noise profile of the environment conducive to the work? |

in transporting oxygen throughout the body (called the "Bohr Effect"). Also, both muscles and nerves can fatigue, leading to lower physical efficiency.

Although this is a greatly simplified picture of the processes that go on inside the body when physical work is performed, it provides the basis for four categories of ergonomic considerations (Table 12.1). Each category is accompanied by questions that can serve as the basis for improvement projects. Biomechanical, physiological, and psychophysical considerations are specific to each individual worker, whereas environmental conditions may be similar for all workers. Although a worker's biomechanical limits will be influenced primarily by gender, age, and muscle characteristics, physiological limits will be determined by physical fitness. Only psychophysical assessments are subjective, but they can also vary depending on an individual's mood and cognitive state.

The biomechanical and psychophysical approaches to MMH have some limitations—in particular, that they do not effectively factor cognitive workload into assessment and design tasks, which can have physical impacts (DiDomenico & Nussbaum, 2011). Consequently, physiological and psychophysical measures provide an essential complement to biomechanical measures during job design and assessment. Ideally, job design and assessment will include measurement approaches across all categories.

Kadir et al. (2019) reviewed all the research in cyber-physical systems and Industry 4.0 to determine how human factors and ergonomics are shifting as a result of technological progress. Like Laudante (2017), they noted that "environment, competitiveness and safety are the main current drivers of innovation of products and process, achievable through the implementation of the current production systems of digitization, simulation of processes and use of advanced digital technologies." Laudante also believes that virtual and augmented reality will provide the mechanism for a "new ergonomic study of the workplace."

## Personal Protective Equipment (PPE) and Wearables

Safety management is an ongoing process. Safety managers control occupational safety and health risks, address vulnerabilities, and strive to continuously improve conditions and encourage workers to comply with safety regulations. Like quality, the work of safety is never finished. Safety is also management led: the directive to achieve a safe, healthy, and secure workplace must be championed by senior and executive management, who set the tone. It is a risk-based practice, and different industries will have different risk appetites and risk profiles.

In addition to audits, inspections, training, and risk management, organizations can protect their workers by providing them with protective gear and clothing. Personal protective equipment (PPE) refers to the robust equipment, clothing, and gear used to protect workers from hazards. PPE includes hard hats and steel-toed boots, safety glasses, latex gloves, respirators, aprons, and sensors. It includes protection for eyes, lungs, hearing, hands, and feet. Wearables provide indirect protection by sensing hazards or issues (e.g., worker in a confined space with expired training, operator using equipment that has not been calibrated) and alerting the worker or triggering an alert or corrective action request.

> In Industry 4.0, wearable tech will provide real-time insights into the safety of each worker, granting us greater insights into how to manage. It will allow us to seize an unprecedented opportunity to proactively identify and mitigate hazards and risks, protecting worker well-being and keeping people alive.
>
> —TAMARA PARRIS, DIRECTOR OF COMMUNITY DEVELOPMENT, SAFEOPEDIA

Podgorski et al. (2017) investigated the next generation of PPE, enhanced with smart materials, internet connections, and wearable sensors referred to as integrated computing technologies (ICTs):

New sensor technologies offer numerous possibilities for the improvement of OSH by means of real-time monitoring of hazardous and strenuous factors, such as noise, exposure to toxic chemical substances, optical radiation and high or low temperature. Furthermore, ICT applications allow facilitating other key functions of OSH management related to hazard identification and risk management. Such functions cover, e.g., monitoring workers' health state by measuring key physiological parameters (i.e., body temperature, heart rate, breathing rate, etc.); monitoring work comfort (e.g., underclothing temperature and humidity, work posture); geographical localisation of workers with regard to other, potentially dangerous objects or high-risk zones; monitoring the current protection level provided by PPE; detecting the end-of-service-life of PPE used by workers; providing warnings to workers in case of emergence of hazardous situations; and the activation of protective systems after exceeding a high-risk threshold value.

At the same time, the role of PPE in the management of the working environment has started to change. Besides being used as a means of passive protection against hazards, PPE items have also started to be used as carriers of sensors for monitoring work environment parameters, worker's health status and his or her location in the workplace space. Another trend has concerned incorporating signalisation systems into the PPE modules, which enabled displaying warnings to the worker, e.g., information on the occurrence of hazards or instructions on how to avoid them. . . . On the other hand, the implementation of new ICT technologies in the working environment leads to significant changes by modifying methods of work and introducing new objects and complex systems that may have functions which are not fully recognised. . . . Such systems may not function according to users' expectations and may be subject to unforeseen failures and consequences.

Podgorski et al. (2017) describe the convergence of these technologies as the Smart Work Environment. However, they do not see it as a panacea for safety but a call to action to develop new methods for risk management that fit the innovation: "As a result of dynamically changing manufacturing processes, workplaces, together with their potential hazards and the environment surrounding workers, will be subjected to frequent fluctuations that are imposed by hardly predictable process variations. New approaches to OSH risk management are therefore needed to sufficiently address the mentioned challenges." The ergonomic design of human work integrated within cyber-physical systems, and the safe design of human-machine interaction, is still nascent.

## Human-Machine Interaction

The most significant shift, though, is that pervasive digital technologies imply that interactions between people and machines will have a greater impact on health and safety than previously. Function allocation (Fitts, 1951; Jordan, 1963; McCauley & Matsangas, 2004) is one technique used to determine the degree of automation that is optimal for a system.

How do you decide which level of automation is right for a particular task? How do you decide what humans should do, and what the system should do? Function allocation separates activities into three groups: tasks that should be done by humans, tasks that can be shared with machines, and tasks that can

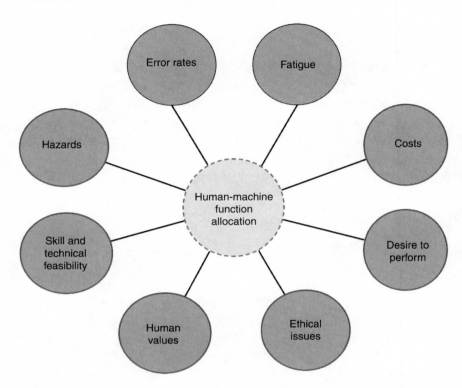

FIGURE 12.3. Considerations for human-machine function allocation (McCauley & Matsangas, 2004).

be fully automated without increasing risk. Next, consider the eight areas in Figure 12.3. This information can also be included in value stream maps.

## CYBERSECURITY

Safety and security (including cybersecurity) can be strongly interdependent; in fact, one can be a precondition for the other:

- **Safety** is associated with accidental risks originating from the system that could result in loss or damage to humans or assets. Ensuring safety means protecting someone or something, both physically and emotionally, from accidental harm.

- **Security** is related to malicious risks (e.g., attacks), which can be accomplished physically (through local access to the system) or electronically (through local or remote access). Security is freedom from the danger or damage that can arise from malicious intent.

This section presents three tools that can help you distinguish between safety and security, and adopt best practices to support cybersecurity from the physical and digital perspectives.

## SEMA Referential Framework

The SEMA Referential Framework (Figure 12.4) illustrates the interrelationship between security and safety. It considers harm in two dimensions: where the harm comes from (S-E for system-environment) and what was intended (M-A for malicious-accidental) (Piètre-Cambacédès & Bouissou, 2013). Because threat actors can gain access to control rooms or critical equipment and plant malware or ransomware, delete files, or otherwise vandalize systems and software, physical security is a prerequisite for cybersecurity (Kriaa et al., 2015). The old practice of "gates, guards, and guns" is no longer sufficient to protect organizations, even though physical controls and access controls may still be needed to protect systems and software (Rogers & Weinstein, 2019).

Each segment of the SEMA diagram is associated with an area of management focus—that is, *what* you should work on to mitigate each safety or security

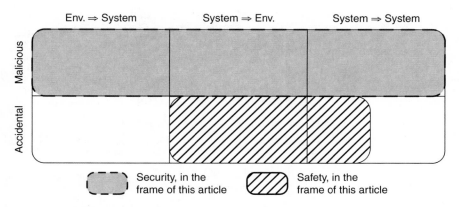

FIGURE 12.4: The SEMA Referential Framework (Piètre-Cambacédès & Bouissou, 2013).

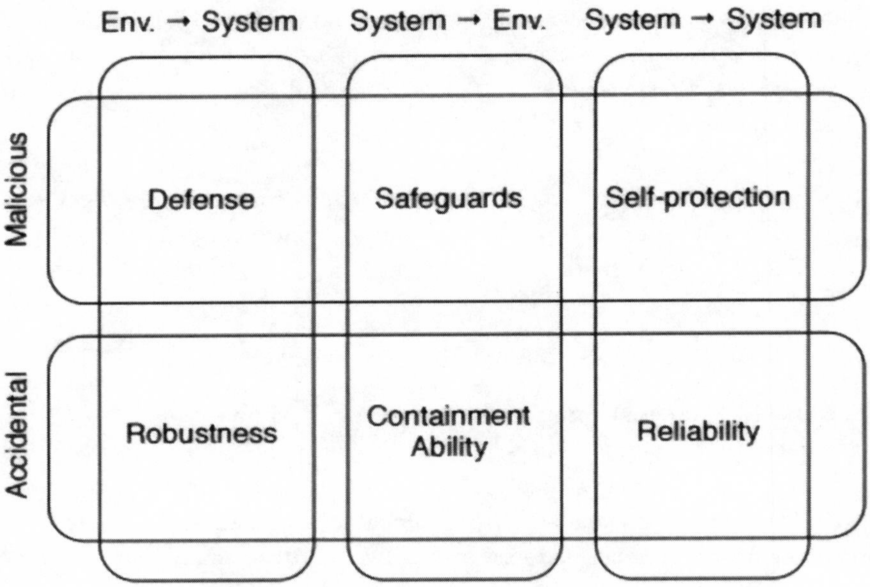

FIGURE 12.5. Recommendations for action in the SEMA Referential Framework (Pietre-Cambacédes & Chaudetb, 2010).

risk (Figure 12.5). The top row specifies actions that can be taken to address cybersecurity. The bottom row shows that systems should be hardened to protect against threats from the environment (e.g., temperature, excess dust), systems should be contained to prevent accidental impact on the environment, and systems should be made more reliable to protect against accidental threats from within. In addition, quality systems can help you identify *how* to mitigate each risk in a consistent and reliable manner.

For most information technology systems, security specialists aim to protect confidentiality, integrity, and availability, in that order. In contrast, priorities for industrial plan operations technology (OT) are completely reversed: safety, availability, integrity, and confidentiality are the concerns in order of criticality (Hahn, 2016). Industrial control systems (ICSs) require special standards and guidance because of the unique nature of OT:

- OT failures tend to have physical consequences, such as damage to production equipment or other assets

- OT security problems often present as maintenance failures or other small process issues, and so are difficult to detect and repair
- OT security can be more difficult to manage, since the systems are often old: life cycle for OT is 10–30 years, while life cycle for IT is 2–10 years
- OT security is also impacted by other factors, including additional network protocols, commands that can't be delayed or blocked (e.g., interrupts), or real-time requirements that preclude scanning or inspecting network traffic

All four of these factors were present in the Stuxnet case. But by far the most serious cybersecurity threat to ICS (or any other business system) is people. "In many situations the potential for accidental or intentional subversion of security by a person will be the system's weakest link" (Boyce et al., 2011). Insider threats from contractors or disgruntled employees (Dalal & Gorab, 2016), security fatigue (Stanton et al., 2016), decision fatigue (Nobles, 2019), groupthink and ignorance (Kelly, 2017), and bad design of processes and interfaces (Nurse et al., 2011) can all pose threats to cybersecurity.

## NIST Cybersecurity Framework (NIST CSF)

The NIST Cybersecurity Framework provides risk-based guidance and a common, technology-neutral language for managing cybersecurity risk. It was designed to complement an organization's preexisting cybersecurity program but can also be used to launch new cybersecurity programs.

The NIST CSF is organized in terms of five functions or practice areas: Identify, Protect, Detect, Respond, and Recover (Figure 12.6). Within each group, there are multiple categories and subcategories. Each subcategory specifies a best practice. Rather than criteria elements or clauses, the NIST CSF is instead

| Identify | Protect | Detect | Respond | Recover |
|---|---|---|---|---|
| • Asset management<br>• Business environment<br>• Governance<br>• Risk assessment<br>• Risk management strategy | • Access control<br>• Awareness and training<br>• Data security<br>• Info protection processes and procedures<br>• Maintenance<br>• Protective technology | • Anomalies and events<br>• Continuous monitoring<br>• Detection processes | • Response planning<br>• Communications<br>• Analysis<br>• Mitigation<br>• Improvements | • Recovery planning<br>• Improvements<br>• Communications |

FIGURE 12.6. NIST CSF.

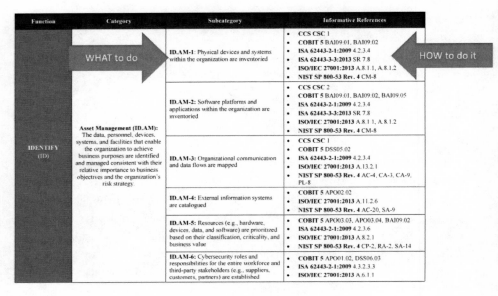

| Function | Category | Subcategory | Informative References |
|---|---|---|---|
| IDENTIFY (ID) | Asset Management (ID.AM): The data, personnel, devices, systems, and facilities that enable the organization to achieve business purposes are identified and managed consistent with their relative importance to business objectives and the organization's risk strategy. | ID.AM-1: Physical devices and systems within the organization are inventoried | • CCS CSC 1<br>• COBIT 5 BAI09.01, BAI09.02<br>• ISA 62443-2-1:2009 4.2.3.4<br>• ISA 62443-3-3:2013 SR 7.8<br>• ISO/IEC 27001:2013 A.8.1.1, A.8.1.2<br>• NIST SP 800-53 Rev. 4 CM-8 |
| | | ID.AM-2: Software platforms and applications within the organization are inventoried | • CCS CSC 2<br>• COBIT 5 BAI09.01, BAI09.02, BAI09.05<br>• ISA 62443-2-1:2009 4.2.3.4<br>• ISA 62443-3-3:2013 SR 7.8<br>• ISO/IEC 27001:2013 A.8.1.1, A.8.1.2<br>• NIST SP 800-53 Rev. 4 CM-8 |
| | | ID.AM-3: Organizational communication and data flows are mapped | • CCS CSC 1<br>• COBIT 5 DSS05.02<br>• ISA 62443-2-1:2009 4.2.3.4<br>• ISO/IEC 27001:2013 A.13.2.1<br>• NIST SP 800-53 Rev. 4 AC-4, CA-3, CA-9, PL-8 |
| | | ID.AM-4: External information systems are catalogued | • COBIT 5 APO02.02<br>• ISO/IEC 27001:2013 A.11.2.6<br>• NIST SP 800-53 Rev. 4 AC-20, SA-9 |
| | | ID.AM-5: Resources (e.g., hardware, devices, data, and software) are prioritized based on their classification, criticality, and business value | • COBIT 5 APO03.03, APO03.04, BAI09.02<br>• ISA 62443-2-1:2009 4.2.3.6<br>• ISO/IEC 27001:2013 A.8.2.1<br>• NIST SP 800-53 Rev. 4 CP-2, RA-2, SA-14 |
| | | ID.AM-6: Cybersecurity roles and responsibilities for the entire workforce and third-party stakeholders (e.g., suppliers, customers, partners) are established | • COBIT 5 APO01.02, DSS06.03<br>• ISA 62443-2-1:2009 4.3.2.3.3<br>• ISO/IEC 27001:2013 A.6.1.1 |

WHAT to do · HOW to do it

FIGURE 12.7. Pointers to guidance in one NIST CSF.

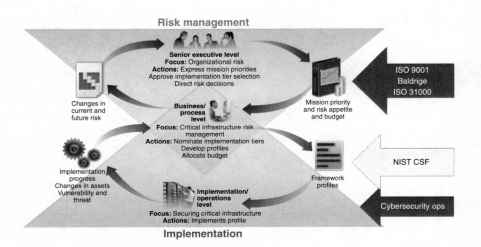

FIGURE 12.8. NIST CSF, risk management, and quality management relationships (adapted from NIST, 2018).

a toolkit of "pointers" to guidance provided by COBIT, the CCS CSC (Center for Cybersecurity Top 20 Critical Security Controls), ISA 62443, ISO/IEC 27001, and NIST SP 800-53 (Figure 12.7). In this sense, it ties together controls and best practices from other authoritative sources to define a holistic reference for cybersecurity practice.

The NIST CSF can be used in conjunction with ISO 31000 (risk management), ISO/IEC 27005 (information technology risk management), and NIST SP 800-395 (information security risk) (Figure 12.8). Together, these provide systematic processes for aligning cybersecurity risk management with process risk management, and continuously improving risk management processes.

### Baldrige Cybersecurity Excellence Builder (BCEB)

The Baldrige Cybersecurity Excellence Builder (BCEB) is a self-assessment tool to help organizations assess how effectively they are applying the NIST CSF. It is derived from the Baldrige Excellence Framework (BEF) and administered by NIST, which supports the Malcolm Baldrige National Quality Award (MBNQA). The BCEB complements an organization's cybersecurity program, connects cybersecurity operations to business results, and is flexible (so can also be used independent of the NIST CSF). For organizations that plan to go through the full process with COBIT or ISO 27001, the BCEB may provide additional information about how to link holistic business practices with cybersecurity (Figure 12.9).

The BCEB is used by senior and executive management to:

- "determine cybersecurity-related activities that are important to business strategy and the delivery of critical services;

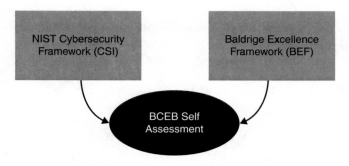

FIGURE 12.9. Relationships between NIST CSF, BEF, and BCEB.

- prioritize investments in managing cybersecurity risk;
- assess the effectiveness and efficiency in using cybersecurity standards, guidelines and practices;
- assess their cybersecurity results; and
- identify priorities for improvement." (NIST, 2016)

The BCEB is a collection of questions focused on conduct and management of cybersecurity operations. There are six process-oriented sections and one results section, corresponding to the criteria sections in the BEF:

- Each answer in the process section should include information about the systematic approach the organization takes to satisfy that element, the manner in which this approach is deployed (broadly across the organization, and deeply, throughout all organizational levels), the feedback and learning process in place to ensure that improvement occurs, and how the element is integrated with other elements.
- Each answer in the results section should report which quantitative metrics are used, and provide recent levels (values) for those metrics. Depending on the maturity of the organization's cybersecurity program, the answer can also include trends to show how those levels have changed, comparisons to indicate whether the measurements are good, and information about how that metric is used in conjunction with other metrics (integration) to ensure that the cybersecurity program is meeting the needs of the business.

The BCEB can help organizations plan, manage, assess, and continually improve cybersecurity operations and risk management. The BCEB should not, however, be used if there is no cybersecurity risk management or cybersecurity operations in place. As guidance, it may be useful for new organizations that want to start developing systematic, repeatable processes for cybersecurity to learn how to design cybersecurity operations that best support customers and business results.

## QUALITY IMPROVEMENT IN ENVIRONMENT, HEALTH, AND SAFETY (EHS)

Where Lean and Six Sigma have been used in conjunction with ergonomics, the methods often involve reviewing records of incidents rather than aiming to anticipate or prevent them. The projects aim to reduce workers' compensation costs, reduce the number of injuries or strains, prevent workplace accidents, reduce perceived stress, demonstrate cost savings in health and safety admin-istration, or prevent workplace deaths.

Ng et al. (2005), for example, aimed to reduce work accidents in the ship-ping industry by studying historical data linking work procedures to falls from cargo containers. They examined the data using Six Sigma tools and a root cause analysis approach, and discovered that lack of concentration at work, disregard of safety regulations, and use of worn-out hooks and slings all con-tributed to threats to worker safety. Nunes (2015) described two case studies, one that reduced physical demands on the workers by almost half while in-creasing productivity slightly, and another that related a reduction in lead time and process efficiency to the redesign of an MMH task.

There will always be trade-offs: even if a task is biomechanically and physi-ologically sound, a person may not feel comfortable doing it; similarly, an indi-vidual may be comfortable performing a task that is physically dangerous. As a result, all considerations should be used together, and an improvement project should never focus on just one. Consequently, a quality improvement process in EHS should ask each of the following questions, in order:

1. Is the job biomechanically sound?
2. Is the job physiologically sound?
3. Is the job psychophysically sound?
4. Is the job environment sound?
5. Is the job environmentally sound?
6. If the job is sound, can you increase the workload or productivity without com-promising worker safety, health, or comfort, or the environment?

Although improvements can target job design or process design, these questions are mainly focused on human-centered job design. To answer them, relevant performance measures are needed (Table 12.2). When collecting this data, you may wish to involve an ergonomist to ensure that each measurement is being collected appropriately. For example, surface electromyography (EMG) requires frequent calibration and skin preparation to ensure validity of measurements.

These performance measures can be easily integrated into Six Sigma prob-lem solving using the Define-Measure-Analyze-Improve-Control (DMAIC) process improvement approach:

- **Define:** Identify goals in terms of one or more ergonomic performance measures, preferably from more than one category of considerations. Create a process map

**TABLE 12.2. Ergonomic variables and data collection resources.**

| Ergonomic performance measure | Data collection | Reference |
|---|---|---|
| Compressive and shear forces on lower back (especially L5/S1 spinal disc) | Univ. of Michigan 3-Dimensional Static Strength Prediction Program.™ | University of Michgan |
| Lifting capacity | NIOSH lifting equation. Capacity changes with lift frequency, tightness of grip, and vertical and horizontal distances traveled. | Waters et al. (1994) |
| Muscle force and %MVC | Surface EMG using components from online stores that supply makerspaces. | Roberts & Gabaldón (2008) |
| Metabolic energy | Break task into subcomponents and assess the energy required for each using the 48 models provided by Garg et al. (1978). | Garg et al. (1978) |
| Heart rate and heart rate variability | Heart rate monitor using components from online stores that supply makerspaces. Make sure that heart rate does not exceed (220–age) and that the difference between resting and working heart rate is 40 beats per minute or less. | Paritala (2009) |
| Core body temperature | Thermometer | Richmond et al. (2015) |
| Physical requirements | Borg's Rating of Perceived Exertion. Ranges from 6 to 20 and is approximately equal to the maximum heart rate divided by 10. | Borg (1982) |
| Cognitive requirements | Effort (1 = easy to 9 = hard), difficulty (1 = easy to 9 = hard), or response time to a secondary task. | DeLeeuw & Mayer (2008) |
| Heat stress | Wet Bulb Globe Temperature. Use three-term version if in the sun, and two-term version if in the shade. | Budd (2008) |
| Cold stress | Required clothing insulation. | Holmér (1994) |

or value stream map. Draw pictures or diagrams to illustrate the motions associated with the task. Describe how the job is split into multiple component tasks.

- **Measure:** Collect historical data to describe the past state of the system in terms of ergonomic performance measures. Determine baseline values for ergonomic performance measures in terms of historical data, anthropometric data, or benchmark data for a particular worker or worker category.

- **Analyze:** Gather data and/or perform sensitivity analysis using the performance measures and techniques described in Table 12.2. Reduce waste by preventing overexertion. Reduce variation in how jobs are conducted, how well employees conform to task requirements (as designed), or how well employees conform to rest requirements. Reduce defects by associating performance measures with incidence of injury or accident. Correlation analysis can be conducted between performance measures, injury reports, and costs.
- **Improve:** Based on the analysis, identify appropriate interventions at the levels of the worker, the environment, and job design.
- **Control:** Identify how the selected interventions will be maintained at the levels of the worker, the environment, and the job.

Note that improvement can involve any combination of factors at one or more levels: the worker, the environment, and the job design. Modifying the worker's clothing, physical fitness, and food and drink intake (on or off the job); controlling or adjusting the environment; or changing the recommended approach to the work itself are all potential improvement strategies.

## QUALITY IMPROVEMENT IN CYBERSECURITY

Quality techniques can be used to improve cybersecurity, which in turn can enhance quality by protecting operational processes from threats. For example, Bishop et al. (2014) suggest that process improvement provides a unique opportunity to directly address one of the most challenging attack vectors in cybersecurity, the insider threat, by strategically identifying monitor points:

Traditional approaches have focused on examining the actions of agents (people with access to the data or resources under consideration). In these approaches the actions themselves are analyzed to find suspicious or unexpected patterns, where agent behavior may be derived from technical logs or external indicators . . . [but] rather than focusing on identifying insider attacks as they occur by monitoring event logs, we use a rigorous, process-based approach to identify places in organizational processes where insider attacks can be successfully launched. In knowing how such attacks might be launched, organizations will often be able to restructure their processes to reduce their exposure to such insider attacks.

This knowledge should give organizations insights that will help them to know where to look first if an attack is suspected, and what steps in the process should be considered for more careful activity monitoring (e.g., using audit logging or video surveillance) or restructuring, e.g., to avoid single points of failure or add redundancy.

Their proposed method uses fault tree analysis and suggests that failure modes and effects analysis (FMEA) may also be promising. In either case, taking the perspective of the process rather than the attacker is the distinctive contribution of this work.

## THE BOTTOM LINE

Environment, health, safety, and quality are all interconnected, a theme that becomes apparent when cyberattacks threaten assets, operations, and worker health and safety. Risk, which emerges from threats, vulnerabilities, and the possible consequences that can emerge from them, should be examined holistically for organizations pursuing digital transformation in the Industry 4.0 era. Guidelines like the SEMA Referential Framework, the NIST CSF, and the BCEB can be used to align quality, risk, and cybersecurity operations in a way that addresses both safety and security.

Because many of the risks are driven by human factors, quality may be compromised unless workers are well suited for their tasks. The human body must be fit for purpose, the brain must be capable of the task at the right time, and training and situational preparation must be provided. In addition to industrial hygiene and safety considerations, organizations should also be on the alert for these issues that can impair workers and threaten quality: fatigue, attention, confusion, environment, training, and situation awareness.

Ergonomic job and safe process design can also protect quality. Biomechanical, physiological, psychophysical, and environmental considerations are all important. But as a result of more pervasive digital and networked technologies, interactions between people and machines will have a greater impact on health and safety in the decades to come.

## REFERENCES

Agrafiotis, I., J. R. Nurse, M. Goldsmith, S. Creese, and D. Upton. (2018). "A taxonomy of cyber-harms: Defining the impacts of cyber-attacks and understanding how they propagate." *Journal of Cybersecurity* 4 (1).

Bishop, M., H. M. Conboy, H. Phan, B. I. Simidchieva, G. S. Avrunin, L. A. Clarke, and S. Peisert. (2014, May). "Insider threat identification by process analysis." In *Security and Privacy Workshops (SPW)*, 251–264. IEEE.

Borg, G. A. (1982). "Psychophysical bases of perceived exertion." *Medical Science of Sports Exercise* 14 (5): 377–381.

Boyce, M. W., K. M. Duma, L. J. Hettinger, T. B. Malone, D. P. Wilson, and J. Lockett-Reynolds. (2011, September). "Human performance in cybersecurity: A research agenda." In *Proceedings of the Human Factors and Ergonomics Society annual meeting* 55 (1): 1115–1119.

Budd, G. M. (2008). "Wet-bulb globe temperature (WBGT)—its history and its limitations." *Journal of Science and Medicine in Sport* 11 (1): 20–32.

Carvalho, Fernando P. (2011). "Environmental radioactive impact associated to uranium production." *American Journal of Environmental Sciences* 7 (6): 547.

Chan, Caroline, Therese S. Hughes, Susan Muldoon, Tim Aldrich, Carol Rice, Richard Hornung, Gail Brion, and David Tollerud. (2010). "Mortality patterns among paducah gaseous diffusion plant workers." *Journal of Occupational and Environmental Medicine*/American College of Occupational and Environmental Medicine 52 (7): 725.

Charter, M. (2016). "Sustainable innovation and design: Future implications and lessons learnt from the last 20 years." Sustainable Innovation & Design 20th International Conference, 9th–10th November 2016. University for the Creative Arts Epsom, Surrey, UK.

Dalal, R. S., and A. K. Gorab. (2016). "Insider threat in cyber security: What the organizational psychology literature on counterproductive work behavior can and cannot (yet) tell us." In *Psychosocial dynamics of cyber security*, 122–140. Routledge.

Dangelico, R. M., and D. Pujari. (2010). "Mainstreaming green product innovation: Why and how companies integrate environmental sustainability." *Journal of Business Ethics* 95 (3): 471–486.

DeLeeuw, K. E., and R. E. Mayer. (2008). "A comparison of three measures of cognitive load: Evidence for separable measures of intrinsic, extraneous, and germane load." *Journal of Educational Psychology* 100 (1): 223.

DiDomenico, A., and M. A. Nussbaum. (2011). "Effects of different physical workload parameters on mental workload and performance." *International Journal of Industrial Ergonomics* 41 (3): 255–260.

Fitts, P. (1951). *Human engineering for an effective air-navigation and traffic-control systems.* Columbus, OH: Ohio State University Foundation.

Foglietta, C., S. Panzieri, and F. Pascucci. (2015). "Algorithms and tools for risk/impact evaluation in critical infrastructures." In *Intelligent monitoring, control, and security of critical infrastructure systems*, 227–238. Berlin: Springer.

Garg, A., Chaffin, D. B., & Freivalds, A. (1982). "Biomechanical stresses from manual load lifting: A static vs dynamic evaluation." *IIE Transactions* 14 (4): 272–281.

Garg, A., Chaffin, D. B., & Herrin, G. D. (1978). Prediction of metabolic rates for manual materials handling jobs. *The American Industrial Hygiene Association Journal* 39 (8): 661–674.

Hahn, A. (2016). "Operational technology and information technology in industrial control systems." In *Cyber-security of SCADA and other industrial control systems*, 51–68. Springer.

Holmér, I. (1994). "Cold stress: Part I—Guidelines for the practitioner." *International Journal of Industrial Ergonomics* 14 (1): 139–149.

Jordan, N. (1963). "Allocation of functions between man and machines in automated systems." *Journal of Applied Psychology* 47: 161–165.

Kadir, B. A., O. Broberg and C. S. da Conceição. (2019). "Current research and future perspectives on human factors and ergonomics in Industry 4.0." *Computers & Industrial Engineering* 137: 1–12.

Kelly, R. F. (2017). *The insider threat to cybersecurity: How group process and ignorance affect analyst accuracy and promptitude.* Monterey, CA: Naval Postgraduate School.

Kopotic, J. D., M. S. Ferri, and C. Buttram. (2013, July). Lessons-learned from D and D activities at the five gaseous diffusion buildings (K-25, K-27, K-29, K-31 and K-33) East Tennessee Technology Park, Oak Ridge, TN 13574. *WM Symposia*, Tempe, AZ.

Kriaa, S., L. Pietre-Cambacedes, M. Bouissou, and Y. Halgand. (2015). "A survey of approaches combining safety and security for industrial control systems." *Reliability Engineering & System Safety* 139: 156–178.

Laudante, E. (2017). "Industry 4.0, innovation and design: A new approach for ergonomic analysis in manufacturing system." *The Design Journal* 20(sup1): S2724–S2734.

McCauley, M. E., and P. G. Matsangas. (2004). *Human systems integration and automation issues in small unmanned aerial vehicles.* http://calhoun.nps.edu/bitstream /handle/10945/746/NPS-OR-04-008.pdf?sequence=1

National Institute of Standards and Technology. (2016, September 15). "NIST releases Baldrige-based tool for cybersecurity excellence." https://www.nist.gov/news-events /news/2016/09/nist-releases-baldrige-based-tool-cybersecurity-excellence

———. (2018, April 16). *Framework for improving critical infrastructure cybersecurity v1.0.* https://nvlpubs.nist.gov/nistpubs/CSWP/NIST.CSWP.04162018.pdf

Ng, E., F. Tsung, R. So, T. S. Li, and K. Y. Lam. (2005). "Six Sigma approach to reducing fall hazards among cargo handlers working on top of cargo containers: A case study." *International Journal of Six Sigma and Competitive Advantage* 1 (2): 188–209.

Nobles, C. (2019, May 21–22). "Establishing human factors programs to mitigate blind spots in cybersecurity." *Proceedings of the Fourteenth Midwest Association for Information Systems Conference.* Midwest Association for Information Systems.

Nunes, I. L. (2015, August). "Ergonomics and Lean Six Sigma integration. A systems approach." In *Proceedings 19th Triennial Congress of the IEA* 9:14. IEA.

Nurse, J. R., S. Creese, M. Goldsmith, and K. Lamberts. (2011, September). "Guidelines for usable cybersecurity: Past and present." In *2011 third international workshop on cyberspace safety and security (CSS)*, 21–26. IEEE.

Paritala, S. A. (2009). *Effects of physical and mental tasks on heart rate variability.* PhD dissertation, Kakatiya University.

Pietre-Cambacédès, L., and M. Bouissou. (2013). "Cross-fertilization between safety and security engineering." *Reliability Engineering & System Safety* 110: 110–126.

Pietre-Cambacédes, L., and C. Chaudetb. (2010). *SEMA: Un référentiel pour éviter équivoques et ambigüités entre sûreté et sécurité.*

Podgorski, D., K. Majchrzycka, A. Dąbrowska, G. Gralewicz, and M. Okrasa. (2017). "Towards a conceptual framework of OSH risk management in smart working environments based on smart PPE, ambient intelligence and the Internet of Things technologies." *International Journal of Occupational Safety and Ergonomics* 23 (1): 1–20.

Richmond, V. L., S. Davey, K. Griggs, and G. Havenith. (2015). "Prediction of core body temperature from multiple variables." *Annals of Occupational Hygiene* 59 (9): 1168–1178.

Roberts, T. J., and A. M. Gabaldón. (2008). "Interpreting muscle function from EMG: Lessons learned from direct measurements of muscle force." *Integrative and Comparative Biology* 48 (2): 312–320.

Rogers, M., and D. Weinstein. (2019, June 9). *Protecting our critical infrastructure in the digital age.* The Hill. https://thehill.com/opinion/cybersecurity/447596-protecting -our-critical-infrastructure-in-the-digital-age

Sarnowski, J. (2019, March 11). "On environment: Tips for the reporting season." *Intelex Community.* https://community.intelex.com/explore/posts/environment-tips -reporting-season

Schein, E. H. (1985). "Defining organizational culture." *Classics of Organization Theory* 3 (1): 490–502.

Schiff, B. (1985). *The proficient pilot.* Wiley.

Shi, J., S. Gardner, K. K. Wheeler, M. C. Thompson, B. Lu, L. Stallones, and H Xiang. (2015). "Characteristics of nonfatal occupational injuries among US workers with and without disabilities." *American Journal of Industrial Medicine* 58 (2): 168–177.

Siva, V., I. Gremyr, B. Bergquist, R. Garvare, T. Zobel, and R. Isaksson. (2016)." The support of quality management to sustainable development: A literature review." *Journal of Cleaner Production* 138: 148–157.

Smith, S. (2014, February). "Thinking lean: Muda, muri, and mura." *Six Sigma Forum* 13 (2): 36–37.

Stanton, B., M. F. Theofanos, S. S. Prettyman, and S. Furman. (2016). "Security fatigue." *IT Professional* 18 (5): 26–32.

Thimm, H. (2017, June). "Using IoT enabled multi-monitoring data for next-generation EHS compliance management systems." In *2017 IEEE International Conference on Environment and Electrical Engineering and 2017 IEEE Industrial and Commercial Power Systems Europe (EEEIC/I&CPS Europe),* 1–6. IEEE.

University of Michigan Center for Ergonomics. 3DSSSP Software. http://c4e.engin .umich.edu/tools-services/3dsspp-software/

Wages, R., S. Markowitz, S. Kieding, S. Griffon, and E. A. Samaras. (2003). *Former worker medical surveillance program at Department of Energy gaseous diffusion plants. Phase I: Needs assessment.* https://www.energy.gov/sites/prod/files/2013/10/f3/Portsmouth _Paducah_K25ProductionWorkers_NeedsAssessment.pdf

Waters, T. R., V. Putz-Anderson, A. Garg, and National Institute for Occupational Safety and Health. (1994). *Applications manual for the revised NIOSH lifting equation,* 94–110. US Department of Health and Human Services, Public Health Service.

Zetter, K. (2014). *Countdown to zero day: Stuxnet and the launch of the world's first digital weapon.* New York: Broadway Books.

Zhang, H. (2019, September 5) Personal Communication.

## CHAPTER THIRTEEN

# VOICE OF THE CUSTOMER (VOC)

Customers are empowered by digital; they have new behaviors and
new expectations that are pushing us to evolve. We have to be more
reactive and more agile. We need to listen to what our customers
expect and also track how our digital environment is evolving.

—BENEDICTE JAVELOT, CHIEF STRATEGY OFFICER
OF ORANGE GROUP

On September 2, 2013, Twitter user Hasan Syed (@HVSVN) had a problem: his father's suitcase had been lost on a British Airways flight. After
an unsatisfactory experience at the service desk (and still no luggage),
he posted a scathing tweet: "Don't fly @BritishAirways. Their customer service
is terrible." Eight hours later, a real person monitoring the airline's account re-
sponded, asking him for more information to hopefully help him find his bag.

But the damage had already been done—the airline had stumbled into an
admission that it wasn't able to find the information he and his father had
given it earlier. Also, the time gap from the initial tweet to the response provided a fertile environment for the tweet to go viral, which it did; there were
76,000 retweets, nearly 10,000 an hour—and still no luggage. Although we're
not sure how (or if) Hasan's case was resolved, since 2013, British Airways has
made tremendous investments in its customer service presence so that issues
like this no longer arise.

Many companies have implemented automated bots to handle these
kinds of publicly accessible online customer service inquiries, but ensuring a
lightning-fast response time comes with a trade-off: appropriateness of the response. Sarcasm, in particular, is hard to detect. Consider these examples from
Rajadesingan et al. (2015):

**User 1:** You are doing great, @MajorAirline! Who could predict heavy travel between #Thanksgiving and #NewYearsEve. And bad cold weather in Dec. Crazy!
**Major Airline 1:** We #love the kind words! Thanks so much.

**User 2:** Ahhh . . . **** reps. Just had a stellar experience w them at Westchester, NY. #CustomerSvcFail

**Major Airline 2:** Thanks for the shout-out Bonnie. We're happy to hear you had a #stellar experience flying with us. Have a great day.

A November 2016 survey by Publicis Groupe DigitasLBi indicated that in addition to the customer service problems, in exchanges like this, 60% of Americans were turned off by the awareness that they were not engaging with a human. This was the case whether the conversational agent was implemented on Twitter, on the company's website, or in other forums. Seventy-three percent said they would not engage with a chatbot after just one bad experience. Even so, more than 60% were open to continue using chatbots if they were able to reliably get the answer or recommendation they were seeking.

By 2020, four out of five organizations that provide customer service will offer one or more chatbots for customer support and to provide information (Faw, 2016). But chatbots and Twitter bots are just one class of emerging technology that can help us connect with our customers and stakeholders, to identify their wants, needs, and expectations—and how they change over time. While digital transformation emphasizes incorporating new technologies to solve old problems, established mechanisms for communicating and uncovering needs are just as important.

This chapter outlines many of these approaches, techniques, and tools and presents them within a conceptual framework you can use to organize a contemporary VoC process. Although explanations of how to use each VoC method are not provided (you will have to consult the primary sources in the tables and references), the information in this chapter will help you create a strategy that incorporates Industry 4.0 and Quality 4.0 technologies into your VoC efforts.

## THE EVOLVING VOICE OF THE CUSTOMER (VOC)

Most organizations are better at speaking to their customers than listening to them. Listening activities are usually bound to motives: upselling products and services, placating customer anger, or gaining market intelligence. Marketing budgets are often dedicated to persuasive advertising, lead generation, and public relations, despite the evidence that poor listening practices contribute significantly to customer attrition and distrust in corporations (Macnamara, 2018).

Proactively addressing customer needs and requirements is a good business decision: it is far more cost-effective to incorporate customer needs when new products are developed, thus reducing the likelihood of waste and rework later. *Voice of the customer* (VoC) refers to the process of identifying, articulating, and prioritizing customer needs, and the results obtained from that process. Understanding the multiple levels on which customer needs are expressed can help organizations produce, improve, and prioritize work to drive quality.

Organizations with mature VoC processes avoid a common pitfall: most effort dedicated to promoting the appearance of being seen listening to customers rather than acting on the insights obtained. Gathering but not using customer data can present a significant waste of resources and loss of opportunities that could have been identified from actionable customer information (Chauhan & Sarabhai, 2018). Overall customer experience is diminished when organizations depend on simplistic trackers to get feedback; while providing ample amounts of data, these approaches rarely uncover the customers' fundamental motivations and behaviors, the true VoC.

Over the past three decades, more than 40 VoC techniques have been described in the research literature and demonstrated in practice. Some are qualitative, and some are quantitative, with newer methods that are primarily digital in nature. Additionally, VoC techniques based on Quality 4.0 technologies like machine learning have recently emerged or become more democratized and accessible.

Today, VoC is data driven and requires infrastructure to handle data. Paul Jarman (2013), CEO of call center firm inContact, explains that understanding customers and how they change requires solid and strategic data management: "[Voice of the customer] is evolving to include all the various interactions that customers have with your organization, including written communications through chat, text/SMS, and emails. . . . The most successful organizations share this data broadly and strategically across other departments, including sales, marketing, and operations."

## A FOUR-STAGE VOC PROCESS

Strategies, tools, and techniques for gathering VoC data and transforming it into actionable intelligence abound. Every VoC process consists of the following stages:

- **VoC Stage I: Identify customer needs.** This stage involves using traditional techniques from marketing and quality management to identify:

  o **IA: Stated needs** that can be directly expressed by the customer
  o **IB: Implied needs** that cannot be expressed, but can be found indirectly
  o **IC: Silent (or hidden) needs** that cannot be expressed or determined indirectly but can be identified by examining context, relationships, and needs expressed by large-scale social and societal trends

- **VoC Stage II: Understand and prioritize needs.** One need can be satisfied in many different ways. In Stage II, methods such as quality function deployment (QFD) and Kano's model can be used to determine specific means to address customer needs, and determine the best order in which they should be satisfied.
- **VoC Stage III: Create meaningful customer experiences.** This stage involves using the insights derived from Stage II to produce business value and engineer exceptional customer experiences.
- **VoC Stage IV: Anticipate future needs.** *Identifying needs that do not currently exist is the basis for innovation.* In this stage, advanced analytical methods (including applied machine learning) can be used to predict future needs, possibly even before customers are aware that those needs exist. In this context, VoC tools can be used to find voices of the future, catalyzing innovation.

To realize the benefits from a VoC process, organizations must design structures and functions to support continuous data collection, analysis, implementation, and learning—blending each of these elements. The case study on VoC architecture later in this chapter illustrates this.

## Stage I: Identify Customer Needs

In the past, VoC data would typically be gathered at regular intervals (daily, monthly, annually) and analyzed. Results could be incorporated into choices and initiatives in strategic plans, and sometimes into more granular action plans. Today, data from many sources (e.g., social media) is being produced continuously so companies need to continuously monitor it, look for leading indicators to project future characteristics, and prepare forward-looking models to anticipate and adapt to shifts in sentiment.

## STAGE IA: IDENTIFY STATED NEEDS

Many organizations have experience capturing stated needs. These methods involve customers explicitly expressing their needs and describing how they would like products and services to meet those needs. The organization has flexibility in determining how the needs will be satisfied, and sometimes this involves further work with the customer or cooperative development (for example, using agile methods).

Table 13.1 lists 24 methods that can be used to gather needs from customers, with references that describe those approaches in greater detail. Most VoC programs triangulate results from several of these methods to construct more detailed and nuanced pictures of customers' stated needs.

## STAGE IB: IDENTIFY IMPLIED NEEDS

Tools used to identify implied needs originate in the fields of psychology, philosophy, ethnography, and data science. This section provides a brief overview of tools for gathering the implied needs that influence product or service quality but that are rarely called out by the methods in Table 13.1. A selection of the 11 techniques in Table 13.2 should be used in conjunction with a selection of the 22 techniques in Table 13.1 to ensure that both stated and implied needs are covered by VoC data collection.

## STAGE IC: IDENTIFY SILENT (OR HIDDEN) NEEDS

Some customer needs cannot easily be expressed by the customer or suggested by the customer indirectly. As a result, it will be difficult to detect these needs using the methods described above. These additional needs are just as critical and must be articulated because they can mean the difference between success and failure over the full life cycle of the product or service.

Many ways to identify hidden or silent needs are made possible by Industry 4.0 technologies and Quality 4.0 approaches. Table 13.3 outlines 11 that can be used to articulate silent or hidden requirements. The methods broadly represent three scenarios: new knowledge made possible by public (or social) data and trends, new knowledge gathered from the customer's products or environments, and new knowledge from combining emerging technologies with established methods from anthropology. Table 13.3 is not a comprehensive list

**TABLE 13.1. VoC techniques for assessing stated needs.**

| Technique | Description | References |
|---|---|---|
| Customer knowledge management | Tracking all activities related to customer characteristics, demographics, and interactions to better understand the customers, considering them to be at the center of the business. | Sachamanorom & Senoo (2016) |
| Observations of customers/ "Lens Model" | Observing customers using a product or service can reveal needs that they may be able to express, but only when prompted by the experience. Talking out loud with a representative of the organization collecting the observational data is recommended. | Griffin & Hauser (1993) |
| Online brand communities | Online portals, usually membership based, that track customer questions and answers allow customers to provide support to one another and allow firms to solicit information about branding, satisfaction, product features, and service elements. | Lee et al. (2014) |
| Surveys and direct elicitation | Surveys use a series of defined questions (and sometimes use predefined answer choices) to provide easily quantifiable feedback to the organization. They can be conducted in person, over the phone, through a web form, or through videoconferencing. Surveys are useful for assessing and monitoring customer preferences and satisfaction, and to evaluate and assess the impact of changes to products or services. | Ding et al. (2011) Hayes (2008) |
| Benchmarking | Benchmarking helps organizations understand how other organizations and market leaders satisfy their customers' needs. Benchmarking allows organizations to study successes and best practices in other organizations, pinpointing places where they can make improvements in their own processes. | Brandt (2018) Cooper (1998) |
| Gemba visits | By going directly to the workplace (gemba), information about what customers want and need can be directly obtained. Because many unsatisfied customers do not complain, direct observation can reveal the source of that dissatisfaction. | González Bosch & Tamayo Enríquez (2005) Cooper (1998) |

(continued)

**TABLE 13.1.** (*continued*)

| Technique | Description | References |
|---|---|---|
| Focus groups/ customer interviews | Focus groups and customer advisory panels allow organizations to spend time with select groups of customers to solicit specific information or engage in brainstorming sessions, and can be conducted in-person or using collaborative technology. | Cooper & Dreher (2010) Hayes (2008) Griffin & Hauser (1993) |
| Brainstorming | Brainstorming, in which ideas and suggestions flow in a less structured way than they do in structured surveys and interviews, is an effective tool that can be used internally as well as externally (with customers). | Cooper & Dreher (2010) Berry & Parasuraman (1997) |
| Social media analysis | Feedback on social media provides timely and unmediated customer insights that can be addressed, actioned, and, if necessary, remedied immediately. | Kohl et al. (2018) Jeong et al. (2017) Trainor et al. (2014) |
| Chat transcripts | Transcripts of chats with service representatives on customer websites provide evidence of product deficiencies, customer difficulties, and how call centers solve (or don't solve) their problems. | Trainor et al. (2014) |
| Web analytics | Web analytics specify the amount of traffic to specific parts of a website, which can provide significant insight into customer priorities and interests. | Croll & Power (2009) |
| Feedback forms | Feedback forms are often provided immediately after customers have an interaction with the organization providing the product or service. | Brandt (2018) Cooper & Dreher (2010) Hayes (2008) |
| E-mail | E-mails can be an excellent source of unstructured feedback from customers who have interacted with the organization. | Brandt (2018) |
| Research results | Many organizations have market research departments that conduct both qualitative and quantitative research, the results of which they regularly share internally. | Gopalani & Shick (2011) Griffin & Hauser (1993) |
| Analyst reports | Research firms, such as Gartner and Forrester, offer market and needs assessments for purchase. | Gopalani & Shick (2011) Griffin & Hauser (1993) |
| Call center/ customer service notes | Customer service agents take notes or make recordings during customer interactions that can provide insights into customer dissatisfaction and product or service defects. | Shaw & Hamilton (2016) Goodman (2006) |

(continued)

**TABLE 13.1.** (*continued*)

| Technique | Description | References |
|---|---|---|
| Suggestion box | The traditional suggestion box, in which customers or employees can place handwritten feedback into a sealed box, remains a valuable way of collecting spontaneous feedback. | Cooper & Dreher (2010) |
| Complaints | Analyzing customer complaints provides an opportunity for an organization to move beyond solving the immediate customer dissatisfaction and to diagnose process or product failures that are producing it. | Goodman (2006) |
| Product cancellation information | Many organizations provide feedback forms requesting details when customers cancel products or services. | Wu (2012) Goodman (2006) |
| Lost deals | Sales teams frequently collect valuable insights from informal conversations with potential customers after failed bids or deals. | Snelgrove (2017) |
| Delphi Method | The Delphi Method presents multiple rounds of questionnaires to subject matter experts. Respondents deliberate on responses during each round until they reach a consensus. | Lee & Huang (2009) |
| Sales meetings, service calls, trade show interactions, communities of enthusiasts | Personal interactions with customers and potential customers can provide valuable anecdotal information that may not make it into a generic customer survey. The drawback is that most of these interactions are not effectively documented or analyzed, therefore limiting the potential impact of the information. | Cooper & Dreher (2010) |
| Willingness to pay | The amount of money a customer is willing to pay for a product, and the minimum amount a person is willing to accept to abandon a product or put up with negative features can provide valuable insight into the financial meaning they attach to needs. | Snelgrove (2017) |
| Warranty data | Warranty data collected during the servicing of warranty claims is a valuable source of product failures and customer dissatisfaction; it suggests the thresholds at which customers believe that their products fail to live up to promises of performance. | Wu (2012) |

**TABLE 13.2. VoC techniques for assessing implied needs.**

| Technique | Description | References |
|---|---|---|
| Observation | Observing a customer as he or she uses a product or service can reveal information about needs that he or she may not be able to articulate directly. | Zultner (1993) Karat et al. (2003) |
| Lead user process | A specific customer or group of customers is selected to actively and iteratively participate in a new product development or continuous improvement process. Originally developed at 3M. | von Hippel (1986) von Hippel et al. (1999) |
| Typology of customer value | Based on the idea that perception of value depends on the interaction between a customer and a product or service, this technique helps an organization identify feelings and beliefs that may be associated with definitions of quality. | Holbrook (1996) |
| Prosumerism and customization | When customers participate (partially or fully) in the creation or improvement of a product or service, or in generating supporting artifacts (e.g., videos, blog posts) for a product, those contributions can yield a valuable source of intelligence about implied needs. | Hartmann (2016) |
| Experience sampling | Prompting a customer on an occasional basis to provide brief information and insights about a product or service (usually using an electronic tool, such as a smartphone) as they are using it can provide surprising insights because the customer has little time to think about their answers. | Larson & Czikszentmihalyi (1983) |
| Repertory grid | A comprehensive method that requires users to identify quality attributes, rate them (one or more times) on a five- or seven-point scale, and categorize them to find signals for what contributes to perception of value in a complex customer experience. | Tan & Hunter (2002) Lemke et al. (2011) Pike (2003) |

(continued)

**TABLE 13.2.** (*continued*)

| Technique | Description | References |
|---|---|---|
| Ergonomic studies | Implicit in a statement of needs is the customer's desire to remain safe, comfortable, and free from bodily harm and injury. Ergonomic studies provide information about how products and services can be designed for safety and comfort, and about requirements that a customer may not be able to verbalize. | Nath et al. (2017) |
| A/B testing | Customers are presented with Option A and Option B and then decide which one they like better. This does not require the customer to explain why a particular option is more desirable. Results are analyzed statistically to determine which option is more effective and desirable. | Kohavi & Thomke (2017) |
| Semantic differential technique | This method asks customers to evaluate the degree to which a product, service, or concept aligns with one or more pairs of descriptive words (e.g., wet/dry, brave/cowardly, confusing/clear). It became one of the cornerstones of kansei engineering decades later. | Snider & Osgood (1969) |
| Kansei engineering | This family of methods seeks to incorporate emotional needs and responses into product design. Semantic differential technique is one method used in kansei engineering. | Huang et al. (2012) Schutte et al. (2014) |
| Zaltman Metaphor Elicitation Technique (ZMET) | This approach asks customers to choose images that reflect their feelings and needs regarding products, services, and concepts, or alternatively, asks them to tell stories about images that reveal unconscious issues. Zaltman's 2003 book was written to help companies apply the ZMET method. | Zaltman & Coulter (1995) Zaltman (2003) |

but rather a representative list, because new technologies are emerging regularly to identify customer needs from new sources.

As the IoT and connected products become more common, VoT (Voice of Things) will become a more significant mechanism for identifying hidden needs. What better way to learn about a customer than to have their purchases report on their needs, interests, and behavior? This may sound far-fetched, but Facebook's software is already surveilling users' browsing habits to inform its advertising network, which stretches far beyond the social media platform itself. The next step is for products to participate in that surveillance.

Interesting legal ramifications may also emerge. For example, the 2010 Supreme Court decision for *Citizens United v. Federal Election Commission* (FEC) was based on a curious assumption: that nonpersons have the right to free speech. Although the case examined the free speech rights of corporations, technically, any object that can "speak" also has that right. The IoT will require us to rethink fundamental questions about how our interests as consumers and stakeholders are represented. For example:

- What will the world look (and feel) like when objects you interact with have a "voice"?
- How will VoC be interpreted when customers *and* the things they own or use have a voice? What if there is a conflict or difference of perspective?
- Will IoT objects have "agency"—that is, the right to represent your needs and interests to other products, services, or companies?

Although there are potential pitfalls, the volume and variety of data from your customers and about your customers will continue to expand. Ethical collection and use of that data must be considered as VoC programs are designed and implemented.

## Stage II: Understand and Prioritize Needs

Collecting VoC data is only the first step of the process. Next, organizations must translate *what* is needed into *how* those needs will be satisfied—and decide which are most important. Stated, implied, and hidden customer needs must be critically examined to understand how requirements should be used to make design choices. In some cases, an organization will know the needs

**TABLE 13.3. VoC techniques for uncovering silent needs.**

| Technique | Description | References |
|---|---|---|
| Sensor surveillance (e.g., IoT) | Sensors embedded into products or the customer's environment (e.g., via the IoT) can provide information about how, when, and why they use products and services. | Radziwill & Benton (2017a) |
| Voice of product (VoP), Voice of things (VoT), Online behavior monitoring | IoT devices, particularly in consumer settings, can provide information about product usage, the user's context of use, and the user's environment. Other products, particularly software products, may be able to provide indirect information about how and why a product or service is used. | Davies (2017) Radziwill & Benton (2017b) |
| Topic modeling | This method takes any collection of unstructured text (e.g., online reviews, social media posts, customer comments) and identifies themes or priorities. | Radziwill (2018) Ko et al. (2018) |
| Sentiment analysis | This method takes any collection of unstructured text (e.g., online reviews, social media posts, customer comments) and compares it with established lexicons that contain word characteristics to determine whether overall sentiment is positive or negative, and/or whether certain emotions are represented more than others over time. | Radziwill (2018) Jeong et al. (2017) |
| Voice of the customer table (VoCT) | This method examines vague or nebulous customer needs to extract true, actionable needs. | Tague (2005) |
| Corporate ethnography | This method involves observing or shadowing customers to inductively build concept maps that explain needs, motivations, and preferences. It requires highly trained researchers to gather and interpret data. | Anderson (2009) Ladner (2014) |
| ISO 26000 Guidance on Corporate Social Responsibility | This method provides best practices for "how to ensure social equity, healthy ecosystems and good organizational governance, with the ultimate objective of contributing to sustainable development" (Frost, 2011). | Hahn (2012) |

(continued)

**TABLE 13.3.** (*continued*)

| Technique | Description | References |
|---|---|---|
| ISO 9241 Ergonomics of Human System Interaction | This method provides guidance on meeting customers' usability needs for hardware and software displays, Interactive Voice Response systems, visual presentation of information, forms, tactile and haptic response, and so on. | Bevan (2009) Bevan et al. (2015) |
| Sensor surveillance (e.g., IoT) | Sensors embedded in products or the customer's environment (e.g., via the IoT) can provide information about how, when, and why they use products and services. | Radziwill & Benton (2017b) |
| Sousveillance | This technique involves customers actively countersurveilling organizations that monitor their needs, desires, and behaviors. | Mann et al. (2002) Levy & Barocas (2018) |
| VoT—anthropomorphized | Adding intelligence to the raw customer-oriented data could create opportunities for greater insights: "Humanizing a connected thing creates the opportunity to obtain feedback from it in the same way we would from a human being, complementing existing human feedback." | Davies (2017) |

of customers and how they should best be satisfied, but will be constrained by feasible or available options for satisfying those needs.

For example, when a patient with a medical condition wants their issue to be resolved, there are usually multiple ways to make it happen. The availability of medical procedures and technology, and the feasibility of performing those procedures on the patient at a particular time, need to be evaluated. Table 13.4 outlines some qualitative and quantitative techniques used to extract meaning from VoC data and prioritize outcomes. Although all methods have been in use for decades by quality professionals and market researchers, the analytical approaches listed here are more powerful thanks to reliable open source software packages that can be used to analyze large volumes of customer data.

**TABLE 13.4.  Examples of methods for analyzing and prioritizing VoC data.**

| Technique | Description | References |
|---|---|---|
| Kano model | A classification technique to help prioritize features of products and services | Kano et al. (1984) |
| Quality function deployment (QFD) | A qualitative tool for examining trade-offs to decide how customer needs should best be met | Chan & Mazur (2017) |
| Analytic Hierarchy Process | A quantitative tool to make complex decisions based on multiple attributes by simplifying the problem into many smaller pairwise comparisons | Saaty (1999) |
| Conjoint Analysis | A quantitative tool to identify the best combination of features based on which ones contribute the most to overall perception of value | Green & Srinivasan (1990) |
| Technique for Order of Preference by Similarity to Ideal Solution | A quantitative tool that compares a set of alternatives based on weights assigned with components of those alternatives | Yoon (1987) Hwang et al. (1993) |

## Stage III: Create Meaningful Customer Experiences

While VoC can be useful for uncovering customer preferences for pricing, product features, or product configurations, a truly insightful VoC process will discover what constitutes a meaningful customer experience. While customers are often rational decision makers, they are also emotional and value pleasurable, beneficial, and educational experiences. Customers (and people in general) also value experiences that allow them to co-create objects and experiences through their interaction with brands and companies (Hwang & Seo, 2016).

Total customer experience describes the end-to-end evolution of social, physical, and emotional realities as a customer moves from the initial stages of awareness to the postransactional "nurture" stage (Hwang & Seo, 2016).

Greater connectedness to opportunities for co-creation means that customer experience has the potential to be a powerful differentiator and competitive edge for today's organizations (Alcántara et al., 2014). Another approach to co-creation is Customer Experience Management, which complements Customer Relationship Management (CRM) by negotiating the gap between customer expectations and customer experience and working to continually enhance customer loyalty (Hwang & Seo, 2016).

A universal way to assess customer experience, however, remains elusive. Some organizations evaluate a customer's experience only as positive or negative, while other more holistic approaches include values such as pleasure, learning, nostalgia, and fantasy. Measurements are made along a continuum that is intended to reflect the highs and lows of everyday life (Hwang & Seo, 2016). Using this baseline, a meaningful experience does not need to be something overwhelming or sublime, but something that honors the needs and personal realities of the customers, or their intentions for enhancing the well-being of the environment or the needs of cultures and social groups (Jensen, 2014). Experience is more than simply solving a customer's immediate problem: it is a dialogue or interaction or exchange that helps that person explore possibilities for living, self-actualization, and building relationships (Jensen, 2014).

VoC is the fundamental exercise that exposes the customer values that can help organizations create those meaningful experiences. Customers adopt products when those products interact with, and are seamlessly interwoven into, the complex social systems that make up the customers' lives. VoC is therefore a critical input to the design process for all products and services.

One of the most valuable ways to create meaning around VoC is storytelling. Storytelling is how people and organizations create common cultures by sharing their knowledge and values to create emotional connections (Beckman & Barry, 2009). Designers can incorporate storytelling into their process by using the VoC toolbox to learn how the customer views the narrative of their lives, the performative actions that might confirm or contradict their interpretation of their experiences, and workarounds that they could consciously or unconsciously incorporate into their lives as they negotiate the complex systems of their experiences on a daily basis. The stories the customer tells, whether overtly or tacitly through their behavior, become the inputs designers use to inspire new and innovative solutions that complement and delight customers (Beckman & Barry, 2009). Designers learn from those stories (and become characters within them) as they find ways to inspire customers to adopt products that blend seamlessly and profoundly with the customer experience. These stories can also become structural narratives that inspire other customers and create new communities and cultural values.

## Stage IV: Anticipate Future Needs

Capturing customer needs and desires is important, but being able to identify the needs that they will have in the future is the basis for innovation. Anticipating future needs, which is an active area of study in market and quality management research, is being driven in large part by Industry 4.0 technologies and Quality 4.0 approaches. This section introduces techniques to forecast or infer future needs. Many of these methods are exploratory and not yet validated, so only a few are summarized.

For example, Gotzamani et al. (2018) used multivariate Markov Chain models to capture the dynamic nature of VoC and how it changes over time and in different contexts. They developed an adaptation of QFD, a qualitative tool to help organizations evaluate trade-offs and select product specifications that will meet customers' needs. Stansfield and Azmat (2017) have started exploring "artificial intelligence infused ISO 16355" to make QFD more responsive to the new data available from the IoT and sensor networks. Trainor et al. (2014) studied "social customer relationship management (CRM)" to see if customers' needs can be predicted or anticipated by the decisions their friends and social contacts are making.

Horizon scanning, the "systematic search for incipient trends, opportunities and constraints that might affect the probability of achieving management goals and objectives" (Sutherland et al., 2011), has also emerged as a paradigm for capturing high-level trends in VoC. Some software packages are now available to manage the collection, organization, and analysis of horizon scanning data that can contribute to understanding VoC. Ernstsen et al. (2018) recommend a three-step horizon scanning process specifically aimed to anticipate disruptive forces that will transform customer needs and requirements. The steps are (1) defining, (2) identifying, and (3) synthesizing. Most significantly, they recommend examining the following resources to anticipate future customer needs:

- Technology reports (e.g., those by analysts including Gartner, IDC, Forrester, McKinsey)

- Industry-specific reports (e.g., those from World Economic Forum, McKinsey)
- Conferences and seminars
- Technology conferences
- Foresight reports (e.g., those by public authorities, governments, U.S. agencies, European Union)

Ernstsen et al. (2018) also recommended that techniques like topic modeling and sentiment analysis be applied to the horizon scanning resources to uncover future needs. The most critical requirement, however, is to embed horizon scanning into product design and development processes to adapt to newly emerging needs as soon as they can be detected.

## Case Study: VoC Architecture at Fuji Xerox

In the late 2000s, Fuji Xerox changed its strategy from "Make and Sell" to "Sense and Respond." The company wanted to be more agile and responsive to changing customer needs, instead of relying on long product development life cycles and the hope that a market would be in place when the product was released. Consequently, a modern overhaul of its VoC program that took into consideration Industry 4.0 technologies and Quality 4.0 values was in order (Sachamanorom & Senoo, 2016).

### STAGE I: IDENTIFY CUSTOMER NEEDS

Fuji identified customer needs according to the three levels and provided labels (VoC 1.0, 2.0, and 3.0) to describe the increase in maturity as new varieties of data were added:

- Stated needs = knowledge *from* customer (VoC 1.0)
- Implied needs = knowledge *about* customer (VoC 2.0)
- Hidden needs = knowledge *discovered* through interactions (VoC 3.0)

Within each of these levels, it identified several elements or mechanisms of the process to draw out the customer needs and desires, the characteristics of the workplace, and the tools required to enable the data collection. This started the process of identifying an architecture (Table 13.5) to make VoC gathering and analysis systematic, repeatable, and robust.

## STAGE II: UNDERSTAND AND PRIORITIZE NEEDS

To synthesize and prioritize needs, Fuji Xerox established (1) an integrated call center/support center to serve as the main contact point for customers; (2) a VoC data collection and management system that is available and accessible to all employees; (3) a website with 24-hour customer support, including human and automated response mechanisms; (4) a process for external market research; and (5) strategic initiatives focused on customer satisfaction survey results. Systematic, repeatable processes were established so that people would know when, how, and why the data would be examined, and how the findings would be incorporated into new product development.

**TABLE 13.5. Fuji Xerox approach to modernizing the VoC program.**

| Needs | Process | Workplace | Tools |
|---|---|---|---|
| Stated needs | Customer satisfaction data, daily interactions (complaints, claims, requests, inquiries through traditional and social media channels), position statements | Integrated customer support center; provide onsite service to customers to gather data | • Customer support<br>• Help desk<br>• Website |
| Implied needs | Focus group market research; cross-division meetings | New product development meeting room; another Fuji development and technology center | • Meeting monitoring devices |
| Hidden needs | Share company information with participants in co-creation lab; ask customers to talk about their business challenges; use facilitators and visualization methods during co-creation sessions | Use open space for discussion instead of closed rooms; incorporate themes into environment to stimulate innovation | • Open-office-board<br>• Anonymous open-office-cards<br>• Co-creation lab |

Source: Adapted from Sachamanorom & Senoo, 2016.

## STAGE III: CREATE MEANINGFUL CUSTOMER EXPERIENCES

To add the dimension of creating meaningful customer experiences, which the company felt it had been missing, the company launched the Co-creation Laboratory in 2010. The lab brings together executives, managers, and decision makers from customer companies in "diversified co-creation sessions" that promote free, open, two-way communication. Results from the sessions are captured in a Co-creation Database that becomes part of the unified VoC data collection and management system accessible to all employees.

There were several Key Performance Indicators (KPIs) assigned to the lab, designed to ensure continuity between understanding and prioritizing needs, and creating and deploying meaningful experiences that led to improved customer satisfaction:

- Number of visitors/number of companies participating in co-creation events
- Number of projects entering the intellectual property secured lab
- Number of projects emerging from the intellectual property (IP) secured lab
- Number of customer/partner co-creations
- Number of projects that became revenue-producing business ideas
- Number of VoC records generated
- Number of new themes introduced in open laboratory
- Number of employees who use the VoC system
- Percentage of executive-level participants
- Percentage of business ideas characterized by "usefulness"
- Net Promoter Score (NPS) to gauge changing customer satisfaction

From 2010 to 2014, NPS improved from −4 (bad) to +35 (very good), as the number of employees actively using the VoC data collection and management system to launch and track co-creation events grew from 1,840 to 4,090. Fuji Xerox was pleased with the definitive results.

## STAGE IV: ANTICIPATE FUTURE NEEDS

By evaluating the performance of the Co-creation Laboratory over the five-year period from 2010 to 2014, Fuji Xerox confirmed that it had been able to capture ideas for innovation in its Co-creation Database. Because the data was available cross-functionally and easy to access, employees from every department of the organization were able to analyze it and provide different insights.

Guiding the innovative approach to modernizing the VoC program at Fuji was the recognition that its internal experts may not be the best people to identify customer needs. "Value which Fuji Xerox believes . . . is essential might not always be the same from their customers' viewpoints." The company wanted to develop a strategy that involved customers more directly in the ideation and design processes. "That is why Fuji Xerox invites their customers, partners and other stakeholders to join the co-creation session and by doing activities together, they can find and create value through the process" (Sachamanorom & Senoo, 2016). The experience of Fuji Xerox shows one example of how traditional and modern methods can be incorporated into a robust VoC system that is responsive to emerging needs while being an exemplar of innovation through VoC.

## THE BOTTOM LINE

This chapter outlined 24 techniques for extracting customers' stated needs, 11 methods for identifying implied needs, and 11 approaches for examining silent needs. To design a VoC program, this proven four-stage process can be applied:

- **VoC Stage I: Identify customer needs.**

  - o **IA: Stated needs** that can be directly expressed by the customer
  - o **IB: Implied needs** that cannot be expressed but can be found indirectly
  - o **IC: Silent (or hidden) needs** that must be determined by context or models

- **VoC Stage II: Understand and prioritize needs.**
- **VoC Stage III: Create meaningful customer experiences.**
- **VoC Stage IV: Anticipate future needs.** *Identifying needs that do not currently exist is the basis for innovation.* Advanced analytical methods can be used to predict future needs, possibly even before customers are aware that those needs exist.

A robust, comprehensive VoC program will incorporate multiple methods from each of the three needs categories (IA, IB, IC) and use techniques like co-creation and storytelling to create and deliver meaningful customer experiences. Developing systematic, repeatable processes for data collection and management, and identifying the requirements for workspaces and tools, can be the basis for an architecture of systems and processes to support the VoC program.

Understanding customer needs and desires is important because it helps organizations develop high-quality products and services now—and helps drive innovation to meet customer needs in the future. For guidance beyond the four-stage process, Brandt (2018) provides insights about how to design systematic, repeatable processes to support a VoC program.

## ACKNOWLEDGMENTS

This chapter is based on the findings published in Freeman and Radziwill (2018).

## REFERENCES

Alcántara, E., M. A. Artacho, N. Martínez, and T. Zamora. (2014). "Designing experiences strategically." *Journal of Business Research* 67 (6): 1074–1080.

Anderson, K. (2009, March). "Ethnographic research: A key to strategy." *Harvard Business Review.* https://hbr.org/2009/03/ethnographic-research-a-key-to-strategy

Beckman, S., and M. Barry. (2009). "Design and innovation through storytelling." *International Journal of Innovation Science* 1 (4): 151–160.

Berry, L. L., and A. Parasuraman. (1997). "Listening to the customer—the concept of a service-quality information system." *MIT Sloan Management Review* 38 (3): 65.

Bevan, N. (2009, August). "What is the difference between the purpose of usability and user experience evaluation methods." In *Proceedings of the Workshop UXEM* (9): 1–4.

Bevan, N., J. Carter, and S. Harker. (2015, August). "ISO 9241-11 revised: What have we learnt about usability since 1998?" In *International Conference on Human-Computer Interaction*, 143–151. Cham, Switzerland: Springer.

Brandt, D. R. (2018). "The current state of corporate voice of the consumer programs: A study of organizational listening practices and effectiveness." *International Journal of Listening*, 1–27.

Chan, C. Y., and G. H. Mazur. (2017, February). "ISO 16355: Modern QFD generated from 50 years of practice." In *Proceedings of the 21st international conference on ISO* (9000).

Chauhan, P., and S. Sarabhai. (2018). "Customer experience management: Evolution and the paradigm shift in marketing." *Business Perspectives* 17 (1): 18–34.

Cooper, R. (1998). "Benchmarking new product performance: Results of the best practices study." *European Management Journal* 16 (1): 1–17.

Cooper, R. G., and A. Dreher. (2010). "Voice-of-customer methods." *Marketing Management* 19 (4): 38–43.

Croll, A., and S. Power. (2009). *Complete web monitoring: Watching your visitors, performance, communities, and competitors.* O'Reilly Media.

Davies, J. (2017). "Three ways to listen to the voice of 'things' in the IoT." *Gartner Reports.* https://www.gartner.com/document/3825264

DigitalLBi. (2016, December 12). "New DigitasLBi Research Shows More than 1 in 3 Americans are willing to make Purchases via Chatbots." https://www.digitas.com /en-us/news/press-releases/new-digitaslbi-research-shows-more-than-1-in-3 -americans-are-willing-to-make-purchases-via-chatbots

Ding, M., J. R. Hauser, S. Dong, D. Dzyabura, Z. Yang, C. Su, and S. P. Gaskin. (2011). "Unstructured direct elicitation of decision rules." *Journal of Marketing Research* 48 (1): 116–127.

Ernstsen, S. K., C. Thuesen, L. R. Larsen, and A. Maier. (2018). "Identifying disruptive technologies in design: Horizon scanning in the early stages of design." In *DESIGN2018-15th International Design Conference.*

Faw, L. (2016, December 13). "What's a chatbot? Most Americans don't know." *Media-Post Agency Daily.* https://www.mediapost.com/publications/article/290971/whats -a-chatbot-most-americans-dont-know.html

Freeman, G., and N. M. Radziwill. (2018). "Voice of the customer (VoC): A review of techniques to reveal and prioritize requirements for quality." *Journal of Quality Management Systems, Applied Engineering, and Technology Management (JoQAT),* 2018 (5). https://joqat.files.wordpress.com/2018/09/fullmanuscript-17.pdf

Frost, R. (2011, March 9). *ISO 26000 social responsibility—the essentials.* ISO.

González Bosch, V., and F. Tamayo Enríquez. (2005). "TQM and QFD: Exploiting a customer complaint management system." *International Journal of Quality & Reliability Management* 22 (1): 30–37.

Goodman, J. (2006, February). "Manage complaints to enhance loyalty." *Quality Progress,* 28–34.

Gopalani, A., and K. Shick. (2011). "The service-enabled customer experience: A jump-start to competitive advantage." *Journal of Business Strategy* 32 (3): 4–12.

Gotzamani, K., A. Georgiou, A. Andronikidis, and K. Kamvysi. (2018). "Introducing multivariate Markov modeling within QFD to anticipate future customer preferences in product design." *International Journal of Quality & Reliability Management* 35 (3): 762–778.

Green, P. E., and V. Srinivasan. (1990). "Conjoint analysis in marketing: New developments with implications for research and practice." *Journal of Marketing,* 3–19.

Griffin, A., and John R. Hauser. (1993). "The voice of the customer." *Marketing Science* 12 (1): 1–27.

Hahn, R. (2012, November). "Standardizing social responsibility? New perspectives on guidance documents and management system standards for sustainable development." *IEEE Transaction on Engineering Management* 59 (4): 717–728.

Hartmann, B. J. (2016). "Peeking behind the mask: Theorizing the organization of consumptive and productive practice moments." *Marketing Theory* 16 (1): 3–20.

Hayes, B. E. (2008). *Measuring customer satisfaction and loyalty: Survey design, use, and statistical analysis methods.* Milwaukee, WI: ASQ Quality Press.

Holbrook, M. B. (1996). "Special session summary customer value, a framework for analysis and research." In K. P. Corfman & J. G. Lynch Jr. (Eds.), *Advances in consumer research* (23): 138–142. Association for Consumer Research.

Huang, Yuexiang, Chun-Hsien Chen, and Li Pheng Koo. (2012). "Products classification in emotional design using a basic-emotion based semantic differential method." *International Journal of Industrial Ergonomics* 42 (6): 569–580.

Hwang, C. L., Y. J. Lai, and T. Y. Liu. (1993). "A new approach for multiple objective decision making." *Computers and Operational Research* 20: 889–899. doi:10.1016/0305-0548(93)90109-v.

Hwang, J., and S. Seo. (2016). "A critical review of research on customer experience management: Theoretical, methodological and cultural perspectives." *International Journal of Contemporary Hospitality Management* 28 (10): 2218–2246.

Jarman, P. (2013, August). "Getting to know your customers through data." *Wired.* https://www.wired.com/insights/2013/08/getting-to-know-your-customers-through-data/

Jensen, J. L. (2014). "Designing for profound experiences." *Design Issues* 30 (3): 39–52.

Jeong, B., J. Yoon, and J. M. Lee. (2017). "Social media mining for product planning: A product opportunity mining approach based on topic modeling and sentiment analysis." *International Journal of Information Management.*

Kano, N., N. Seraku, F. Takahashi, and S. Tsuji. (1984). "Attractive quality and must-be quality." *Hinshitsu: The Journal of the Japanese Society for Quality Control* 14 (2): 39–48.

Karat, C. M., C. Brodie, J. Karat, J. Vergo, and S. R. Alpert. (2003). "Personalizing the user experience on IBM.com." *IBM Systems Journal* 42 (4): 686–701.

Ko, N., B. Jeong, S. Choi, J. Yoon. (2018). "Identifying product opportunities using social media mining: Application of topic modeling and chance discovery theory." *IEEE Access* 6, 1680–1693.

Kohavi, R., and Thomke. (2017). "Online controlled experiments and A/B tests." In C. Sammut & G. Webb (Eds.), *Encyclopedia of machine learning and data mining.* Springer.

Kohl, C., M. Knigge, G. Baader, M. Böhm, and H. Krcmar. (2018). "Anticipating acceptance of emerging technologies using Twitter: The case of self-driving cars." *Journal of Business Economics*, 1–26.

Ladner, S. (2014). *A guide to doing ethnography in the private sector.* Routledge.

Larson, R., and M. Czikszentmihalyi. (1983). "The experience sampling method." In H. T. Reis (Ed.), *New Directions for Methodology of Social and Behavioral Sciences* 15: 41–56.

Lee, H., J. Han, and Y. Suh. (2014). "Gift or threat? An examination of voice of the customer: The case of MyStarbucksIdea.com." *Electronic Commerce Research and Applications* 13 (3): 205–219.

Lee, Y. C., and S. Y. Huang. (2009). "A new fuzzy concept approach for Kano's model." *Expert Systems with Applications* 36 (3): 4479–4484.

Lemke, F., M. Clark, and H. Wilson. (2011). "Customer experience quality: An exploration in business and consumer contexts using repertory grid technique." *Journal of the Academy of Marketing Science* 39 (6): 846–869.

Levy, K., and S. Barocas. (2018). "Privacy at the margins: Refractive surveillance: Monitoring customers to manage workers." *International Journal of Communication* 12: 23.

Macnamara, Jim. (2018). Toward a theory and practice of organizational listening. *International Journal of Listening* 32: 1–33.

Mann, S., J. Nolan, and B. Wellman. (2002). "Sousveillance: Inventing and using wearable computing devices for data collection in surveillance environments." *Surveillance & Society* 1 (3): 331–355.

Nath, N. D., R. Akhavian, and A. H. Behzadan. (2017). "Ergonomic analysis of construction worker's body postures using wearable mobile sensors." *Applied Ergonomics* 62: 107–117.

Pike, Steven. (2003). "The use of repertory grid analysis to elicit salient short-break holiday destination attributes in New Zealand." *Journal of Travel Research* 41 (3): 315–319.

Radziwill, N. M. (2018, October). "Let's get digital: The many ways the fourth industrial revolution is reshaping the way we think about quality." *Quality Progress.*

Radziwill, N. M., and M. C. Benton. (2017a). "Design for X (DfX) in the internet of things (IoT)." *Journal of Quality Management Systems, Applied Engineering, & Technology Management (JoQAT), 2017* (1).

———. (2017b). "Quality in chatbots and intelligent conversational agents." *Software Quality Professional* 19 (3).

Rajadesingan, A., R. Zafarani, and H. Liu. (2015, February). "Sarcasm detection on Twitter: A behavioral modeling approach." In *Proceedings of the Eighth ACM International Conference on Web Search and Data Mining*, 97–106. ACM.

Saaty, T. L. (1999). *Decision making for leaders.* 3rd edition. Pittsburgh, PA: RWS Publications

Sachamanorom, W., and D. Senoo. (2016, June). "Voice of the customer through customer cocreation: The case of Fuji Xerox Japan." In *PACIS*, 147.

Schutte, S. T. W., J. Eklund, J. R. C. Axelsson, and M. Nagamachi. (2014). "Concepts, methods and tools in Kansei Engineering." *Theoretical Issues in Ergonomics Science* 5 (3): 214–231.

Shaw, C., and R. Hamilton. (2016). "How to move to the next level of customer experience." In *The Intuitive Customer*, 161–191. London: Palgrave Macmillan.

Snelgrove, T. (2017). "Creating, calculating and communicating customer value: How companies can set premium prices that customers are willing and able to pay." In *Innovation in pricing*, 248–260. Routledge.

Snider, J. G., and C. E. Osgood. (Eds.). (1969). *Semantic differential technique; a sourcebook.* Aldine Pub. Co.

Stansfield, K. E., and F. Azmat. (2017, March). "Developing high value IoT solutions using AI enhanced ISO 16355 for QFD integrating market drivers into the design of IoT offerings." In *Communication, Computing and Digital Systems (C-CODE), International Conference*, 412–416. IEEE.

Sutherland, W. J., S. Bardsley, L. Bennun, M. Clout, I. M. Côté, M. H. Depledge, L. V. Dicks, A. P. Dobson, L. Fellman, E. Fleishman, D. W. Gibbons, A. J. Impey, J. H. Lawton, F. Lickorish, D. B. Lindenmayer, T. E. Lovejoy, R. M. Nally, J. Madgwick, L. S. Peck, . . . A. R. Watkinson. (2011)." Horizon scan of global conservation issues for 2011." *Trends in Ecology and Evolution* 26 (1): 10–16.

Tague, N. R. (2005). *The quality toolbox.* 2nd edition. Milwaukee, WI: Quality Press.

Tan, F. B., and M. G. Hunter. (2002). "The Repertory Grid Technique: A method for the study of cognition information systems." *MIS Quarterly* 26 (1): 39–57.

Trainor, K. J., J. M. Andzulis, A. Rapp, and R. Agnihotri. (2014). "Social media technology usage and customer relationship performance: A capabilities-based examination of social CRM." *Journal of Business Research* 67 (6): 1201–1208.

von Hippel, E. (1986). "Lead users: An important source of novel product concepts." *Management Science* 32 (7): 791–805.

von Hippel, E., S. Thomke, and M. Sonnack. (1999). "Creating breakthroughs at 3M." *Harvard Business Review* 77: 47–57.

Wu, S. (2012). "Warranty data analysis: A review." *Quality and Reliability Engineering International* 28 (8): 795–805.

Yoon, K. (1987). "A reconciliation among discrete compromise situations." *Journal of Operational Research Society* 38: 277–286. doi:10.1057/jors.1987.44.

Zaltman, G., and R. H. Coulter. (1995). "Seeing the voice of the customer: Metaphor-based advertising research." *Journal of Advertising Research* 35 (4): 35–50.

Zaltman, G. (2003). *How customers think: Essential insights into the mind of the market.* Harvard Business Press.

Zultner, R. E. (1993). "TQM for technical teams." *Communications of the ACM* 36 (10): 79–91.

## CHAPTER FOURTEEN

# ELEMENTS OF A QUALITY
# 4.0 STRATEGY

The future will be driven by humans collaborating with other humans
to design work for machines that creates value for other humans.
—GREG SATELL, AUTHOR OF *CASCADES AND MAPPING INNOVATION*

I n a quiet valley in eastern West Virginia, tucked between two ridges just north-east of Snowshoe Mountain, a giant satellite dish looms over the hilly horizon. At 485 feet (148 meters), it's taller than the Statue of Liberty in New York City, and almost as imposing as the 555-foot (169-meter) Washington Monument in the District of Columbia. Mobile phones, microwave ovens, and baby monitors are prohibited here in the National Radio Quiet Zone. Radio frequency interference, which is kryptonite to this larger-than-life machine, must be battled at all costs.

As you may have suspected, the Green Bank Telescope (GBT), named for the town that hosts it, is no ordinary dish. First of all, it receives electromag-netic signals from outer space (rather than transmitting them). The GBT has such a vast collecting surface that it can detect the presence of tiny amino acids light-years away. Astronomers use it to collect tiny amounts of energy from planets, galaxies, pulsars, and star-forming regions to help them understand how astronomical bodies form, evolve, and die (Figure 14.1).

The GBT itself is a marvel of engineering: it is the world's only fully steer-able single dish radio telescope. It took the U.S. National Science Foundation (NSF) 12 years to design, build, and commission it at a cost of $95 million. The 2.3-acre surface of the parabolic dish is built from 2209 individual motorized panels, each of which can be optimally controlled. This way, the antenna can maintain its perfect parabolic shape, even when it's pointed toward the hori-zon and the force of gravity on the 17 million pounds of steel is most intense. Despite constructing a unique and irreproducible instrument, the NSF chose to defund the GBT in 2016. Fortunately, it would continue to be supported

FIGURE 14.1. The GBT in West Virginia (Green Bank Observatory [GBO], 2019).

primarily by private grants and international sources; by 2019, however, the long-term fate of the GBT was still up in the air (Scoles, 2016).

Even though the GBT is not a manufacturing facility, the physical and electronic infrastructure that collects, consolidates, amplifies, and interprets the signals from space is similar to a supervisory control and data acquisition (SCADA) control system. It should thus not be surprising that in the early 2000s, the GBT was a platform for improvements that would now fall under the Quality 4.0 banner—projects leveraging technologies for connectedness, intelligence, and automation to improve quality and performance.

All high-impact initiatives start with at least one compelling business need. In this case, there were two: observing was expensive for the astronomer (the customer), and lost observing time due to equipment issues or weather was expensive for the observatory (the business).

When the telescope started receiving research astronomers in 2002, most would travel long distances to get to Green Bank. In a typical week, in addition to researchers from U.S. universities, it wasn't uncommon to host observers who had flown in from Russia, the Netherlands, Chile, and Canada. The observers paid for their lodging at the on-site dormitory and meals in the full-time cafeteria from their research funding. A typical trip might cost $3000 for a single astronomer. Once onsite, there was no guarantee that you would actually get to run your observation. Radio signals are sensitive to water vapor in the atmosphere, and West Virginia is not a desert. In addition, if there was a problem with any component on the telescope during your scheduled observing time, you may not be able to gather your data. At a cost to taxpayers of $5,800 an hour, neither of these scenarios was particularly good.

I worked at the GBO during this time, so I had the privilege of being part of developing solutions to these problems. First, travel time and costs could be eliminated if the astronomers could "drive the telescope" from their home computer instead of the computer in the telescope's control room. Additionally, we could reduce or remove uncertainty in slotting observations if we monitored the weather in real time with our own sensors, created algorithms to decide whether a particular observation was viable, and then dynamically rescheduled astronomers based on best conditions. This would also help us respond much more effectively to equipment issues. Because astronomers could connect to the telescope from home, adjusting to a shift in schedule would no longer be quite so challenging. The remote observing and dynamic scheduling projects, which would take nearly three years to complete (but were highly successful), were thus born.

The lessons I learned from these projects can be applied to developing strategies for Quality 4.0 today. First, establish one to three measurable goals for improving quality or performance. Next, identify value propositions to support those goals. Finally, identify strategic initiatives (and execute them by building thematically related action plans) that combine work systems and technologies for enhancing connectedness, intelligence, and automation. This progression, with estimates of the results that were presented to audit and review panels, is shown in Table 14.1.

The terms *digital transformation, Industry 4.0,* and *Quality 4.0* are interconnected. Digital transformation refers to the process of incorporating digital technologies for connection, intelligence, and automation, with or without particular quality and performance goals as drivers. Beneficial digital shifts can be catalyzed in any industry, from healthcare to finance, retail, service, or software. Industry 4.0, based on research papers written between 2011 and 2019, specifically refers to digital transformation in manufacturing industries like automotive, aviation, chemicals, defense, medical devices, and pharmaceuticals. Precision agriculture, intelligent transportation, and various terms with "4.0" appended (like Tourism 4.0, Healthcare 4.0, and HR 4.0) also appear in the literature and are used to refer to the digital transformation of those domains using Industry 4.0 and related emerging technologies.

**TABLE 14.1. Early, ad hoc development of a Quality 4.0 strategy in 2004.**

| Quality and performance goals (*what* to achieve) | Value propositions (*how* benefits will be delivered) | Strategic initiatives (*how* to engage and act) | Results (*what* to achieve) |
|---|---|---|---|
| Reduce time and cost of travel for astronomer | Enable a new business model to democratize access to the telescope<br><br>Make it logistically possible for people to remote observe by implementing dynamic scheduling | Remote observing project | Cost reduced from $3,000 to $0 per observer; savings to customers of $600,000–$800,000 per year |
| Reduce lost time/ money due to weather issues and/ or equipment failure | Augment human intelligence to choose better times to observe, thus increasing the quality of observations<br><br>Increase speed and quality of decision making | Dynamic scheduling project | Approximately 1,500 reclaimed hours of time per year at $5,800/hour; total increase in science value delivered ~$8.7 million per year |

A Quality 4.0 strategy is an Industry 4.0 strategy or digital transformation strategy where quality and performance goals are front and center. While Industry 4.0 research may be limited to how cyber-physical systems impact discrete manufacturing and process industries, Quality 4.0 is for all industries. It addresses the interplay among people, processes, and new technologies that enhances connectedness, intelligence, and automation.

This chapter develops the three requirements for a Quality 4.0 strategy—that is, a digital transformation (or Industry 4.0) strategy driven by quality and performance. To identify them, we

- revisit the quality goals associated with digital transformation, in terms of quality costs;
- review drivers for digital transformation: *why* companies pursue it;
- describe digital transformation and Industry 4.0 stories from old and new organizations; and
- summarize existing models for digital transformation strategy and Industry 4.0 maturity.

The resulting approach is applicable to every industry and provides guidance for developing initiatives and action plans to realize success.

## THE VISION FOR QUALITY 4.0

In the Industry 4.0 era, the goals for quality and performance improvement are the same as they have been since the early and mid-1900s. Quality planning, control, and improvement activities are performed in order to:

- achieve conformity,
- meet requirements and specifications,
- reduce variation,
- reduce and/or prevent defects,
- reduce waste and rework,
- eliminate non-value-adding activity,
- prevent human error,
- improve productivity,
- improve efficiency,
- improve the effectiveness of people and resources,
- improve usability and customer experience, and
- drive innovation. (Evans, 2013)

Only two things are changing with the fourth industrial revolution: (1) the amount of data is increasing, and (2) quality goals can be accomplished faster and more completely because emerging technologies are becoming more powerful and accessible.

As an example, the earlier a problem is detected, the less it costs and the lower the impact. We reduce failures by catching them in advance (by applying controls for prevention, mistake-proofing, or implementing early warning systems), catching them as they occur (via controls for detection), or designing them out of the system so they can't happen in the first place. Imagine having an intelligent software system connected to multiple sources of data from a production process in real time. When the data starts to indicate that a failure is imminent, instead of the operator interpreting the data and intervening, the system knows it's about to happen and takes action before an out-of-control action plan is required. This saves time, money, and effort—and recovery from the failures is unnecessary, because they aren't able to occur.

DeFeo (2018) explains these dynamics in terms of quality costs. Each of the columns in Figure 14.2 represents a company's total revenue. The first column on the left of the figure shows that revenue covers operating costs, and whatever is left over is profit. The second column shows that part of the total cost of operations is the Cost of Quality (CoQ), which represents all the activities an organization engages in to prevent (or recover from) quality issues and quality events.

CoQ has four categories: costs to prevent quality problems (e.g., design, mistake-proofing, or planned maintenance and calibrations), costs for appraisal (e.g., assessments, audits, management reviews), costs of internal failures that occur before they impact the customer, and costs of external failures that do ultimately impact the customer.

When siloed or struggling companies start to implement quality management programs, the amount of time and effort spent on prevention and appraisal usually increases so that internal and external failures begin to fall, as shown in the rightmost column in Figure 14.2. After two to three years, the distribution shifts: external failures decrease greatly, while internal failures decrease at a lesser rate; prevention and appraisal activities decrease as efficiencies are discovered. The increase in quality maturity means that costs are lower and profits are higher.

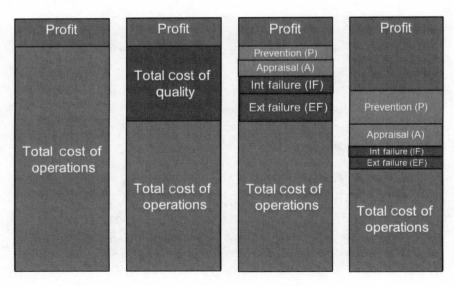

FIGURE 14.2. Distribution of quality costs shifts with digital maturity (Adapted from DeFeo, 2018).

Imagine, though, the level of quality maturity that will be possible once a complete digital transformation has occurred and is being continuously renewed. Failures do not happen, because prevention activities are perfect and self-correcting based on new data that comes in. Appraisal costs are nearly nothing, because operations teams are empowered to respond to assessments and monitor data in real time, audits are performed automatically by the software, and management reviews are just a formality. Some prevention activities are required, like regular calibration and asset maintenance, but with predictive maintenance in place these costs are minimized as well. In addition to optimizing profit potential (Figure 14.3), additional revenue may be generated by making other areas of the business more effective and efficient, like environment, health, and safety (EHS).

Examples of how quality performance goals shift in response to Industry 4.0 and digital transformation are provided in Table 14.2. Although this table is not comprehensive, it does provide representative examples of how quality

FIGURE 14.3. Distribution of quality costs when the vision of Quality 4.0 is realized.

**TABLE 14.2. How quality and performance goals shift in Quality 4.0.**

| Traditional quality | Quality 4.0 |
| --- | --- |
| Reduce total costs of operations by eliminating waste and non-value-adding activity | Minimize total costs of operations by eliminating waste and non-value-adding activity |
| Reduce internal and external failures | Eliminate internal and external failures |
| Reduce costs of appraisal by streamlining audits, reviews, and related activities | Minimize or eliminate costs of appraisal by automating audits and reviews, and detecting issues before they occur |
| Reduce variation | Minimize variation where appropriate; create variation where appropriate (e.g., innovation) |
| Use Theory of Constraints, Pareto analysis, or cost/benefit analysis to determine resource allocation and risk reduction priorities | Automated intelligent systems provide guidance for optimally allocating resources, choosing the highest value corrective actions, and selecting the most risk-reducing actions based on the organization's current risk appetite and priorities |
| [No Analog] | Increase profit by identifying and capturing new opportunities for growth |

practice can become deeper, more complete, and more powerful by adding the connectedness, intelligence, and automation of Quality 4.0.

## DRIVERS FOR DIGITAL TRANSFORMATION

The essence of digital transformation is *establishing new ways to create value.* For example, in an article in *Forbes,* Newman (2018) explains that connected consumers, customized experiences, empowered employees, optimized production, and connected products are the desired outcomes driving digital transformation. Although his article focuses on digital transformation in Industry 4.0, these drivers of change and technology adoption apply across all industries.

Why do companies start a digital transformation journey? New companies, like Uber and Lyft, are "digital natives" that have based their business models on the availability of pervasive networked digital technologies. Established enterprises, on the other hand, see opportunities for growth and greater customer satisfaction, and fear falling behind and losing competitive advantage if they don't take action. Small and medium-sized companies are somewhere in between, with many recognizing the value of digital and trying to figure out how and where to make the highest-impact investments.

Liere-Netheler et al. (2018) studied 67 research articles on digital transformation and found six organizational drivers, five external drivers, and a workforce driver. The organizational drivers were process improvement, workplace improvement (e.g., improving safety, ergonomics, or usefulness), vertical integration, horizontal integration, management support, and cost reduction. External drivers were responding to customer demands, streamlining the supply chain, enhancing innovation, responding to market pressures, and improved compliance with laws and regulations. The workforce driver was employee support (providing digital systems to help employees better perform tasks).

## Organizational Drivers

Organizational drivers are goals that the organization would like to achieve for itself as a result of digital transformation initiatives. These are typically related to strategy development, leadership and governance practices, workforce development and operations, and emphasize operational efficiency. The World Economic Forum (WEF, 2015) identified the earliest organizational drivers as creating new revenue streams, reducing operations costs, and optimizing asset utilization, followed closely by improving sustainability, worker productivity, safety, and customer experience.

Machadoa et al. (2019) held a workshop with seven companies in the aviation, machining processes, plastic packaging, heavy vehicles, and automobile industries. This study uncovered several drivers in operations: improving safety, improving data quality, improving products and processes, enhancing efficiency, assessing machine conditions, collecting and using project data, controlling and stabilizing processes, closing resource gaps, increasing speed and time-to-value, and making it possible to analyze systems by simulation. They found that low digital maturity—in particular, lack of technology and data literacy in the front-line workforce—negatively impacted the success of digital transformation initiatives in the companies that participated.

More drivers for digital transformation can be found by exploring the research specific to paradigms for thinking about manufacturing, like "smart manufacturing" and "sustainable manufacturing." The results from Lu et al. (2016) in Table 14.3 summarize the findings across these paradigms. Most of the drivers

**TABLE 14.3. Manufacturing paradigms.**

| Manufacturing paradigm | Drivers | Enablers |
|---|---|---|
| Smart | Enhance productivity through connectedness, interoperability, intelligence, and collaboration across the supply network; enhance decision making for energy and resource efficiency and accelerating innovation | • Pervasive digitization<br>• Connected devices<br>• Connected supply chain<br>• Advanced sensors<br>• Advanced analytics |
| Lean | Eliminate waste | • Process monitoring<br>• Resource leveling<br>• Workflow optimization<br>• Real-time monitoring |
| Flexible | Adapt to changes in production volume, process, and types | • Modular design for interoperability<br>• Service-oriented architecture |
| Sustainable | Conserve energy and natural resources, and enhance human safety | • Advanced materials<br>• Processes designed for sustainability |
| Digital | Reduce production cost and time-to-value | • Model-based engineering<br>• Product life-cycle management (PLM) |
| Cloud | Improve maintainability and make it possible to focus on core competencies by outsourcing infrastructure and services | • Cloud computing<br>• Internet of Things (IoT)<br>• Virtualization<br>• Advanced analytics |
| Intelligent | Adapt to changing environments and process requirements; optimize asset utilization | • Artificial intelligence (AI)<br>• Advanced sensing & control<br>• Optimization |
| Holonic | Make changes dynamically and continuously to adapt to changing requirements and environment | • Multiagent systems<br>• Decentralized control |
| Agile | Respond quickly to customer needs and market changes | • Collaborative/concurrent engineering<br>• Supply chain management (SCM)<br>• PLM |

Source: Adapted from Lu et al., 2016.

should be familiar to quality professionals. What has changed in the era of Industry 4.0 and digital transformation is that the technologies available to realize quality and performance goals are more plentiful, powerful, and accessible.

## External Drivers

Sometimes organizations pursue digital transformation because they feel external pressure from customers, stakeholders, or markets (WEF, 2015). More effectively responding to market volatility, shifting client expectations, and new competitors that are not as firmly entrenched in a market can also be motivating factors (Ismail et al., 2017). This includes enhancing the richness of interactions among employees, suppliers, and customers; providing a seamless customer experience across all channels and touch points; ensuring immediacy and availability of information; enhancing transparency and visibility across the business ecosystem; and providing self-service mechanisms wherever it is feasible.

Enhanced engagement and customer service are not the only motivators. Legal and regulatory changes (for example, more stringent requirements for labeling and provenance in the food and beverage industry) may also compel organizations to adopt digital strategies. Alternatively, input from customer advisory groups or user committees can provide the push for change (Stentoft et al., 2019).

## Workforce Drivers

Finally, organizations may pursue digital transformation because they want to improve employee experience, employee engagement, or both. This means supporting better communication and collaboration between humans and machines, both within organizations and across organizational boundaries, to enhance productivity and make tasks easier to accomplish (Lu et al., 2016). Improving communication and access to resources internally can also enable better customer experiences (WEF, 2015; Lu et al., 2016).

Retention of employees is another challenge that can be addressed through effective digital transformation initiatives. Many employees desire flexibility in their working environment and demand that tools be available to enable their success, so digitally enabled companies are often more attractive employers. Engineers and developers, in particular, tend to gravitate to new environments where they can learn about cutting-edge technologies. Finally, demands from a younger workforce to advance sustainability goals and corporate social responsibility can also push organizations to engage in digital transformation (Ismail et al., 2017).

## CASE STUDIES: FROM DRIVERS TO DIGITAL OUTCOMES

Quality 4.0 strategies emphasize real-time access to information and visibility into processes, intelligent decision support, and enhanced communication among people, systems, and machines. In contrast with pure digital transformation or Industry 4.0 strategies, when quality is the central element, specific quality and performance goals are emphasized more than the digital transformation itself. The following cases show how other companies have approached these initiatives, to what degree they focused on quality and performance goals, and the relative degrees of success (or failure) experienced.

### Audi: Big Data Analytics

Automotive manufacturing is being upended by Industry 4.0. In addition to competitive pressure not related to connected, intelligent, automated technologies (like stylish electric vehicles, and the charging infrastructure to make them practical), autonomous driving, pay-as-you-go services like Uber and Lyft, and revolutions in predictive maintenance are changing the nature of auto ownership. This impacts the viability of traditional auto manufacturers, which are required to respond to maintain competitiveness in unique and differentiating ways.

At German manufacturer Audi AG and its subsidiary Lamborghini S.p.A., the senior leadership team recognized that its best chance was to compete on insights from data, which it already had, scattered around the organization. At the same time, the team was well aware that a data transformation would require a more profound shift: "Adopting and assimilating big data analytics requires structural, capability, cultural, and procedural transformations across the entire organization. First, issues concerned with data access, data ownership, and joint analytics projects become important, and their resolution often requires organizational transformations" (Dremel et al., 2017).

With this in mind, Audi planned a three-stage, five-year evolutionary process. In the first two years, it would obtain analytics capabilities from partners, pursue reporting and descriptive analytics, and engage in one-off problem solving driven by business needs. In years three and four, the organization would begin bringing the skills and capabilities in-house as it introduced more advanced analytics and began prioritizing analytics initiatives across business units. Finally, starting in the fifth year, Audi would implement analytics as an internal service, gradually introducing predictive analytics and integrating the insights into strategic planning as well as operations.

The gradual advancing-enabling-leveraging approach helped the company successfully shift to a data-driven culture and break down silos. Although Audi's inaugural Industry 4.0 strategy was not specifically tuned to quality and performance goals, it still produced demonstrable change.

## Media Corp: Connected Products and Services

The anonymous, century-old Nordic publishing and broadcasting business Media Corp launched a digital transformation initiative in 2016 to maintain its leading position in several market segments. Plagued by the strategic challenges of decreasing revenue and low profitability, it recognized the need to identify new sources of revenue and decrease costs while responding to changing customer needs and increasing pressures from the market. New entrants to the market, which were not limited by a workforce trained primarily in traditional publishing and broadcasting, were also threats to Media Corp.

The company's response was to take an incremental approach. It dedicated 3%–5% of revenue to digitizing current products and services, and hired a chief digital officer and chief transformation officer to manage the technological and process perspectives, respectively. The new leaders empowered business unit directors to develop concepts for connected products and services that would leverage the new governance structures. The incremental approach helped Media Corp maintain its market position, but it was not bold enough to substantially improve its financial position (Hyvönen, 2018).

## Nikon: Real-Time Measurement

Nikon is one of the few companies that specifically discuss their Quality 4.0 strategy. Its approach, announced in 2018 and described by the tagline "digital, automated, connected," focuses on real-time measurement: improving and automating measurement systems, automating inspections, and centralizing the results electronically. By digitizing and connecting as much as possible, its goal is to shorten the time it takes to make decisions about production processes, thus improving availability and productivity (Cutting Tool Engineering, 2018). Results and outcomes are not yet available.

## Sweden: Near-Real-Time Dementia Monitoring

Initiated in 2013 and continuing through 2017, this digital transformation project was launched by a consortium of healthcare providers and universities in Sweden that were interested in improving the safety and quality of long-term residential dementia care. The performance of the system was evaluated through regular workshops with nurses, care providers, and family members where incremental results could be compared with ongoing risk assessments in a collaborative setting where providers could co-create new processes to most effectively make use of the technology.

The technology consisted of 67 monitoring systems that linked IoT-based door sensors and bed monitors, middleware, and a cloud-based portal. Care providers could access the data from computers, tablets, or smartphones, and alerts were sent by text message if movement outside nominal patterns was detected. The system was adaptable so that it could respond to changes in patients' needs, behaviors, and dementia progression.

From year to year, demonstrable benefits in patient safety and quality care were observed during the regular workshops. Adoption of the technology was enhanced by the co-creation aspect of the project, with relatives of the patients driving interest and support.

This Quality 4.0 project successfully achieved its qualitative goals for improving safety and care. In addition, "the care providers became experienced innovators . . . [and took] calculated risks and experimented with the technology in contrast to previous reports from implementation of monitoring technology in residential care" (Dugstad et al., 2019, p. 366). This was a marked improvement from previous projects attempting to implement digital monitoring technology in healthcare.

## Kaiser Permanente: Connected Customer Engagement

In the early 2010s, the nonprofit healthcare consortium Kaiser Permanente chose to shift from a volume-based model to a value-based model. To make this happen, its strategy incorporated digital elements to help patients gain access to care, care providers, and information about their condition from any device. Its quality and performance goals were to (1) increase the quantity and quality of communications channels between healthcare providers and patients, (2) identify best practices for personal outreach and supporting compliance to medical protocols, like taking medication according to instructions, and (3) deepen engagement among patients, providers, and others who share similar interests.

These overall goals were enacted through three initiatives, respectively: enhancing online communications, investing in predictive and prescriptive analytics, and using social media to create care circles. Its "Generation 2" digital services platform, a cloud-based portal, was released in mid-2014 combining the results of the initiatives. By 2016, the company reported the industry's highest NPS for customer satisfaction, with 70% of customers actively engaged with the new online platform (Sebastian et al., 2017). The company attributes its success to a clear vision, well-defined initiatives, and strong, consistent support from executives.

## LEARNING FROM SUCCESS

What these examples reveal is that digital transformation and Industry 4.0 initiatives are similar, even though only some are driven specifically by quality and performance goals. Every organization that embarks on a digital transformation journey, though, wants to be successful.

To identify a prescription for success, researchers have (1) developed readiness and maturity models based on the real experiences of companies that have started their transformations, and (2) performed meta-analyses to find patterns and trends. These models (referred to as digital maturity models, digital readiness models, Industry 4.0 readiness models, or Industry 4.0 maturity models in the research) can be used to self-assess and identify gaps between current and desired digital and Industry 4.0 capabilities. Because no models have been designed specifically for Quality 4.0 across industries yet, this section aims to outline the common elements in these models and meta-analyses that support quality and performance goals.

### Readiness and Maturity Models

To find themes and patterns, Schumacher et al. (2019) compared and contrasted the main Industry 4.0 assessment models that have been developed and used in Europe:

- IMPULS Industrie 4.0 Readiness (Lichtblau et al., 2015)
- Empowered Implementation Strategy for Industry 4.0 (Lanza et al., 2016)
- Industry 4.0/Digital Operations Self Assessment (PricewaterhouseCoopers, 2016)
- Connected Enterprise Maturity Model (Rockwell Automation, 2014)
- Industry 4.0 Reifegradmodell (Jodlbauer, 2016)

By systematically examining the research literature in both German and English that informed these models, and applying concept mapping, they found nine themes. Those themes were translated into an assessment tool that was validated in two enterprises, where additional feedback was provided to ensure that the model was reliable. The themes Schumacher et al. (2019) found, which demand focused attention during any digital transformation efforts, were the following:

- **Strategy**—implementation road map, available resources, new business models
- **Leadership**—executive commitment, management competencies, central coordination
- **Customers**—customer data management, digitization of sales and services
- **Products**—individualization, customization, connected products, integrated into systems
- **Operations**—decentralized, interdisciplinary, use of modeling and simulation
- **Culture**—knowledge sharing, open innovation, openness to technology
- **People**—openness, interest, autonomy, competencies, learning
- **Governance**—adoption of standards, security, privacy, intellectual property
- **Technology**—openness to mobile, cloud, machine-to-machine, and other enablers

The elements in each of these themes reflected competencies or activities that were observed across multiple models or case studies. Although this collection is not prescriptive, it does cover facets of digital transformation and Industry 4.0 initiatives that are very often encountered.

## Meta-Analyses of Digital Transformation

The meta-analysis technique scours volumes of research studies to uncover patterns, in this case the elements that are common to digital transformation strategies in Industry 4.0 and other markets. Ross et al. (2019) found five building blocks of digital transformation: an operational backbone (digital systems supported by systematic, repeatable processes), shared customer insights, a digital platform (or single source of truth for organizational knowledge assets), an accountability framework (with clear ownership, roles, and responsibilities), and an external developer platform (so that suppliers, partners, and collaborators can interact directly with a company's digital systems, without requiring a human in the loop).

Sony and Naik (2019) examined 68 research papers on digital transformation in Industry 4.0 to identify six key themes: Strategy, Digitization of the Organization, Digitization of the Supply Chain, Smart Product and Services, Employee Adaptability, and Top Management Commitment. Their work called out an important element of Industry 4.0—that it is not just about production or operations, but encompasses every aspect of a business and extends beyond the business to examine its role within (and impact on) society.

Basl and Doucek (2019) found six areas of emphasis in Industry 4.0 assessment: Technologies, Human Resources, Strategies, Processes, Data, and Security. They also looked at relationships between the organization and nature (e.g., resource utilization and sustainable practices), the organization and local community support, the organization and its customers, the organization and its value chain partners (e.g., collaboration and coordination), and the organization and society as a whole (for example, by supporting dignified work conditions). Each of these conditions, the authors explain, should be considered when constructing an Industry 4.0 strategy.

Many researchers have attempted to develop frameworks for ideation and execution of Industry 4.0 initiatives. For example, Trotta and Garengo (2019) identify five areas to focus on: Strategy, Technology, Production, Products, and People. Specific recommendations include adding a digital officer to the leadership team, moving manufacturing operations to the cloud, enabling automated data exchange between machines, embedding sensors in products, gathering product usage data automatically, assessing individual attitudes toward emerging technologies, and "implementing big data." With the exception of the last item, which will differ in scope and details between companies and across industries, the recommendations are actionable and generalizable.

## Recipe for a Digital Foundation

Combining the results from the models and meta-analyses, a holistic view of what is required for digital transformation success emerges:

- **Product—Smart Products and Services**

  - o Leveraging connectedness of people, machines, and data
  - o Embedding intelligence into products and systems
  - o Implementing appropriate levels of automation

- **Process—Organizational Backbone** to align people, processes, and technologies

  - o Clear, consistent executive commitment
  - o Framework for translating strategic objectives into action plans
  - o Clear roles, responsibilities, and accountability
  - o Standard work/descriptions of work processes
  - o Value stream maps for production
  - o Workforce capability and capacity development
  - o Framework for continuous organizational learning
  - o Effective communication channels between customers, suppliers, collaborators, workforce, and leaders

- **Data—Digital Platform** (single source of truth)

  - o Knowledge repository and digital services for workforce
  - o Knowledge repository and digital services across supply network
  - o Data platform for sharing and managing critical data, and ensuring data integrity
  - o External developer platform for shared work and collaborative innovation
  - o Cybersecurity and data management infrastructure
  - o Availability of mobile, machine-to-machine, and prototyping tools

## THE BALDRIGE EXCELLENCE FRAMEWORK (BEF)

Fortunately, there's no need to develop an entirely new framework for Industry 4.0 or Quality 4.0. Guidance for developing and maintaining an Organizational Backbone and Digital Platform is embodied in the Baldrige Excellence Framework (BEF), maintained by the U.S. National Institute of Standards and Technology (NIST, 2019), which can be extended to accommodate the new requirements of quality-driven digital transformation.

The elements in the models and meta-analyses covered in the previous section are mapped to the Baldrige criteria sections in Table 14.4. This shows how projects, initiatives, and capabilities can be aligned with the Organizational Backbone recommended by several of the studies.

**TABLE 14.4.  Key elements of digital transformation/Industry 4.0 mapped to Baldrige Criteria.**

| BEF criteria element | Assessment items | References |
|---|---|---|
| Organizational Profile: *Products and Services* | • Use of cloud technology<br>• Product individualization<br>• Flexibility of product characteristics<br>• Availability of information about product use<br>• Data processing in products<br>• Internet connection to/info exchange with products<br>• Digital compatibility and interoperability<br>• IT services attached to physical products<br>• Legal protection for digital products and services<br>• Embedding sensors in products<br>• Gathering product usage data automatically<br>• Leveraging voice of things (VoT) | Schumacher et al. (2019);<br>Sony & Naik (2019)<br><br><br><br><br><br><br><br>Trotta & Garengo (2019) |
| Organizational Profile: *Assets* | • Technology for information exchange<br>• Decentralized information storage<br>• Sensors for data collection<br>• Resource utilization and sustainable practices | Schumacher et al. (2019)<br><br><br>Basl & Doucek (2019) |
| Leadership | • Financial resources to support new technologies<br><br>• Centralized coordination of initiatives<br>• Effective two-way communication about incorporation of new technologies<br>• Provision of resources and access to appropriate competencies<br>• Accountability framework, with clear roles, responsibilities, and decision-making processes<br>• Add one or more members to the leadership to be solely responsible for digital transformation | Schumacher et al. (2019)<br><br><br><br>Wolf et al. (2018)<br><br>Ross et al. (2019)<br><br>Trotta & Garengo (2019) |
| Strategy | • Road map for integration of new technologies<br>• Risk assessment for new technologies<br>• Holistic thinking and networking across the enterprise; connected processes<br>• Guide resource allocation and drive decisions about capital investments | Schumacher et al. (2019)<br>Wolf et al. (2018)<br><br>Sebastian et al. (2017) |

*(continued)*

**TABLE 14.4.** (*continued*)

| BEF criteria element | Assessment items | References |
|---|---|---|
| Customer Focus | • Platform for VoC and shared customer insights<br>• Openness to new technology<br>• Digitization of customer contact<br>• Customer integration in product development<br>• IT collaboration for product development/evolution<br>• Use of customer-related data | Ross et al. (2019)<br>Schumacher et al. (2019) |
| Measurement, Analysis, and Information Management | • Digital platform and master data management (single source of truth)<br>• Digital information management processes<br>• Automated data collection<br>• Automated provision of information<br>• Individualization of information<br>• Digital process visualization<br>• Simulation/exploration of future scenarios | Ross et al. (2019);<br>Basl & Doucek (2019)<br>Schumacher et al. (2019) |
| | • Active knowledge management programs<br>• External developer platform (so that suppliers, partners, and collaborators can participate in co-creation of value and innovation) | Wolf et al. (2018)<br>Ross et al. (2019);<br>Basl & Doucek (2019) |
| | • Identification and leverage of dark data | Sundarraj & Natrajan (2019) |
| | • Enabling automated data exchange between machines | Trotta & Garengo (2019) |
| Workforce Development | • Workforce segment competence with information and communication technologies<br>• Training/retooling on digital competencies<br>• Flexible work arrangements<br>• Autonomy of workers at all levels<br>• Experience with interdisciplinary work<br>• Openness to continuous learning<br>• Incentive systems that cross organizational boundaries | Schumacher et al. (2019) |
| | • Motivation, encouragement, and time to try out new things; support for employee adaptability | Wolf et al. (2018);<br>Sony & Naik (2019) |
| Operations | • Integrated computing/edge computing in machines<br>• Integrated computing/edge computing in tools<br>• Use of additive manufacturing<br>• Use of robotics<br>• Autonomy of machines<br>• Digitization of communication with partners<br>• Information exchange between machines<br>• Remote control of machines<br>• Automated quality control<br>• Collaboration of humans and robots<br>• Mobile devices for business processes<br>• Mobile devices for operations processes | Schumacher et al. (2019);<br>Basl & Doucek (2019) |
| | • Agile methods<br>• Operations moved to the cloud; as much as possible outsourced to focus on core competencies | Wolf et al. (2018)<br>Trotta & Garengo (2019) |

## THE BOTTOM LINE

A Quality 4.0 strategy is an Industry 4.0 strategy or digital transformation strategy where quality and performance goals are front and center. Since 2010, there have been thousands of research publications on digital transformation and Industry 4.0 that document case studies and lessons learned. They provide insight into the three elements that are required for a quality-driven digital transformation to be successful:

- **Smart Products and Services** with connectedness, intelligence, and automation
- **Organizational Backbone** to coordinate people, processes, and technologies
- **Digital Platform** (single source of truth for data, information, and sharing)

Jeanne Ross, principal research scientist for MIT's Center for Information Systems Research, has identified a similar structure in her research. She describes the first part as identifying and delivering viable digital offerings, while digitization incorporates the latter two. Digitization promotes "disciplined adoption of appropriate standardized business processes to ensure reliability, predictability, security, and visibility into customer interactions." She encourages organizations to consciously address both digital and digitization initiatives for optimal success (Ross et al., 2019). In the next chapter, I bring these concepts together in a practical, actionable playbook for quality-driven digital transformation success.

## REFERENCES

Basl, J., and P. Doucek. (2019). "A metamodel for evaluating enterprise readiness in the context of Industry 4.0." *Information* 10 (3): 89.

Cutting Tool Engineering. (2018, July 12). "Nikon opens up about its strategic focus on Quality 4.0." https://www.ctemag.com/news/industry-news/nikon-opens-about-its-strategic-focus-quality-40

DeFeo, J. A. (2018, March 28). "The smart factory, Industry 4.0, and quality." Juran Institute. https://www.youtube.com/watch?time_continue=4&v=z4-R4YZ_A08

Dremel, C., J. Wulf, M. M. Herterich, J. C. Waizmann, and W. Brenner. (2017). "How AUDI AG established big data analytics in its digital transformation." *MIS Quarterly Executive* 16 (2).

Dugstad, J., T. Eide, E. R. Nilsen, and H. Eide. (2019). "Towards successful digital transformation through co-creation: A longitudinal study of a four-year implementation of digital monitoring technology in residential care for persons with dementia." *BMC Health Services Research* 19 (1): 366.

Evans, J. R. (2013). *Quality & performance excellence: Management, organization, & strategy*. 7th edition. Mason, OH: Cengage Learning.

Green Bank Observatory. (2019). GBT - NRAO/GBO image #945. Image Gallery. https://greenbankobservatory.org/media/image-gallery/#jp-carousel-645

Hyvönen, J. (2018). *Strategic leading of digital transformation in large established companies—a multiple case-study.* Aalto University Learning Centre. https://aaltodoc.aalto.fi/handle/123456789/30110

Ismail, M. H., M. Khater, and M. Zaki. (2017, November). "Digital business transformation and strategy: What do we know so far" [Working paper]. Cambridge Service Alliance. https://cambridgeservicealliance.eng.cam.ac.uk/resources/Downloads/Monthly%20Papers/2017NovPaper_Mariam.pdf

Jodlbauer, J. (2016, November). "Reifegradmodell I 4.0. FH OÖ, Institut für Intelligente Produktion, Steyr." https://www.wko.at/site/tip/RGM-WK-NOe.pptx

Lanza, G., P. Nyhuis, S. M. Ansari, T. Kuprat, and C. Liebrecht. (2016). "Befähigungs-und Einführungsstrategien für Industrie 4.0." *ZWF Zeitschrift für wirtschaftlichen Fabrikbetrieb* 111 (1–2): 76–79.

Lichtblau, K., Volker Stich, R. Bertenrath, M. Blum, M. Bleider, A. Millack, K. Schmitt, E. Schmitz, and M. Schröter. (2015). "Industrie 4.0 Readiness. IMPULS-Stiftung for mechanical engineering, plant engineering, plant engineering, and information technology." https://industrie40.vdma.org/documents/4214230/26342484/Industrie_40_Readiness_Study_1529498007918.pdf/0b5fd521-9ee2-2de0-f377-93bdd01ed1c8

Liere-Netheler, K., S. Packmohr, and K. Vogelsang. (2018). "Drivers of digital transformation in manufacturing. The digital supply chain of the future: Technologies, applications and business models." In *Proceedings of the 51st Hawaii International Conference on System Sciences (HICSS)*, 3926–3935.

Lu, Y., K. C. Morris, and S. Frechette. (2016). "Current standards landscape for smart manufacturing systems." National Institute of Standards and Technology, *NISTIR, 8107*, 39.

Machadoa, C. G., M. Winrotha, D. Carlssonb, P. Almströma, V. Centerholtb, and M. Hallin. (2019). "Industry 4.0 readiness in manufacturing companies: Challenges and enablers towards increased digitalization." *Procedia CIRP* 81, 1113–1118.

National Institute of Standards and Technology (NIST). (2019). "Baldrige Excellence Framework (Business/Nonprofit): Proven leadership and management practices for high performance." https://www.nist.gov/baldrige/publications/baldrige-excellence-framework/businessnonprofit

Newman, D. (2018, June 12). "Four digital transformation trends driving Industry 4.0." *Forbes.* https://www.forbes.com/sites/danielnewman/2018/06/12/four-digital-transformation-trends-driving-industry-4-0

PricewaterhouseCoopers. (2016). "The Industry 4.0/digital operations self assessment tool." https://i40-self-assessment.pwc.de/i40/landing/

Rockwell Automation. (2014, July). *The connected enterprise maturity model.* Publication CIE-WP002-EN-P. https://literature.rockwellautomation.com/idc/groups/literature/documents/wp/cie-wp002_-en-p.pdf

Ross, J. W., C. M. Beath, and M. Mocker. (2019). *Designed for digital: How to architect your business for sustained success.* MIT Press.

Schumacher, A., T. Nemeth, and W. Sihn. (2019). "Roadmapping towards industrial digitalization based on an Industry 4.0 maturity model for manufacturing enterprises." *Procedia CIRP* 79: 409–414.

Scoles, S. (2016, October 7). "What happens when a space observatory goes rogue." *Wired*. https://www.wired.com/2016/10/happens-space-observatory-goes-rogue/

Sebastian, I., J. Ross, C. Beath, M. Mocker, K. Moloney, and N. Fonstad. (2017). "How big old companies navigate digital transformation." *MIS Quarterly Executive* 16 (3): 197–213.

Sony, M., and S. Naik. (2019). "Key ingredients for evaluating Industry 4.0 readiness for organizations: A literature review." *Benchmarking: An International Journal*.

Stentoft, J., K. W. Jensen, K. Philipsen, and A. Haug. (2019, January). "Drivers and barriers for Industry 4.0 readiness and practice: A SME perspective with empirical evidence." In *Proceedings of the 52nd Hawaii International Conference on System Sciences*.

Sundarraj, M., and R. M. Natrajan. (2019). "A sustainable method to handle dark data in a smart factory." *Software Quality Professional* 21 (4).

Trotta, D., and P. Garengo. (2019, March). "Assessing Industry 4.0 maturity: An essential scale for SMEs." In *2019 8th International Conference on Industrial Technology and Management (ICITM)*, 69–74. IEEE.

Wolf, M., A. Semm, and C. Erfurth. (2018, June). "Digital transformation in companies—challenges and success factors." In *International Conference on Innovations for Community Service*, 178–193. Springer.

World Economic Forum. (2015, January). *Industrial internet of things: Unleashing the potential of connected products and services*. https://www.accenture.com/t20150527T205433Z__w__/us-en/_acnmedia/Accenture/Conversion-Assets/DotCom/Documents/Global/PDF/Dualpub_8/Accenture-Industrial-Internet-of-Things-WEF-Report-2015.pdfla=en

## CHAPTER FIFTEEN

# PLAYBOOK FOR TRANSFORMATION

The fundamentals of leadership—being the kind of leader you would want to have—won't change. But much of leadership is contextual, and in the digital world, how you exercise leadership will change.

—BILL TROY, FORMER CEO, ASQ

E very other chapter in this book started with a story. But this chapter is different. It's going to be about *your story*—your journey to lead your organization to improved operational efficiency; enhanced environment, health, and safety outcomes; and new product lines and business models. Although this chapter will help make it easier, your journey will not be easy. Digital transformation, like managing quality, risk, health, and safety, is a never-ending process that can get bumpy.

If you've ever taken a trip on an airplane, you've probably encountered turbulence there too. Although turbulence can be frightening for passengers, it's usually nothing to worry about, and pilots don't mind at all. They're accustomed to "light chop," the most common way to describe those minor bumps, and take action only if it interferes with maintaining a steady altitude. If the turbulence is severe, though, the pilot has to protect the aircraft from damage. Structural failures can occur if the pilot flies the plane through areas where forces exceed the design limits of the aircraft.

When confronted with a potentially dangerous situation, there are some options. The pilot can go around the turbulence, but only if there's information available about the conditions ahead, which is usually obtained from other pilots. The other option is to change altitude to a level where the airflow has different characteristics.

If neither of these options is viable, there's one more action that can be taken: slow the aircraft to the "turbulence penetration speed" or maneuvering speed—the ideal speed for navigating through bad air, one that assumes a heavy load on the wings. Some pilots even believe that the safest speed is *even slower* than the recommended speed:

> Most of the time, turbulence is only a mild annoyance, but occasionally, it can attack with a vengeance, assaulting the aircraft unmercifully. For those inside, this can lead to fatigue, nausea, and injury. Deterioration of vision also may be experienced, because turbulence excites an airplane's natural vibrations, making it almost impossible for a pilot to read the instruments . . . [for many reasons] the slower the better.
>
> A pilot is not likely to jeopardize safety unless he manhandles the controls, is close to the ground, or is flying an airplane with undesirable stall characteristics. . . . When an aircraft is flown into severe or extreme turbulence, gust loads are punishing and potentially destructive. Unfortunately, many pilots compound the problem by rapidly jerking and shoving the controls in an effort to maintain a reasonably level attitude. The effect of this, however, is to create maneuvering loads that combine with gust loads to make the total G load greater than necessary. Although a pilot understandably is filled with anxiety (and possibly fear) at such a time, he must make every effort not to contribute to the hazard. The controls should be moved deliberately yet smoothly. (Schiff, 1985)

Although it may be the pilot's instinct, more control is not the solution. Fearful attempts to get a handle on the situation can compound the problem, even to the point of damaging or destroying the aircraft and endangering the lives inside.

When there's no way to avoid turbulence by shifting your strategy, you do have the option of adjusting your speed. "We can't slow down; there's too much to do!" you might hear from your workforce or your leaders. You might be saying it yourself. And despite the very real pressure that people feel, and the desire to get to growth goals or revenue targets quickly, even well-intentioned attempts to harness control can backfire. The solution is to take it slowly and focus on building the habits and discipline that provide the foundation for every technology you leverage.

*Transformation* means shifting from one form or experience to another, or becoming something entirely new; it does not have to be rapid. The caterpillar becomes a butterfly. Baking a cake or a casserole transforms raw ingredients into a final product. A mathematical transformation turns numbers or models into entirely new ones. Learning is perhaps the most profound transformation, as it helps people see themselves and the world in new ways, exposing opportunities for growth and improvement.

When the concept of transformation emerged in the business literature in the 1990s, there were two research streams: business transformation and IT-enabled transformation. Business transformation consisted of four constructs: business process reengineering and restructuring (to obtain efficiencies), renewal through employee empowerment, and regeneration of the business concept itself. IT-enabled transformation was all about reframing a company's view of itself, restructuring for agility and flexibility, revitalizing the value chain by horizontal and vertical integration, and empowerment of people (Ismail et al., 2017).

Digital transformation, in contrast, is not focused just on a company's internal processes. The digital technologies used to make operations more efficient and effective can also be used to improve, enhance, or transform the customer's experience. In addition, they can be used to build robust supplier networks (and even broader ecosystems) that support near-real-time visibility, communication, and collaboration. Always-on information networks challenge the old concepts of seams and transitions that kept workers and companies siloed (ASQ, 2015).

## PLANNING FOR DIGITAL TRANSFORMATION

How should you transform? Table 15.1 brings together everything you've learned in this book to create a quality-driven road map for digital transformation. Even though many of the concepts have been drawn from Industry 4.0 and manufacturing, they are applicable across industries.

## STEP 1: WHO ARE YOU?

Before you begin any transformation or major change, it is important to have a clear understanding of your organization. What makes your company unique and compelling? How do you go to market and compete for business? What advantages, challenges, and risks will influence your interest and ability in pursuing Industry 4.0 and digital transformation objectives and initiatives? It is important to answer all these questions (in writing, preferably) because they can provide an anchor for your workforce during a time of disruption and change.

The Organizational Profile section of the BEF is one approach that can guide you through this process of anchoring. NIST (2019) explains that it is impor-

**TABLE 15.1. Steps in a quality-driven digital transformation.**

| Step | Guiding question | Tools | References |
|---|---|---|---|
| 1 | **Who** are you and **how** does your organization work? (create an organizational profile) | **1. Organizational Profile** (e.g., from Baldrige Excellence Framework (BEF): <br>• Product offerings <br>• Delivery mechanisms <br>• Mission, vision, values, and culture <br>• Workforce profile <br>• Assets <br>• Regulatory environment <br>• Organizational structure <br>• Customer and stakeholder profile <br>• Suppliers, partners, and collaborators <br>• Competitive environment <br>• Strategic challenges and advantages <br>• Processes for evaluating performance <br>**2. Organizational Backbone**: Work systems—learning and communications framework <br>**3. Single Source of Truth**: Data quality, management, and governance processes | NIST (2019); Ross et al. (2019) |
| 2 | **Why** do you want to transform? (establish strategic orientation and quality/ performance goals) | Select an **orientation** and set targets for the **quality and performance goals** you want to achieve: <br><br>1. Internal and external process optimization <br>2. Customer interface and experience improvement <br>3. New ecosystems and value networks <br>4. New business models: smart products and services | Ismail et al. (2017) Ibarra et al. (2018) Figure 15.1 Table 15.2 |
| 3 | **How can** you transform? (establish value propositions and brainstorm initiatives) | Determine viable initiatives by: <br>• Examining **entities and processes** <br>• Examining **degrees of connectedness, intelligence, and automation** | Table 15.3 Table 15.4 |
| 4 | **How should** you transform? (prioritize initiatives) | Evaluate each potential initiative in terms of: <br>• **Magnitude**—anticipated impacts on customers, stakeholders, employees, society, environment | Radziwill (2018) |

(continued)

**TABLE 15.1.** (*continued*)

| Step | Guiding question | Tools | References |
|---|---|---|---|
| | | • **Opportunity**—how well the initiative captures opportunities and addresses intelligent risks<br>• **Deployment**—whether sufficient workforce capability, capacity, and available assets exist to advance the initiative | |
| 5 | **How much** should you transform? (make buy-build-partner decisions) | Examine **strategic orientation** with respect to technology, **value creation, structural changes,** and **financial aspects**. | Hess et al. (2016) |
| 6 | How will you **measure** success? (determine KPIs) | Identify **metrics** for each initiative to determine if you are winning and if you should make adjustments. | Kotarba (2017) |

tant because it can help you quickly call out areas where "conflicting, little, or no information is available, [using] these topics for action planning." The Organizational Profile establishes a context and process for formulating a shared understanding of how your organization operates, from which you can identify gaps.

Kendall and Bodinson (2016) discuss another compelling reason to invest in an Organizational Profile before you launch a transformation:

> Dr. Katherine Gottlieb, CEO of Southcentral Foundation (a Baldrige Award health care recipient in 2011), described exploring the Baldrige Criteria with her VPs and the board in 2003. "We knew it would cost us time, cost us money. But we came to an agreement that this [using the Baldrige framework] would be the tool to drive systematic change." She also found that developing the Organizational Profile—the five-page description of what an organization does, how it functions, and its competitive environment—was a driving force for gaining leadership buy-in from the VPs and the managers as they provided input and feedback to document who Southcentral Foundation was.

In addition to providing a basis for shared understanding across the workforce, the Organizational Profile can also serve as a tool for buy-in and change management support. Keep in mind, though, that this step is not intended to constrain your vision. Kane et al. (2015) recommend using the approach taken by spice manufacturer McCormick & Company, which set an audacious goal in its digital transformation strategy process that it would be able to personalize and customize spice blends to meet local and regional tastes around the globe. Even though the vision wasn't technologically feasible when the company dreamed of it, "McCormick [now] uses FlavorPrint to recommend recipes to its consumers. But the vision is much bolder. McCormick thinks of FlavorPrint as the Pandora of flavorings, which has prompted the organization to see itself as a *food experience company* rather than a purveyor of spices."

## STEP 2: WHY TRANSFORM? WHY NOW?

If your organization has committed to or started a digital transformation journey, you've already decided that you want to be different in some way. Have you articulated exactly what you're aiming for? Digital transformation mechanisms developed by Ismail et al. (2017) and Ibarra et al. (2018) can help you identify the best focus (Figure 15.1).

Figure 15.1 was developed by examining how sustainability could serve as a driver for growth and innovation through digital transformation. Process optimization is the first step, but as processes mature, new interfaces can be provided to customers for co-creation of products and services. When companies move value creation ecosystems and supply networks online, all partners can benefit from greater information sharing. Beyond networks and ecosystems, ideation of smart products and services can create entirely new markets. Using

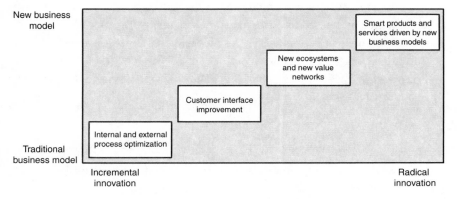

FIGURE 15.1. Mechanisms for digital transformation. Adapted from Ibarra et al. (2018).

this model requires that you use your Organizational Profile to choose where you want to focus, based on your strategic advantages, strategic challenges, and organizational capabilities.

This may require that you drill down a little deeper. Research shows that strategy, and not technology, determines successful digital transformation initiatives (Kane et al., 2015). This is why so many transformation initiatives that focus on implementing a specific technology fail, like the $50 million data science horror story in Chapter 7.

> Technology is increasingly not about the tech itself but how that tech best serves the end user. User-centric products have always been important but I don't think the expectations have ever been this high in our modern world of consistently newer and shinier things. Making a difference in someone's quality of life is more important than finding someone to impress. Users are demanding and deserving of more. The best technologies and services will be those that focus on what they can do to make users' lives easier, more efficient, happier. The tables have turned really, and with end-user focus being such a natural part of what quality assurance practitioners live for, that makes it a very exciting time to be in the quality profession.
>
> —AUSTIN S. LIN, TECHNICAL PROGRAM MANAGER, GOOGLE, ASQ CHAIR (2020)

Setting quality and performance goals up front, with a clear understanding of why those goals are important for your business, can make the difference between success and failure. Once you have identified your focus area, evaluate the quality and performance goals in Table 15.2 to establish why your organization is motivated to pursue that focus. Finally, why now? For each focus area and quality goal, establish a case for why that pursuit is timely.

To supplement Table 15.2 and to provide more examples, Table 15.3 outlines additional ways from the research to think about how to frame quality and performance goals.

## STEP 3: HOW CAN YOU TRANSFORM?

Once you know why you are pursuing a digital transformation initiative, and have identified quality and performance goals for each target mechanism (process optimization, customer interface and experience, new ecosystems, new

**TABLE 15.2. Quality goals associated with digital transformation mechanisms.**

| | 1. Efficiency and Effectiveness | 2. Customer Value | 3. Connective Intelligence | 4. Innovation |
|---|---|---|---|---|
| Transformation mechanism | Internal and external process optimization | Customer interface and experience improvement | New ecosystems and value networks | New business models: smart products and services |
| Degree of innovation | Incremental | Moderate | Significant | Radical |
| Business model | Traditional | Incremental | Innovative | New |
| Quality and performance goals | Improve product or process quality; meet specifications<br><br>Standardize to support digitalization<br><br>Reduce cost, waste, defects, rework, and Cost of Quality (CoQ)<br><br>Enhance workforce capability/capacity<br><br>Improve environment, health, or safety outcomes and conditions<br><br>Improve leadership, governance, and ethics<br><br>Improve ability to meet legal, financial, or compliance requirements<br><br>Improve ability to realize strategic objectives | Increase customer satisfaction<br><br>Reduce customer dissatisfaction<br><br>Builder stronger long-term relationships with customers<br><br>Enhance decision-making capability<br><br>Improve ability to realize strategic objectives<br><br>Enable co-creation of value to accelerate new product development<br><br>Align online and off-line experiences to provide customer consistency | Improve product or process quality<br><br>Increase customer satisfaction<br><br>Reduce customer dissatisfaction<br><br>Enhance decision-making capability<br><br>Improve leadership, governance, or ethical outcomes<br><br>Improve financial results<br><br>Enable co-creation of value to accelerate new product development | Bring products and services to market faster<br><br>Bring better products and services to market<br><br>Standardize to support digitalization<br><br>Grow<br><br>Diversify the business or portfolio<br><br>Provide greater benefits to society<br><br>Improve financial results<br><br>Improve ability to realize strategic goals |

**TABLE 15.3. Quality goals associated with digital transformation mechanisms.**

| 1 | 2 | 3 | 4 | How | Reference |
|---|---|---|---|---|---|
| X | X | X | X | Increased cooperation, information sharing, and collaboration using digital channels | Katsamakas (2014) |
| X | X | X |   | Software-intensive co-creation of value | Zimmerman et al. (2016) |
| X | X |   |   | Building of deep relationships through long-term repeated interactions based on trust | Katsamakas (2014) |
| X | X |   |   | Products and services designed with "virtual visitor" in mind | Pryor (2016) |
| X |   |   |   | All new services interoperable, modular, and reusable to promote greater agility and flexibility | Pryor (2016) |
|   | X |   |   | Digital by default: new products and services leverage digital in appropriate ways while "those who can't use digital services are not excluded" | Pryor (2016) |
|   | X |   |   | Expanded digital interactions (e.g., creating, eliminating, and consuming content at will) | Lanzolla & Anderson (2008) |
|   | X |   |   | Expanded digital distribution of products, information, and services | Lanzolla & Anderson (2008) |
|   | X |   |   | Easy and convenient: online self-service to the greatest extent possible to give customers and stakeholders optimal agency | Pryor (2016) |
|   |   | X |   | Expanding digital reach to overcome physical barriers | Lanzolla & Anderson (2008) |
|   |   | X |   | Specialized coinvestments with partners stimulate collaboration and deepen relationships | Katsamakas (2014) |
|   |   | X |   | Customers have digital experiences with products and services (e.g., enrich interactions at all stages of the customer journey) | Pryor (2016) |

business models, and smart products) that is important to you, it's time to start thinking more about specifics.

One fun way to come up with digital transformation objectives is to brainstorm. Mixing and matching descriptors, entities, and processes like those listed in Table 15.4 (which provides representative, but not exhaustive, lists) can provide many ideas.

## Brainstorming

Choose one descriptor, plus an entity, process, or both (preferably as part of a group or team exercise), and see where it leads you. For example, here are

**TABLE 15.4. A brainstorming grid to convert quality and performance goals to initiatives.**

| Descriptors | Entities | Processes |
|---|---|---|
| Connected: | Person: | Planning |
| • Electronic | • Agent | Exploring |
| • Digitized | • Customer | Designing |
| • Integrated | • Intelligent agent | Improving |
| • Remote | • Operator | |
| | • Worker | |
| Intelligent: | | |
| • Smart | Place: | Executing: |
| • Predictive | • Factory | • Monitoring |
| • Prescriptive | • Office | • Control |
| | • Site | • Maintenance |
| Automated: | | |
| • Augmented | Entity: | Compliance: |
| • Autonomous | • Asset | • Auditing |
| | • Building | • Managing |
| | • City | |
| | • Data platform | |
| | • Data repository | |
| | • Energy | |
| | • Environment | |
| | • Equipment | |
| | • Health | |
| | • Pollutants | |
| | • Product | |
| | • Process | |
| | • Safety | |
| | • Software system | |
| | • Vehicle | |
| | • Waste | |

some speculative combinations determined by randomly choosing from the above lists:

- **Integrated customer planning**—horizontal integration of systems to better anticipate services and offerings through all stages of the customer journey
- **Remote vehicle control**—providing the ability to navigate vehicles through small, dangerous, or unsafe places, increasing capabilities while protecting workers

- **Intelligent asset management**—using machine learning to anticipate threats and vulnerabilities to high-risk assets
- **Predictive safety design**—using simulation or analyzing incidents and near misses to design safer work processes
- **Autonomous process planning**—using intelligent systems and automation technologies to dynamically plan and execute operational processes

## Contextualization

After brainstorming, it's time to think about the ideas you generated in the context of your entire organization. Table 15.5 presents multiple ideas for how you can realize your objectives, and illustrates the relationships you will have to consider in doing so.

All organizations should have an organizational backbone in place, supported by solid data management, to coordinate people, processes, and technologies. ISO 9001, Baldrige, and lean management can all be used for this, because they provide a framework for translating strategic objectives into actions, clear roles, responsibilities, and accountability; standard descriptions of work and value delivery; and approaches to learning and communication.

Similarly, a solid digital platform is a prerequisite for digital transformation initiatives. This includes a knowledge repository and digital services for the workforce and the supply network, a platform for sharing and managing critical data and ensuring data integrity, and tools for shared work and collaborative innovation within and beyond your organization.

The digital platform supports connectedness among people, machines, and data. The extent of connectedness can be thought of in terms of these categories, based on Monostori (2014):

- **Manual process**—processes are managed ad hoc, on paper or on spreadsheets
- **Digitization**—some processes support electronic data entry, retrieval, and visualization
- **Horizontal integration**—some systems can exchange information across functional areas of the organization (e.g., sales, marketing, product)
- **Vertical integration**—information and material flows connect sensors, control systems, and business systems
- **Connected work systems**—information and materials flow horizontally and vertically; organization can anticipate, adjust, and adapt

**TABLE 15.5. Extent of connectedness, intelligence, and automation with enabling technologies.**

**Smart products and services:**

Online/offline convergence, cradle to grave product life cycle, product as agent

| Capability | Connectedness | | Intelligence | | Automation | |
| --- | --- | --- | --- | --- | --- | --- |
| **Type** | **Digitization** | **Integration** | **Descriptive/ Diagnostic** | **Predictive/ Prescriptive** | **Augmentation** | **Autonomy** |
| Extent of capability [top = high to bottom = low] | • Connected work systems<br>• Horizontal integration<br>• Vertical integration<br>• Connected/remote process<br>• Digital process (digitization)<br>• Manual process | | • Machine intelligence<br>• Augmented collective intelligence<br>• Collective intelligence<br>• Augmented intelligence<br>• Human intelligence | | • Complete autonomy<br>• Partial reports (if asked)<br>• Machine reports (if asked)<br>• Machine reports action<br>• Human aborts<br>• Human approves<br>• Machine selects<br>• Machine suggests<br>• Machine gives options<br>• Machine executes/augments | |
| Enabling technologies connect us to people, machines, data | • Social media<br>• Mobile devices<br>• Cloud computing<br>• Wearables<br>• Augmented reality<br>• Virtual reality<br>• Enterprise software systems (CRM, ERP, MES, QMS, EHS, EHSQ, PLM, etc.)<br>• 4G and 5G | | • Big data<br>• Descriptive analytics<br>• Diagnostic analytics<br>• Predictive analytics<br>• Prescriptive analytics<br>• Artificial intelligence<br>• Machine learning<br>• Advanced/smart materials | | • Additive manufacturing/3D printing<br>• Robotics<br>• Robotic Process Automation (RPA)<br>• Internet of things (IoT)<br>• Industrial internet of things (IIoT)<br>• Edge computing<br>• Blockchain (automates data integrity)<br>• Modeling and simulation<br>• Driverless vehicles and drones | |
| Support systems in digital platform | • Cybersecurity (protect data, assets, process integrity) | | • Environment, health, and safety (EHS) systems (protect people and communities) | | • Process framework (protect process integrity; do not automate a bad process!) | |

**Organizational Backbone:** work systems, learning framework, communications framework, APIs
**Single Source of Truth:** data quality, management, and governance

Digitization is the easiest way to connect people with data, taking it out of manila folders and off desktops and personal data directories to benefit multiple people. Systems integration is another way to enhance connectedness, because when systems can exchange data and information, it is often easier for people to find and use it.

The extent of intelligence can be represented in terms of how people and machines collaborate to generate accurate and actionable insights:

- **Human intelligence**—relying on the understanding and interpretations of one person
- **Augmented intelligence**—intelligent systems that provide assistance and guidance to add to the understanding and interpretations of one person
- **Collective intelligence**—relying on the wisdom of many (e.g., crowdsourcing)
- **Augmented collective intelligence**—using data and intelligent systems to add to, improve on, or leverage the knowledge of the crowd
- **Machine intelligence**—intelligent systems that can monitor, control, and respond to processes and changes, improving on the understanding and interpretations of one or more people

Finally, the extent of automation can be evaluated using the Sheridan and Verplank (1978) framework, where the simplest form of automation involves a machine performing a task that the human has completely specified in advance, and continuing on to complete autonomy, where the machine just executes on its own—it does not provide any insight into its choices or behavior.

## Gap Analysis

At this point, you should have a list of initiatives you would like to consider. For most of these efforts, it is unlikely that your organization will be starting from scratch. Figure 15.2 can help you compare the as-is state of your area of interest with the to-be (desired) state. An accurate assessment of the gap will help you determine whether pursuing the initiative is feasible.

For example, consider a company that manages all of its processes and quality events on paper and in Excel. Its performance goal is to reduce costs without compromising safety. It identified an integrated management system for handling quality and safety events as a potential digital transformation opportunity.

To examine options from the spectrum of possibilities, consider the degrees of connectedness, intelligence, and automation for this particular company using Figure 15.2. At present, it has a digitized but not connected process for handling quality events and incidents, and it relies on human intelligence to administer and manage its work processes. There is no automation, but the people involved in these processes say that they would be open to guidance from an intelligent system (for example, knowing which corrective actions they should work on first). The distinction between the current state and the possible future state is shown as follows:

- **As-is:** digitized process, human intelligence, no automation
- **To-be:** connected process (shared software systems), augmented intelligence, machine gives options or selects options for what to work on next (but human controls the work)

The gap analysis process can be used to support group discussions that can call out exactly how much digitization, integration, intelligence, or automation is appropriate to achieve defined quality and performance goals. There is a caveat though: it is not always best to move up all three scales, nor is it best to jump straight to the top of any of the scales. Every project or initiative will

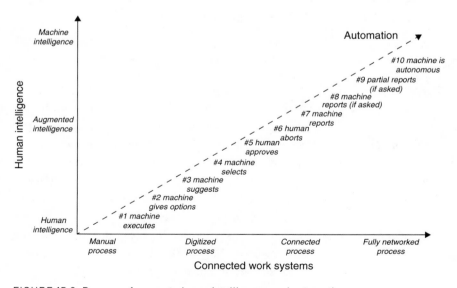

FIGURE 15.2. Degrees of connectedness, intelligence, and automation.

have its own context, and what is appropriate and desirable in one case may be inappropriate, dangerous, infeasible, or unethical in another. Not all processes should be governed by connected, autonomous work systems driven by machine intelligence (Stormont, 2008).

## STEP 4: HOW SHOULD YOU TRANSFORM?

By the time you get to step 4, you should have a list of potential initiatives that match the strategy and quality goals you identified in step 2. Before moving forward, prioritize the possibilities. Evaluate each one in terms of the following:

- **Magnitude**—the extent of anticipated impacts on customers, stakeholders, employees, society, and the environment (both positive and negative)
- **Opportunity**—how well the initiative affirms strategic advantages, responds to strategic challenges, captures opportunities, and/or addresses intelligent risks
- **Deployment**—whether sufficient workforce capability, capacity, and available assets exist (or can be obtained) to advance the initiative

Your organization may have additional factors to incorporate into the assessment and prioritization process. The most important outcome of this step is to have a prioritized list your organization can use to make decisions about resourcing and adjustments to work plans if necessary.

## STEP 5: HOW MUCH SHOULD YOU TRANSFORM?

Digital transformation is a team sport, and it is rare for any organization to shoulder the entire burden of a transformation initiative. It is possible to buy capabilities (e.g., contractors, consultants, companies), build or develop capabilities, or partner to obtain capabilities, and these options are widely used. Before committing to specific initiatives, defining project plans, and dedicating resources, think about how much of the job your organization should do itself.

Hess et al. (2016) provide an excellent overview of the key decisions that need to be made at this step, which are summarized in Figure 15.3. There are four categories of questions: strategic use of technology (two questions), means of value creation (three questions), whether structural changes will be required (four questions), and financial considerations (two questions). Although these

| Tech | IT is an/a: | Enabler | | | Supporter | | |
|---|---|---|---|---|---|---|---|
| | Company ambition: | Innovator | | Early adopter | | Follower | |
| Value | Digital diversification: | Online sales channels | In product/service delivery | Digitalization of products/services | Content platforms | | Extended business |
| | Product type: | Paid | Freemium | | Ad supported | | Referrals |
| Structure | Leadership: | CEO | | Group CEO | Chief data officer (CDO) | | Chief information officer (CIO) |
| | Transformation activities are: | Integrated into organization | | | In separate organizational unit | | |
| | Operational changes: | Products/services | | Internal business processes | | Skills and capabilities | |
| | Build competencies: | Internally | | Through partnerships | Mergers and acquisitions | | External sourcing |
| Finance | Financial pressure: | Low | | Medium | | High | |
| | Financing source: | Internal | | | External | | |

FIGURE 15.3. Key decisions for a digital transformation strategy. Adapted from Hess et al. (2016).

authors focus on digital transformation in media companies, their recommendations translate well to many other industries.

## Strategic Use of Technology

Because digital transformation initiatives are technology intensive, companies must determine what role information technology will play in their overall strategy, and whether it is one of their core capabilities. This can help you choose the right technologies. For example, a legal firm that uses information technology to support operations but does not have aspirations to be a technology innovator or early adopter should not build its own blockchain-based system for document management. In contrast, a software company that considers itself a technology innovator may want to take this step.

## Value Creation

Because digital technologies can impact a company's business model, it will be important to examine what those changes might be and how they might impact revenue generation. First, Hess et al. (2016) recommend looking at the digital environment of your products and services. Are digital channels just being used to enhance sales and distribution, or do they provide a value-add to existing products and services? Do they define completely new offerings?

Next, what are the revenue models associated with the enhanced or new offerings? Although there are more possibilities, these authors call out paid digital content, "freemium" content (some free teaser material or services, with paid products, memberships, or enhanced service agreements for customers who want more), ad sales, and product referrals as opportunities. Finally, how do the new digital offerings transform the scope of the business? For example, a company whose revenues are dominated by product sales may find new service-based revenue channels as a result of its digital transformation efforts.

## Structural Changes

Digital transformation may also require building new internal or external relationships, or growing new capabilities internally. Changes to the organizational structure may be needed to communicate executive commitment, establish authority for key decisions, and create spaces for new relationships to form. Hess et al. (2016) recommend deciding who has overall responsibility for the digital transformation and ensuring that they are in a position where authority and accountability are clear.

Next, leadership must decide whether the transformation activities will be separate from other functional areas or tied to existing functional areas (e.g., by setting up matrix management structures). Whether the emphasis is on product development, process optimization, or new skills development is also important to identify, because this will impact which leaders need to be part of the transformation effort. Also, the way in which competencies will be developed must be identified. This will be tied to buy-build-partner decisions.

## Financial Considerations

Finally, the organization should evaluate whether there is financial pressure that needs to be addressed, and whether financial support will come from internal or external sources. This will depend on the age and size of the company (a new startup will be under different pressures than an established enterprise) and the forces motivating the transformation effort (e.g., growth vs. the need to gain operational efficiencies).

## STEP 6: HOW WILL YOU MEASURE SUCCESS?

Are you winning? Are you *done*? Choosing the right measures for success will not only guide the progress of your initiatives but also provide you with a clear indication of what benefits your organization has obtained from its digital transformation efforts. Key Performance Indicators (KPIs) should be tied to the quality and performance goals you outlined in step 2.

For example, if you want to reduce operations costs, how much do you want to reduce them by—and by when? If you want to make work processes more efficient, do you want to save time, eliminate non-value-adding steps, use less energy, or do all the above? Who is impacted, what is impacted, and when they are impacted will all be valuable indicators of whether progress is taking place or whether adjustments should be made.

### EXAMPLE

In Chapter 14, I shared a Quality 4.0 strategy—a digital transformation strategy guided by quality and performance goals—that was developed by a national research laboratory in 2004. The details of that approach are repeated in Table 15.6. The only elements that are not apparent are steps 1 and 5, but the activities from both of those steps were indeed undertaken. For step 1, the organization made sure that its wiki, which contained information about work processes and a digital repository with information for staff, customers, and partners, was accurate and up to date. It had robust processes in place for cybersecurity, and effective data management and governance, which was externally reviewed once every two years.

For step 5, strategic use of technology definitely played a role in the decision to build the new digital systems internally. At a national lab, technological innovation to support scientific discovery is a prerogative, so building is typically preferred to buying or partnering, unless specialized capabilities need to be sourced from a university that performs basic research. No changes in value creation or organizational structure were required, and financing was internal.

### THE BOTTOM LINE

Evidence of digital transformation is everywhere. You, in fact, may not even have a physical book in your hands right now but may be reading a digital copy

**TABLE 15.6. Early, ad hoc development of a Quality 4.0 strategy in 2004 (from Chapter 14).**

| Step 2: Quality and Performance Goals | Step 3: Value Propositions | Step 4: Prioritized Strategic Initiatives | Step 6: Results |
|---|---|---|---|
| (*what* to achieve) | (*how* benefits will be delivered) | (*how* to engage and act) | (*what* to achieve) |
| **Orientation** Enhance customer interface and experience by: | *In order to reduce time and cost of travel, we need to:* | Remote observing project | Cost reduced from $3,000 to $0 per observer; savings to customers of $600,000–$800,000 per year |
| **Quality and Performance Goal—** Reducing time and cost of travel for astronomer | • Enable a new business model to democratize access to the telescope <br> • Make it logistically possible for people to remote observe by implementing dynamic scheduling | | |
| **Orientation** Internal process optimization (to make it possible to enhance customer interface and experience) | *In order to reduce lost time, we need to:* | Dynamic scheduling project | Approximately 1,500 reclaimed hours of time per year at $5,800/hour; total increase in science value delivered ~$8.7 million per year |
| **Quality and Performance Goal—** Reduce lost time/$ due to weather issues and/ or equipment failure | • Augment human intelligence to choose better times to observe, thus increasing the quality of observations <br> • Increase speed and quality of decision making | | |

on your computer or tablet. Successful digital transformation initiatives place quality and performance front and center, and leaders ensure that the right foundations, processes, capabilities, and habits are in place to bring about the desired results.

> Effective Quality Management is all about process discipline. Blending the emerging tools, process discipline, and a robust closed loop approach is beyond effective—it is powerful and insightful. It is taking real time and predictive data and turning it into actionable information. Just focusing on the latest tools is damaging. A fool with a tool is still a fool, and all you have is a bunch of data you don't know what to do with.
> —JASON GUZEK, DIRECTOR OF QUALITY AND FORTIVE BUSINESS SYSTEM, INDUSTRIAL SCIENTIFIC CORPORATION

This chapter presented a six-step playbook for digital transformation planning that incorporates people, process, and technology:

- Step 1—**Who** are you and how does your organization work? (Create an organizational profile)
- Step 2—**Why** do you want to transform, and why now? (Establish strategic orientation and quality/performance goals)
- Step 3—**How *can*** you transform? (Establish value propositions and potential initiatives)
- Step 4—**How *should*** you transform? (Prioritize initiatives)
- Step 5—**How *much*** should you transform? (Make buy-build-partner decisions)
- Step 6—**How** will you measure success? (Determine KPIs)

Following these steps will produce an actionable Quality 4.0 strategy—that is, a digital transformation strategy driven by quality and performance. While Industry 4.0 research may focus on cyber-physical systems for manufacturing and related industries, Quality 4.0 is for all industries. It addresses the interplay among people, processes, and new technologies that enhance connectedness, intelligence, and automation.

Good luck!

# REFERENCES

American Society for Quality. (2015). *2015 future of quality report: Quality throughout.* https://tinyurl.com/p22skhb

Hess, T., C. Matt, A. Benlian, and F. Wiesböck. (2016). "Options for formulating a digital transformation strategy." *MIS Quarterly Executive* 15 (2).

Ibarra, D., J. Ganzarain, and J. I. Igartua. (2018). "Business model innovation through Industry 4.0: A review." *Procedia Manufacturing* 22: 4–10.

Ismail, M. H., M. Khater, and M. Zaki. (2017, November). *Digital business transformation and strategy: What do we know so far.* Cambridge Service Alliance.

Kane, G. C., D. Palmer, A. N. Phillips, D. Kiron, and N. Buckley. (2015). "Strategy, not technology, drives digital transformation." *MIT Sloan Management Review and Deloitte University Press* 14: 1–25.

Katsamakas, E. (2014). "Value network competition and information technology." *Human Systems Management* 33: 7–17.

Kendall, K., and G. Bodinson. (2016). *Leading the Malcolm Baldrige way: How world-class leaders align their organizations to deliver exceptional results.* New York: McGraw-Hill Professional.

Kotarba, M. (2017). "Measuring digitalization—key metrics." *Foundations of Management* 9 (1): 123–138.

Lanzolla, G., and J. Anderson. (2008). "Digital transformation." *Business Strategy Review* 19 (2): 72–76.

Monostori, L. (2014). "Cyber-physical production systems: Roots, expectations and R&D challenges." *Procedia CIRP, 17,* 9–13. http://www.sciencedirect.com/science/article/pii/S2212827114003497

National Institute of Standards and Technology. (2019). *Baldrige Excellence Framework (Business/Nonprofit): Proven leadership and management practices for high performance.* https://www.nist.gov/baldrige/publications/baldrige-excellence-framework/businessnonprofit

Pryor, M. (2016, April 6–9). "Digital strategy in evolution: Issues and responses emerging from the project to develop a digital transformation strategy for Museum Victoria." MW2016: Museums and the Web, Los Angeles, CA.

Radziwill, N. M. (2018, November 18–19). "Designing a Quality 4.0 strategy and selecting high impact initiatives." Quality 4.0 Summit, ASQ, Dallas, TX.

Ross, J. W., C. M. Beath, and M. Mocker. (2019). *Designed for digital: How to architect your business for sustained success.* MIT Press.

Schiff, B. (1985). *The proficient pilot.* Wiley.

Sheridan, T. B., and W. L. Verplank. (1978). *Human and computer control of undersea teleoperators.* Cambridge: Massachusetts Institute of Technology, Man-Machine Systems Lab.

Stormont, D. P. (2008, July). "Analyzing human trust of autonomous systems in hazardous environments." In *Proc. of the Human Implications of Human-Robot Interaction workshop,* 27–32. AAAI.

Zimmermann, A., D. Jugel, K. Sandkuhl, R. Schmidt, C. Schweda, and M. Möhring. (2016). "Architectural decision management for digital transformation of products and services." *Complex Systems Informatics and Modeling Quarterly* (6): 31–53.

# INDEX OF STATISTICAL AND MACHINE LEARNING MODELS

This appendix provides a brief description of several common machine learning algorithms. The purpose of this section is to help you understand, at a high level, what types of problems these methods are used to solve. These are simple examples using very small datasets, neither of which you will find in the real world. Machine learning practice is messy and requires hours upon hours of data cleaning and data wrangling.

Some of the algorithms in this appendix are just statistical or probabilistic methods that are commonly applied to large datasets—approaches that would be unwieldy if attempted without a computer. Questions about whether some of these approaches (e.g., linear regression) are "real" machine learning methods are routinely posed on Stack Overflow and Twitter—in general, data scientists who are first and foremost programmers say yes, while data scientists who are statisticians say no. The difference seems to lie in how each community uses the models. While programmers are concerned with the ultimate performance of the model (e.g., how well it predicts values or classifies observations), statisticians are more focused on estimating the characteristics of the parameters inside the model to understand why relationships are in place and why variation is occurring.

All the examples in this section are fully reproducible and use prepackaged datasets that come with packages from the R Statistical Software. To run them yourself, sign up for an account at https://rstudio.cloud, and after you launch that application, type the code that appears in the examples below. *Do not type the leading caret > if you see it.* After lines of code that start with the caret character, the results are displayed below the code. To reproduce the example, type everything but the leading caret.

## ALGORITHMS AND EXAMPLES

### Artificial Neural Networks

Neural networks are a supervised machine learning method that can be used to perform prediction and classification tasks. A neural network must be trained with many prior observations. These observations contain one or more inputs (each of which becomes a node in the input layer), and the model is created with the intention of generating one or more outputs (nodes in the output layer). In the following example, we will predict one number (housing price) based on 13 other numbers. For a complex task like predicting whether an integer appears in an image, the input layer would have one node for each pixel in an image in the training set, and the output layer would consist of ten nodes, each representing a digit from zero to nine.

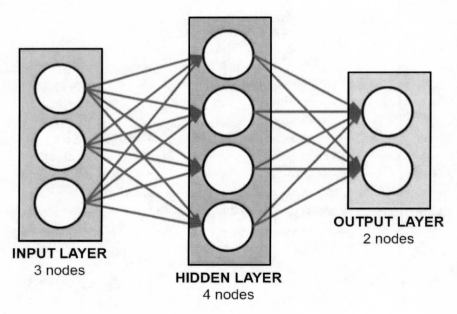

```
install.packages("mlbench")
library(mlbench)
install.packages("nnet")
library(nnet)
```

```
data(BostonHousing)
attach(BostonHousing)

> head(BostonHousing)
     crim zn indus chas   nox    rm   age    dis rad tax ptratio      b lstat medv
1 0.00632 18  2.31    0 0.538 6.575 65.2 4.0900   1 296    15.3 396.90  4.98 24.0
2 0.02731  0  7.07    0 0.469 6.421 78.9 4.9671   2 242    17.8 396.90  9.14 21.6
3 0.02729  0  7.07    0 0.469 7.185 61.1 4.9671   2 242    17.8 392.83  4.03 34.7
4 0.03237  0  2.18    0 0.458 6.998 45.8 6.0622   3 222    18.7 394.63  2.94 33.4
5 0.06905  0  2.18    0 0.458 7.147 54.2 6.0622   3 222    18.7 396.90  5.33 36.2
6 0.02985  0  2.18    0 0.458 6.430 58.7 6.0622   3 222    18.7 394.12  5.21 28.7

# create and train the neural network—scale by 50 so that the network converges
my.nnet <- nnet(medv/50 ~ ., data=BostonHousing, size=6, decay=0.01)

# use the neural network to make predictions
my.pred <- predict(my.nnet)*50

# load the special neural network (nnet) plotting function
fxn.url <- "https://gist.githubusercontent.com/fawda123/5086859/raw/cc1544804d5027d82b
70e74b83b3941cd2184354/nnet_plot_fun.r"
source(fxn.url)

plot(my.nnet, pos.col='darkgreen', neg.col='darkblue', alpha.val=0.7,
     rel.rsc=15, circle.cex=10, cex=1.4, circle.col="lightgray")
```

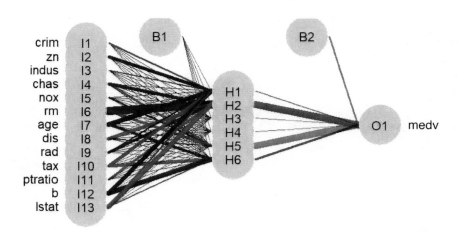

**What this plot shows:** This shows the structure of the neural network we created, but not the output. There are 13 input variables on the left, which are input nodes I1 through I13 (the gray circles are large and partially cover each other). In the middle, there is a hidden layer with 6 nodes, H1 to H6. On top, there are two "bias terms," B1 and B2, that each add a constant to the equation that describes the neural network. Finally, there is one variable in the output layer, O1. This is the median housing value, medv.

Now, compare the predictions from the neural network with the actual values in the training data:

```
plot(medv/50, fitted(my.nnet), main="Predicted vs. Actual")
abline(0,1) # plot a diagonal line through the origin with a slope of 1
```

**Predicted vs. Actual**

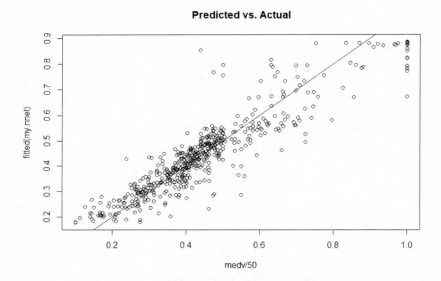

**What this plot shows:** Actual values are on the x-axis, and predicted values are on the y-axis. Most of the values fall close to the diagonal line where the predicted value equals the actual value, so this neural network may be a good prediction model.

## C5.0

C5.0 is an algorithm used to build a decision tree or rule set. It uses information entry to determine the appropriate questions, meaning that it finds the most general questions first and then eventually generates more narrow questions to distinguish between groups.

```
install.packages("C50")
library(C50)
data("iris")
```

```
> iris
             Sepal.Length      Sepal.Width      Petal.Length      Petal.Width      Species
1                    5.1              3.5               1.4              0.2       setosa
2                    4.9              3.0               1.4              0.2       setosa
3                    4.7              3.2               1.3              0.2       setosa
4                    4.6              3.1               1.5              0.2       setosa
5                    5.0              3.6               1.4              0.2       setosa
. . . # plus 145 more observations

model <- C5.0(Species ~ ., data=iris)
plot(model)
```

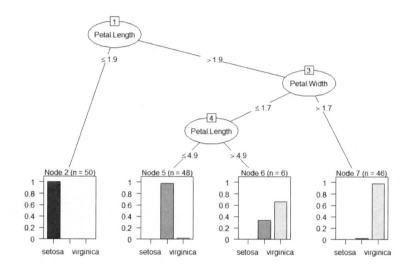

**What this plot shows:** You are observing an iris flower out in the wild. To determine what species it belongs to, check the petal length. If it is less than 1.9 cm, there is a nearly 100% chance it is from species Setosa. If the petal length is greater than 1.9 cm, check the petal width. If the petal width is greater than 1.7 cm, there is a very high chance it is from species Virginica. Otherwise, check the petal length. If petal length is less than 4.9 cm, there is a very high chance it is from species Versicolor (unlabeled). If petal length is greater than 4.9 cm, there is a 70% chance it is species Virginica and a 30% chance it is Versicolor.

## Decision Trees

Decision trees are used to visualize the interconnected rules used to make a complex decision, and (sometimes) display the likelihood of making each choice at each step. In most cases we don't know the rules up front, but instead have examples that we use to figure out what those rules are. The classification and regression tree (CART) approach works with categorical and numerical data and tolerates outliers well.

For example, if we have 150 measurements of the sepals and petals of four different species of iris, we can construct a decision tree that finds the patterns.

```
install.packages("caret")
install.packages("e1071")
install.packages("rattle")
library(caret)
library(e1071)
library(rattle)

data("iris")

> iris
     Sepal.Length    Sepal.Width    Petal.Length    Petal.Width    Species
1            5.1            3.5             1.4            0.2     setosa
2            4.9            3.0             1.4            0.2     setosa
3            4.7            3.2             1.3            0.2     setosa
4            4.6            3.1             1.5            0.2     setosa
5            5.0            3.6             1.4            0.2     setosa
. . . # plus 145 more observations

model <- train(Species ~ ., method="rpart2", data=iris, tuneGrid=data.
frame(maxdepth=2))

> model
CART

150 samples
  4 predictor
  3 classes: 'setosa', 'versicolor', 'virginica'

No pre-processing
Resampling: Bootstrapped (25 reps)
Summary of sample sizes: 150, 150, 150, 150, 150, 150, . . .
Resampling results:

  Accuracy Kappa
  0.9396151 0.9084115

Tuning parameter 'maxdepth' was held constant at a value of 2

> fancyRpartPlot(model$finalModel)
```

**What this plot shows:** You are observing an iris flower out in the wild. To determine what species it belongs to, measure the petal length. If the petal length is less than 2.5 cm, the iris belongs to species Setosa. Otherwise, measure the petal width. If it is less than 1.8 cm, it belongs to species Versicolor. If it is greater than 1.8 cm, it belongs to species Virginica.

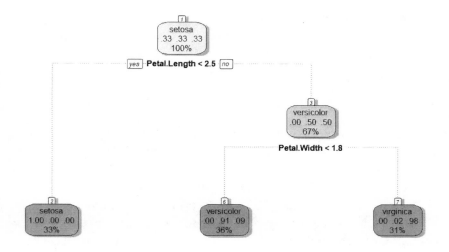

## Deep Learning

Deep neural networks are an extension of neural networks that can be configured with many different types of hidden layers. They are "deep" because they can also have a multitude of hidden layers, making the depth between input and output layers great. Because it can be a challenge to get your runtime environment set up to do deep learning yourself, I do not include an example on RStudio Cloud.

Here is an example of a deep neural network configured using the keras package. There is an art associated with defining layers and ordering of layers in a deep network. This portion of an example comes from Kisler (2018):

```
model <- keras_model_sequential()

# add layers
model %>%
  layer_conv_2d(filter=32, kernel_size=c(3,3), padding='same',
      input_shape=c(img_width, img_height, channels)) %>%
  layer_activation('relu') %>%

  # Second hidden layer
  layer_conv_2d(filter=16, kernel_size=c(3,3), padding='same') %>%
  layer_activation_leaky_relu(0.5) %>%
  layer_batch_normalization() %>%

  # Use max pooling
  layer_max_pooling_2d(pool_size=c(2,2)) %>%
  layer_dropout(0.25) %>%
```

```
# Flatten max filtered output into feature vector and feed into dense layer
layer_flatten() %>%
layer_dense(100) %>%
layer_activation('relu') %>%
layer_dropout(0.5) %>%

# Outputs from dense layer are projected onto output layer
layer_dense(output_n) %>%
layer_activation('softmax')
```

The only type of layer in a regular neural network is the "dense" (or "fully connected") layer, where a linear combination of inputs and weights is formed. Deep neural networks have many additional types of layers. The example above has additional layer types, including *convolution* layers, which scan the inputs for notable features; *activation* layers, which interpret the numerical results from dense and convolution layers and decide how to handle them; *normalization* layers, which shift the mean of their inputs to zero and standard deviation to 1; and *flatten* layers, which reshape the data structure before the next step. A deep learning network can be considered a massive pipeline, where inputs are transformed and manipulated and interpreted at many steps.

## Dynamic Time Warping

If you hear a song you recognize, it usually doesn't matter whether the song is played fast or slow—you will still recognize it. Dynamic time warping is an algorithm that measures the alignment between two sequences of numbers. It is useful in clustering problems when you are trying to figure out whether several plots of some variable over time are similar to each other. This example uses the dtw package from Giordino (2009). Find out more at Sobolewska (2019).

```
install.packages("dtw")
library(dtw)

t=seq(0,10,0.1)
y=sin(t)# create a sine wave
z=sin(2*t)# create a sine wave with a longer wavelength

# plot them
plot(t, z, type="l", xlab="time", ylab="Sine wave")
points(t ,y, type="l", xlab="time", ylab="Sine wave", col="red", lwd=2)

# plot a time warping solution
plot(dtw(y, z, k=TRUE), type="two", off=1, match.lty=2, match.indices=20)
```

**What this plot shows:** It shows the points on the shorter sine wave that match up to the points on the longer sine wave in the best possible way.

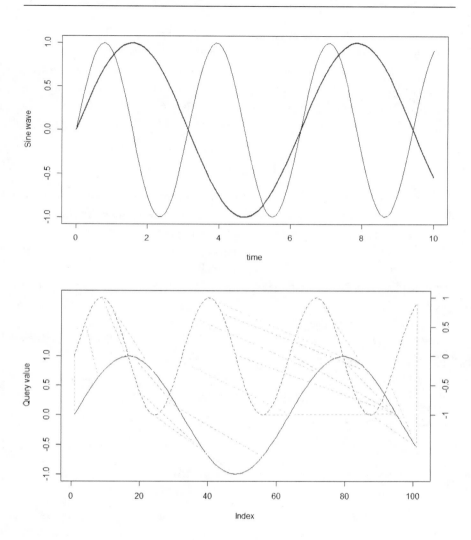

## Expert Systems

Although very popular in the 1970s and 1980s and one of the first success stories in AI, expert systems have been eclipsed by other, more powerful methods, especially those in machine learning. Building the expert system involves defining facts and rules, and then the inference engine generates new facts based on the combination of facts and rules it has available. An expert system typically has a user interface, an inference engine, and a knowledge base. Prolog,

JESS, CLIPS, and PyCLIPS are examples of specially designed packages with built-in inference engines; the user interface is the command line.

Here's how an expert system works. The user enters Fact #1: *Alex is the son of Mary* and Fact #2: *Mary is the daughter of John* into the knowledge base. The user also enters one Rule: *The parent of a parent is a grandparent.* The inference engine in the expert system generates a new fact: *John is the grandparent of Alex.*

## K-Means Clustering

K-means clustering is an unsupervised machine learning task for knowledge discovery. It does not generate a model but rather creates a new variable (cluster membership) that can be used to understand differences between observations. Thus, it cannot be used for prediction: the algorithm only generates the clusters; it does not tell you what those clusters represent or why they are important. People have to manually inspect and interpret clustering solutions to determine what they mean.

There are many different algorithms for clustering. This example uses the k-means approach to see whether we can cluster 150 iris observations into three species: Setosa, Versicolor, and Virginica (Enhance Data Science, 2017). The processing requires only one line of code. This example uses a package called GGally to plot fancy results that are easier to interpret.

```
install.packages("ggplot2")
install.packages("GGally")
library("ggplot2")
library("GGally")

data("iris")

> iris
     Sepal.Length    Sepal.Width    Petal.Length    Petal.Width    Species
1           5.1            3.5            1.4            0.2    setosa
2           4.9            3.0            1.4            0.2    setosa
3           4.7            3.2            1.3            0.2    setosa
4           4.6            3.1            1.5            0.2    setosa
5           5.0            3.6            1.4            0.2    setosa
 . . . # plus 145 more observations

clusters <- kmeans( iris[,1:4], 3 )              # create the clustering solution
iris$cid <- as.factor(clusters$cluster)          # create a new column w/cluster ID
ggpairs(iris, columns=1:5, mapping=aes(color=cid))  # plot the cluster solution
```

**What this plot shows:** Starting with the plot in the upper left, this shows us that sepal length may be able to help us distinguish between the three species, but continuing diagonally down and to the right, sepal width definitely cannot

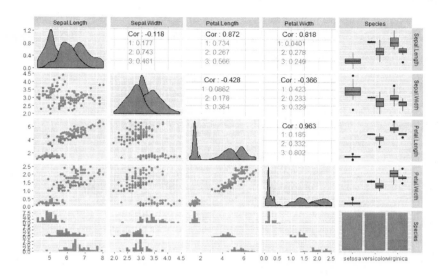

(all the peaks are too close together). Going down and to the right one more time, we see that it should be easy to distinguish Setosa using just petal length and petal width, because its values are far smaller than the other two categories. In the bottom right, we see how well the clustering solution grouped our 150 observations. Setosa were perfectly identified, and Versicolor were almost perfectly identified. It was more difficult to characterize Virginica, and around a quarter of observations were inaccurately assigned to the Versicolor cluster.

## K-Nearest Neighbors

The supervised machine learning algorithm k-nearest neighbors classifies a new observation into a group based on the group membership of its nearest neighbors. The value of k represents the number of neighbors you consult to determine which group to assign the new observation to. "Nearest" can be determined by the Euclidean distance ("as the crow flies") or other distance metrics like Manhattan (distance using only perpendicular streets and avenues).

This example comes from data in Lantz (2019) about characteristics of fruits, vegetables, and proteins. There are two quantitative variables (sweetness and crunchiness), rated from 1 (not sweet or crunchy) to 10 (very sweet or crunchy), one categorical label (ingredient), and one categorical variable representing the classification of the food item (food.type). The goal is to use the first seven observations as the "neighbors" and then classify the last three observations based on that information.

```
> foods <- read.csv("https://raw.githubusercontent.com/NicoleRadziwill/Data-for-R
-Examples/master/foods-lantz-ch3.csv",header=TRUE)
```

```
> foods
     ingredient  sweetness   crunchiness    food.type
1         apple        10            9          fruit
2         bacon         1            4        protein
3        banana        10            1          fruit
4        carrot         7           10      vegetable
5        celery         3           10      vegetable
6        cheese         1            1        protein
7         grape         8            5          fruit
8    green bean         3            7      vegetable
9          nuts         3            6        protein
10       orange         7            3          fruit
```

A plot shows that fruits, vegetables, and proteins are nearly distinct from one another when comparing them in terms of sweetness and crunchiness:

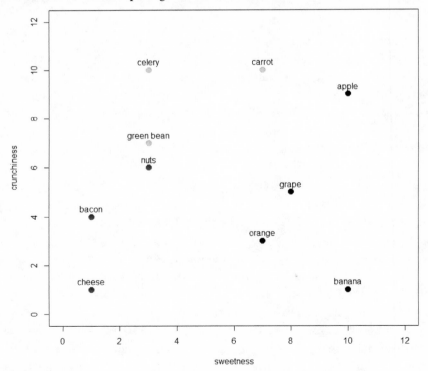

```
attach(foods)
plot(sweetness, crunchiness,pch=16,cex=1.5,xlim=c(0,12),ylim=c(0,12),col=food.type)
with(foods, text(crunchiness~sweetness, labels=ingredient,pos=3))

install.packages("class")
library(class)
train <- foods[1:7,2:3]       # the first 7 observations are our training set
test <- foods[8:10,2:3]       # the last 3 observations will test the classifier
train.labels <- foods[1:7,4]  # here are the correct answers for the first 7 cases
```

The classifier tells us that the last three observations (green bean, nuts, and orange) are a vegetable, protein, and fruit, respectively. (The last line just tells us the labels for all groups.) The confusion matrix at the end provides another way to show this perfect classification. Not all classifiers perform this well, obviously.

```
> knn(train, test, train.labels)
[1] vegetable protein fruit
Levels: fruit protein vegetable

> table(knn(train, test, train.labels), foods[8:10,4])

                  fruit         protein   vegetable
    fruit           1              0          0
    protein         0              1          0
    vegetable       0              0          1
```

## Latent Dirichlet Allocation (LDA)

Latent Dirichlet allocation (LDA) is an unsupervised machine learning algorithm that examines a collection of documents (called a corpus), looking for words that describe a set number of topics. It is a statistical model that looks for how words are distributed within each topic, and also how topics are distributed across the documents. This example finds six topics within a corpus containing 2246 publications from the U.S. Associated Press:

```
install.packages("topicmodels")
library(topicmodels)
data(AssociatedPress)

> AssociatedPress
<<DocumentTermMatrix (documents: 2246, terms: 10473)>>
Non-/sparse entries : 302031/23220327
Sparsity            : 99%
Maximal term length : 18
Weighting           : term frequency (tf)

> model.6 <- LDA(AssociatedPress, k=6)
> terms <- as.data.frame(topicmodels::terms(model.6,20),stringsAsFactors=FALSE)
> terms # (they're truncated at 9, but go to 20 . . .)
```

|   | Topic 1 | Topic 2 | Topic 3 | Topic 4 | Topic 5 | Topic 6 |
|---|---------|---------|---------|---------|---------|---------|
| 1 | percent | soviet | police | two | i | company |
| 2 | million | government | people | military | bush | federal |
| 3 | year | president | two | people | new | court |
| 4 | billion | united | i | officials | dukakis | million |
| 5 | market | party | state | air | president | new |
| 6 | new | states | years | miles | people | last |
| 7 | prices | political | city | iraq | years | case |
| 8 | stock | union | death | army | campaign | department |
| 9 | last | minister | found | fire | time | judge |

## Logistic Regression

Logistic regression uses an S-curve to classify observations into groups. In the case of binary logistic regression, discussed in this example, we use the value on the x-axis to determine the probability of group membership (the y-axis). One group is represented by the lower tail at y = 0 and the other group at y = 1.

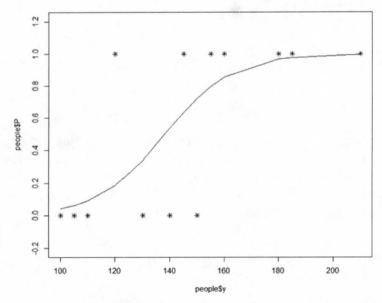

This example is on a simple dataset of fictitious people. We know their height, weight, gender (0 is female and 1 is male), and age. We'd like to predict gender (a binary categorical variable, in this particular problem) from the other variables.

```
x1 <- c(60,69,72,75,66,64,62,68,68,70,56,66,65,64) # height (in)
x2 <- c(22,28,32,40,25,45,30,32,45,38,22,26,34,35) # age (years)
```

```
y <- c(130,150,180,210,155,100,110,140,160,185,105,120,155,145) # weight (lbs)
z <- c(0,0,1,1,1,0,0,0,1,1,0,1,1,1) # gender (0=female; 1=male)
log.model <- glm(z ~ x1 + x2 + y, family=binomial(link='logit'))

> summary(log.model)

Call:
glm(formula = z ~ x1 + x2 + y, family = binomial(link = "logit"))

Deviance Residuals:
Min                 1Q      Median        3Q       Max
-1.3947            -0.4578   0.1228    0.5212    2.0548

Coefficients:
                Estimate      Std.    Error z value    Pr(>|z|)
(Intercept)     -8.14993   15.21065           -0.536      0.592
x1              -0.09286    0.31172           -0.298      0.766
x2               0.07025    0.13239            0.531      0.596
y                0.08725    0.05475            1.594      0.111
```

**What this table shows:** The values in the bottom right column tell us whether each predictor is significant. To be significant, the value should be (at the very least) less than 0.05. None of the values are tiny, so none of our predictors are helpful for trying to determine gender. So let's try another model, only this time we'll try weight as a predictor.

```
log2.model <- glm(z ~ y,family=binomial(link='logit'))

> summary(log2.model)

Call:
glm(formula = z ~ y, family = binomial(link = "logit"))

Deviance Residuals:
Min                 1Q      Median        3Q       Max
-1.5992            -0.4199   0.1443    0.6482    1.8353

Coefficients:
                Estimate      Std.    Error  z value   Pr(>|z|)
(Intercept)    -11.20409    6.03350  -1.857    0.0633      .
y                0.08104    0.04250   1.907    0.0565      .
—
Signif. codes: 0 '***' 0.001 '**' 0.01 '*' 0.05 '.' 0.1 ' ' 1
```

**What this table shows:** The values in the bottom right column, as above, tell us whether each predictor is significant. To be significant, the value should be (at the very least) less than 0.05. The values are right on the edge of that threshold, so this is a tough case. Weight is possibly a significant predictor of gender, but we would have to collect a larger sample to know for sure.

## Naive Bayes Classification

Bayesian classification depends on the concept of conditional probability. Given that you know one condition is true, what is the probability that a second thing is also true? You can think about this in the context of a weather forecast. If you know that skies are overcast today, what is the probability of rain? If you know the skies are clear today, what is the probability of rain? It shouldn't be too hard to understand why the probability in the first case will be much higher than in the second case.

In addition, for a Bayesian classifier, we need to be able to provide a general idea about how many items are expected to be in the groups after we classify them. In the weather case, we might already know that, on average, 10% of our days will be rainy and 90% will be sunny (perhaps we live in Arizona). These are called prior probabilities. This method is often used for building utilities like spam filters—if the word *Viagra* appears in an e-mail message, there's a high probability that it's spam; in general, system administrators will know what proportion of incoming e-mails are spam versus not spam (the prior probabilities). The more accurate estimates we can provide for our priors, the better (in general) the classifier should be.

In this example, we observe four e-mails and classified them as spam or not spam ("ham"). In addition, we determined whether the word *Viagra* appeared in each of them, and built a training set from that information:

```
train <- data.frame(class=c("spam","ham","ham","ham"),
        viagra=c("yes","no","no","yes"))
> train
    class   viagra
1    spam      yes
2     ham       no
3     ham       no
4     ham      yes
```

Bayes theorem, the cornerstone on which this method is based, can help us calculate the probability that an e-mail is spam if it contains the word *Viagra*.

$$P(Email\ is\ Spam \mid Email\ has\ "Viagra"\ in\ it) = \frac{P(Viagra|Spam)\ P(Spam)}{P(Any\ Email\ has\ "Viagra"\ in\ it)}$$

This would be $P = (1 \times 0.25/0.50) = 0.50$ or 50%.

```
library(e1071)
classifier <- naiveBayes(class ~ viagra, train)

> predict(classifier, test, type="raw")

        ham    spam
[1,]    0.5    0.5
```

Just as we calculated earlier, the probability that an e-mail is spam if it contains the word *Viagra* is 50%. Although this result isn't that interesting because it's easy to calculate manually, you can try it with larger datasets by adding more terms to the data.frame at the beginning of this example.

## Principal Component Analysis (PCA)

If a dataset has hundreds or thousands of predictors (i.e., columns in a spreadsheet), that's a potential problem for two reasons. First, it makes machine learning algorithms computationally expensive, meaning you need more time or money (or both) to get results. Second, even if you can get the machine learning algorithm to run, using so many predictors can get you into a terrible situation called *overfitting*, where you're modeling the noise instead of the signal. Principal Component Analysis (PCA) is one statistical technique that can be used to reduce large datasets to only the most important linear combinations of their significant predictors. No example is provided since our test datasets are all small.

## Q-Learning

Reinforcement learning "learns" an optimal path through a system by trying lots of paths, each of which has a reward associated with it, and keeping score. These algorithms gain information by interacting with the environment, and gain feedback using a reward signal. An agent interacts with its environment and takes actions, and these actions impact the state of the environment (e.g., increases or decreases the agent's score). The best path is the one that, after a multitude of attempts, yields the best score. This method is based on Markov decision processes (a model in which the state of the system depends only on its previous state).

Q-learning is one of several algorithms for reinforcement learning. It chooses the best sequence of steps to maximize a reward, given that you don't have a model for how to get from start to finish, only a record of what the reward is when you move between individual steps. For example, McCullock (2012) uses Q-learning to determine the quickest way out of a building (that is, the quickest way to "Room 5"):

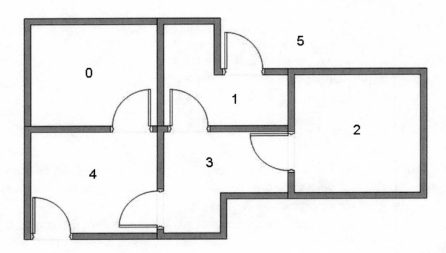

To do Q-learning, we first need to set up a matrix of rewards. We will assign a −1 if "you can't get there from here," a 0 if the next room is not the target state, and 100 if it is the target state. Rows represent "from" and columns represent "to." The top left value, −1, is the "reward" when going from Room 0 to Room 0 (staying in place). Since this gets you no closer to your goal, it is assigned a −1. The bottom rightmost number, which is in row 5 and column 5, represents the reward when you are in Room 5 and you stay in Room 5 (the target).

```
source("https://raw.githubusercontent.com/NicoleRadziwill/
R-Functions/master/qlearn.R")
r    <- c(-1,  -1,  -1,   -1,    0,   1,
         -1,  -1,  -1,    0,   -1,   0,
         -1,  -1,  -1,    0,   -1,  -1,
         -1,   0,   0,   -1,    0,  -1,
          0,  -1,  -1,    0,   -1,   0,
         -1, 100,  -1,   -1,  100, 100)

R <- matrix(r,  nrow=6,  ncol=6)

>q.learn(R,10000,alpha=0.1,gamma=0.8,tgt.state=6)
          [,1]       [,2]       [,3]       [,4]       [,5]        [,6]
[1,]   0.00000    0.00000    0.00000    0.00000   79.99977     0.00000
[2,]   0.00000    0.00000    0.00000   63.99920    0.00000    99.99980
[3,]   0.00000    0.00000    0.00000   63.99949    0.00000     0.00000
[4,]   0.00000   79.99947   51.19934    0.00000   79.99964     0.00000
[5,]  63.99940    0.00000    0.00000   63.99935    0.00000   100.00000
[6,]  64.19567   79.99373    0.00000    0.00000   79.99598    99.99323
```

**What this table shows:** Start at the top left of the matrix, in row zero (which is labeled [1,]). The goal is to start in a room (a row) and find the biggest reward that Q-learning has found, then go there. Continue to the row that is recommended, and do the same process over. A policy is a "path" through the states of the system, so if you:

- Start at 0: Choose 4 (80), then from 4 choose 5 (100) to get outside
- Start at 1: Choose 5 (100)
- Start at 2: Choose 3 (64), then from 3 choose 1 or 4 (80); from 1 or 4 choose 5 (100)
- Start at 3: Choose 1 or 4 (80), then 5 (100)
- Start at 4: Choose 5 (100)
- Start at 5: Stay at 5 (100)

Reinforcement learning has given us optimal paths to get outside, no matter where we start.

## Random Forests

A random forest is a supervised classification method that, because it is an ensemble method, generates a multitude of random decision trees and then picks the one that best predicts the observations in the training set. Typically, cross-validation is used, where you split the training data into a training set (used to build the random forest) and a test set (used to evaluate its performance). In this example, we will build a random forest from the entire iris dataset.

```
install.packages("randomForest")
library(randomForest)

> my.rf <- randomForest(Species~., data=iris, ntree=100, proximity=TRUE)
> my.rf

Call:
 randomForest(formula = Species ~ ., data = iris, ntree = 100, proximity = TRUE)
               Type of random forest: classification
                     Number of trees: 100
No. of variables tried at each split: 2

        OOB estimate of  error rate: 4.67%
Confusion matrix:
           setosa   versicolor   virginica   class.error
setosa         50            0           0          0.00
versicolor      0           46           4          0.08
virginica       0            3          47          0.06
```

**What this table shows:** This is a confusion matrix that shows us how well the random forest predicted which observation belonged to which species. Most of the observations were correctly classified: only seven observations confused Virginica with Versicolor. Setosa were predicted perfectly, which matches the conclusion from the "K-Means Classification" section above.

## Support Vector Machine (SVM)

Support vector machines (SVMs) can be used for both regression (predicting numbers) and classification (predicting group membership). For SVM classifiers, they create a hyperplane boundary between categories in the training data (much like the simplest neural networks create an equation splitting a plane into sections, each of which forms a group). In this example, the goal is to predict the type of glass (1 through 6) by its chemical composition:

```
library(e1071)
library(mlbench)
data(Glass, package="mlbench")

> head(Glass)
        RI     Na    Mg    Al     Si     K    Ca  Ba    Fe  Type
1  1.52101  13.64  4.49  1.10  71.78  0.06  8.75   0  0.00     1
2  1.51761  13.89  3.60  1.36  72.73  0.48  7.83   0  0.00     1
3  1.51618  13.53  3.55  1.54  72.99  0.39  7.78   0  0.00     1
4  1.51766  13.21  3.69  1.29  72.61  0.57  8.22   0  0.00     1
5  1.51742  13.27  3.62  1.24  73.08  0.55  8.07   0  0.00     1
6  1.51596  12.79  3.61  1.62  72.97  0.64  8.07   0  0.26     1

index <- 1:nrow(Glass)
testindex <- sample(index, trunc(length(index)/3))
testset <- Glass[testindex,]
trainset <- Glass[-testindex,]
svm.model <- svm(Type ~ ., data = trainset, cost = 100, gamma = 1)
svm.pred <- predict(svm.model, testset[,-10])

> table(pred = svm.pred, true = testset[,10])
      true
pred    1   2   3   5   6   7
   1   17   6   3   0   0   0
   2    6  16   2   1   1   5
   3    1   1   2   0   0   0
   5    0   0   0   2   0   0
   6    0   0   0   0   1   0
   7    0   0   0   0   0   7
```

**What this table shows:** This is a confusion matrix that compares the values predicted by the SVM with the actual values from the test set. The numbers that appear along the diagonal have been correctly classified (17, 16, 2, 2, 1, 7). The classifier had the most difficulty with glass type #2. It incorrectly put observations in this group 15 times and correctly classified observations into this group 16 times.

## Sentiment Analysis

Comparing words, word frequencies, and word positions within a document can provide insight into the sentiments or emotions in a text document. This is possible because sentiment analysis packages compare the words with pre-defined lexicons that serve as an interpretation guide. Although these packages are excellent at basic analysis, they are not good at detecting sarcasm.

For this example, I analyzed the entire first chapter of this book to make sure it's setting the right tone for your reading experience.

```
install.packages("tidyverse")
library(tidyverse)
install.packages("syuzhet")
library(syuzhet)
chapter1 <- "CHAPTER 1 QUALITY 4.0 AND THE FOURTH INDUSTRIAL REVOLUTION Any suffi-
ciently advanced technology is indistinguishable from magic.—Arthur C. Clarke Covering
140,000 miles, the privately owned and maintained U.S. rail network transported nearly
1.1 million carloads of freight in 2017, and again in 2018. Rail transportation car-
ries coal to power plants, food to . . ." # only I copied in the ENTIRE first chapter.

c1.sent <- get_sentences(chapter1)
c1.tokens <- get_tokens(c1.sent, pattern="\\W")
sentiment <- get_sentiment(c1.tokens)
plot(sentiment, type="l", main="Chapter 1",
        xlab="Narrative Time", ylab="Emotional Valence")
```

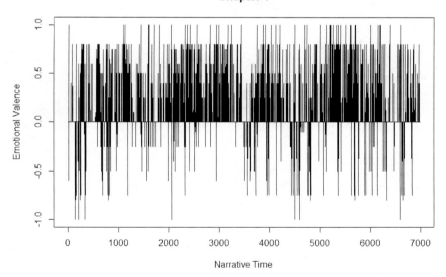

**Chapter 1**

**What this plot shows:** Time moves from left to right, from the beginning of Chapter 1 until the end. Measurements above the line show positive emotional valence, while those below the line show negative emotional valence. Although I kept the majority of the chapter positively oriented, it does look like I started out with a negative or concerning story. This is true.

You can also use lexicons to determine the percentage of the story or narrative that emphasizes or suggests certain emotions. I also ran Chapter 1 through this process, using the NRC emotion lexicon that compares texts with eight basic emotions:

```
barplot(
    sort(colSums(prop.table(nrc_data[, 1:8]))),
    horiz = TRUE,
    cex.names = 0.7,
    las = 1,
    main = "Emotions in Chapter 1",
        xlab="Percentage"
    )
```

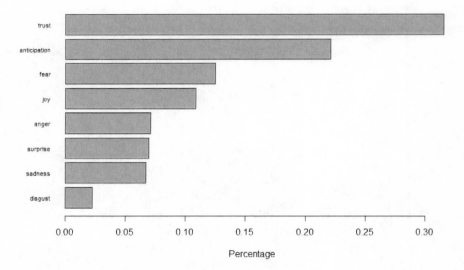

**What this plot shows:** You can be the judge about whether this sentiment analysis accurately represents Chapter 1 of this book.

## REFERENCES

Enhance Data Science. (2017, April 30). *R Basics: K-means with R*. http://enhancedatascience .com/2017/04/30/r-basics-k-means-r/

Giorgino, T. (2009). "Computing and visualizing dynamic time warping alignments in R: The dtw package." *Journal of Statistical Software* 31 (7): 1–24. doi:10.18637/jss.v031.i07

Kisler, D. (2018, June 18). "R vs Python: Image classification with Keras." Data Science Plus. https://datascienceplus.com/r-vs-python-image-classification-with-keras/

Lantz, B. (2019). *Machine learning with R*. 3rd edition. London, UK: Packt Publishing.

McCullock, J. (2012). "Q-Learning step-by-step tutorial." http://mnemstudio.org/path-finding-q-learning-tutorial.htm

R Core Team (2019). "R: A language and environment for statistical computing." R Foundation for Statistical Computing. https://www.R-project.org/

Sobolewska, E. (2019). "Dynamic time warping (DTW) as a mean to cluster time series." https://rpubs.com/esobolewska/dtw-time-series

# DIGITAL TRANSFORMATION PLANNING CHECKLIST

This appendix provides a checklist for the recommendations in Chapter 15.

| Step | Guiding question | Category | Item | Done? |
|------|------------------|----------|------|-------|
| 1 | Who are you? How does your organization work? | Organizational Profile | Product offerings defined | |
| | | | Product and service delivery mechanisms defined | |
| | | | Mission, vision, values, culture defined | |
| | | | Workforce profile defined | |
| | | | Assets identified and categorized by risk profile | |
| | | | Regulatory environment articulated | |
| | | | Organizational structure supports transformation goals | |
| | | | Customer and stakeholder profiles complete | |
| | | | Supplier, partner, collaborator stakeholder profiles complete | |
| | | | Competitive environment defined | |
| | | | Strategic challenges and advantages identified | |
| | | | Performance improvement system defined and functional | |

*(continued)*

| Step | Guiding question | Category | Item | Done? |
|---|---|---|---|---|
| | | Organizational Backbone | Demonstrate clear, consistent executive commitment | |
| | | | Establish framework for translating strategic objectives into actions | |
| | | | Articulate clear roles, responsibilities, and accountability | |
| | | | Document standard work/ descriptions of work processes | |
| | | | Define value stream maps as appropriate | |
| | | | Develop workforce capability and capacity plan | |
| | | | Establish framework for continuous organizational learning | |
| | | | Effective two-way communication channels between customers, suppliers, collaborators, workforce, and leadership | |
| | | Digital Platform | Establish knowledge repository and digital services for workforce | |
| | | | Establish knowledge repository and digital services across supply network | |
| | | | Implement data management and governance (e.g., a data platform for sharing and managing critical data, and ensuring data integrity) | |
| | | | Establish an external developer platform for shared work and collaborative innovation | |

*(continued)*

| Step | Guiding question | Category | Item | Done? |
|------|------------------|----------|------|-------|
| | | | Establish robust cybersecurity practices | |
| | | | Make mobile, machine-to-machine, and prototyping tools available as necessary | |
| 2 | Why do you want to transform? | | Define strategic orientation | |
| | | | Identify quality and performance goals | |
| 3 | How can you transform? | | Establish value propositions | |
| | | | Brainstorm target entities and processes | |
| | | | Examine degrees of connectedness, intelligence, and automation | |
| 4 | How should you transform? | | Assess each opportunity in terms of magnitude, deployment, and other factors important to your organization | |
| | | | Prioritize opportunities | |
| 5 | How much should you transform? | | Examine strategic orientation with respect to technology, value creation, structural changes, and financial aspects | |
| | | | Make buy-build-partner decisions | |
| 6 | How will you measure success? | | Identify metrics for each initiative to determine if you are winning, if you should make adjustments, and how you will communicate benefits to stakeholders | |

# GLOSSARY OF TERMS

## A

**A/B testing.** An experiment where two choices are randomly presented to users or prospects to determine which one is more effective or desirable. *See also* **multiarmed bandit**.

**additive manufacturing.** Building a three-dimensional object from a CAD model in a way that iteratively adds layers, rather than milling a part out of a solid block of raw material.

**agent.** Someone or something that performs an action on behalf of someone or something else—for example, real estate agents, insurance agents, lawyers, home assistants (Siri, Alexa).

**agile methodology.** Approaches for managing software development that have grown to be applied in other disciplines, like sales, marketing, and services. Agile methodologies are iterative and exploratory, depend on close interactions with customers, and are sometimes pull-based. Includes Scrum, Dynamic Systems Development Method (DSDM), kanban, lean, Crystal, and SaFE.

**American Society for Quality (ASQ).** An international professional organization with the mission "to empower people, communities, and organizations of the world to achieve excellence through quality."

**analytics.** Using math, statistics, and/or machine learning to generate and communicate insights about business processes and entities. *See also* **descriptive analytics, diagnostic analytics, predictive analytics,** *and* **prescriptive analytics**.

**anomaly detection.** Identification of data points, rare events, or observations that deviate from the normal characteristics of behavior of a dataset.

**application programming interface (API).** A mechanism for programmers to talk directly to a software system, online service, or data platform. Facilitates connectedness and automation, and supports process optimization.

**artificial general intelligence (AGI).** A long-term goal for artificial intelligence research in which a machine has the capacity to understand or learn in a way similar to humans. *See also* **strong AI.**

**artificial intelligence.** The ability of machines to demonstrate any aspect of human intelligence, such as pattern recognition, language understanding, reasoning, planning, perception, or emotions.

**artificial neural network (ANN).** *See* **neural network.**

**attack surface.** The sum total of places in a software system or infrastructure where an unauthorized "attacker" can gain access to or interact with the data in an environment. An attack surface should be minimized by design.

**autocorrelated process data.** Outcomes or measurements from the same process, produced in different time periods, that are related to one another. Common in production environments and chemical operations.

**autoencoder.** A type of neural network with a hidden layer for encoding and an output layer for decoding, whose purpose is to match the input it was provided.

**automata.** A concept dating back to the 1940s to analyze behaviors of complex systems, where individual agents (or cells) change state based on their interactions with one another. *Also called* **cellular automata.**

**automotive SPICE.** *See* **SPICE.**

# B

**bag of words.** An approach in unsupervised machine learning where text is broken into individual words and analyzed based on frequency, association, and proximity.

**Baldrige Cybersecurity Excellence Builder (BCEB).** A guidebook and self-study road map (and companion to the Baldrige Excellence Framework) to help organizations assess and address cybersecurity risks, threats, and vulnerabilities and their relationship to strategy, operations, and business results.

**Baldrige Excellence Framework (BEF).** A guidebook and self-study road map to help organizations reach their goals and improve their competitiveness by defining and linking strategy, operations, and business results. Administered by the Baldrige Performance Excellence Program (BPEP) at the U.S. National Institute of Standards and Technology (NIST).

**big data.** Extremely large or heterogeneous datasets or data streams that are difficult to analyze, especially in real time.

**bitcoin.** A cryptocurrency built on a public, permissionless, decentralized blockchain infrastructure. Its value does not depend on a central government, bank, administrator, or national currency.

**blockchain.** A series of groups of records, typically describing events or transactions, that cannot be altered because they have been linked together using a cryptographic hash of the previous group of records, the timestamp of the current hash, and newly stored data. *See also* **hyperledger.**

**body of knowledge (BoK).** An outline of the key topics in a discipline and their relationships to each other, often used to organize knowledge requirements for certification and education.

## C

**cellular automata.** *See* **automata.**

**classification and regression trees.** Predictive modeling approaches used to present decision-making processes visually.

**cloud computing.** Delivery model where infrastructure, servers, software, and compute resources can be obtained over the internet without the need to install or support the hardware that runs those services.

**clustering.** An unsupervised machine learning technique that finds patterns in observations that consist of one of more variables.

**community detection.** Algorithms that examine characteristics of nodes and connections in a network to determine which should be grouped together.

**computer-aided design (CAD).** Using computers to build, modify, or evaluate 2D or 3D structures and technical documents; the digital analog of drafting.

**computer-aided engineering (CAE).** Using simulation and modeling to explore dynamic aspects of engineering design, including fluid dynamics, and stress and reliability analysis.

**computer-aided manufacturing (CAM).** Using software to control operations technology (OT), in particular, CNC (computer numerical control) machines for rendering parts from electronic drawings. *See also* **operations technology (OT).**

**computer-integrated manufacturing (CIM).** The practice of using computers to monitor, control, and automate a production process. This acronym emerged in the 1980s and is considered a precursor to Industry 4.0, which is expected to (at least partially) realize the goals of CIM.

**confusion matrix.** $2 \times 2$ grid that describes the performance of a classifier algorithm in terms of true positives, false positives, true negatives, and false negatives.

**Control Objectives for Information and Related Technology (COBIT).** A collection of best practices and controls for IT management and IT governance. Administered by ISACA (Information Systems Audit and Control Association).

**convergence.** When multiple disparate ideas or technologies come together and are replaced by a common concept, single technology, or shared platform or infrastructure.

**coordinate measuring machine.** Specialized industrial robot that is used to test the geometry of a part or assembly against design specifications or intent.

**corpus.** A collection of documents. In machine learning, the data structure used for text analysis algorithms like latent Dirichlet allocation (LDA) and sentiment analysis.

**cost of quality (CoQ).** Costs associated with preventing quality problems, evaluating management systems to anticipate or proactively address quality problems (appraisal), responding to errors and failures that emerge prior to the product or service reaching the customer, and responding to errors and failures experienced by customers.

**cross-validation.** A technique for validating a model where some of the training data is withheld and not used to train the model, but instead is used as "new data" to simulate the process of the model encountering data it has never seen before.

**customer relationship management (CRM).** A software system used to manage information about customers throughout the sales funnel and customer life cycle.

**cyberinfrastructure.** First introduced by the U.S. National Science Foundation (NSF), this refers to research and commercial environments that provide innovative and advanced high-performance computing services.

**cyber-physical systems (CPS).** Smart systems that NIST describes as "co-engineered interacting networks of physical and computational components" (Monostori, 2014). CPSs have a cyber (connected) part and a physical (tangible) part.

**cybersecurity.** The practice of protecting computers, networks, and data from threats and malicious attacks.

# D

**dashboard.** An information management tool that presents key performance indicators (KPIs) to the people who need them for decision making. Most organizations have many dashboards.

**data as a service (DaaS).** Cloud service where (usually clean) data is served over a network, often after receiving and processing data that is owned by or used by the customer.

**data cleaning.** The preprocessing steps taken before any data science modeling or programming activity to remove spurious data, straighten out formatting issues, and select or create the most significant features and/or reduce dimensionality to increase the value of the resulting models.

**data integrity.** The practice of maintaining all the data quality dimensions for a dataset or repository over the full life cycle of the data.

**data lake.** An unstructured storage repository of an organization's raw data, intended to provide visibility and access but not necessarily structure.

**data quality dimensions.** Characteristics that define data quality for a particular user or organization (e.g., accuracy, completeness, credibility, objectivity, conciseness, traceability, timeliness, redundancy, and validity).

**data science.** An interdisciplinary activity that encompasses all the activities required to support data-driven decision making, from collecting or obtaining data, to building and optimizing models, to generating insights, to delivering value and broader impacts.

**data warehouse.** A structured repository to gather and organize a company's multiple sources of data, often for unified reporting.

**data wrangling.** Like data cleaning, but with more of a struggle and a greater chance of being abandoned due to pain. *See also* **data cleaning**.

**decision tree.** A data structure that breaks down complex predictions or classifications into multiple choices or probabilistic results. *See also* **random forest**.

**deep learning.** Technique in machine learning where neural networks with multiple layers, some specialized (e.g., convolution, pooling), are applied to perform complex tasks like image recognition.

**deep neural network.** Data structure used to support deep learning. *See also* **deep learning**.

**descriptive analytics.** Metrics that describe things that have already happened or are in the process of happening. Includes both levels and trends; typically displayed on dashboards.

**diagnostic analytics.** Metrics used to identify causal relationships, determine the most significant variables or features, or uncover root causes. Can reveal why anomalies or patterns have emerged.

**digital native.** New companies, like Uber and Lyft, whose initial business models were based on the availability of pervasive, networked digital technologies.

**digital transformation.** A strategic, pervasive shift driven by customer and business requirements that leverages digital technologies to promote new business models, new sources of value, and the organizational change to realize them.

**digital twin.** A software replica of a physical system that can be manipulated with software in the same way that a physical system can be. Changes to the physical system can be explored by running simulations that use the digital twins.

**digitalization.** New business models and value streams opened up by digital technologies.

**digitization.** Changing from analog to digital; state where some organizational processes are supported by electronic data entry, retrieval, and visualization.

**discrete event simulation (DES).** A modeling technique that tracks entities as they flow through locations in a system where they are sometimes provided service by operators. DES is useful for modeling queuing systems, routing systems, and service systems.

**distributed control system (DCS).** The full state-driven "brain" of an industrial control system. DCS is an active system that is engaged with field controllers and reports results back to a SCADA system.

**DMAIC (Define, Measure, Analyze, Improve, Control).** A methodology associated with Six Sigma to guide process improvement projects.

# E

**Earley algorithm.** A recipe for separating and analyzing words and concepts. *See also* **natural language processing (NLP)**.

**edge computing.** Technique where incoming sensor data is processed onboard before a subset or calculated value is sent to a more centralized hub.

**ensemble methods.** Techniques that combine the results from many models to generate a best-fit or consensus model.

**enterprise resource planning (ERP).** A collection of software systems to manage the key resources for an organization, including (but not limited to) finances, people, information technology, and product offerings.

**environment, health, and safety (EHS).** A discipline that seeks an integrated treatment of environmental impact and occupational health and safety due to common hazards. Also refers to software systems that support these capabilities.

**ergonomics.** The study of work and human physical and cognitive performance.

**exoskeleton.** A mechanical (and often robotic) augmentation to a worker's body that helps him or her accomplish fine, difficult, or hazardous tasks.

**expert systems.** An approach to artificial intelligence in which a body of knowledge is represented by facts and rules, and an inference engine is applied to generate new facts. For example: Fact #1: *Alex is the son of Mary.* Fact #2: *Mary is the daughter of John.* Rule: *The parent of a parent is a grandparent.* New fact generated by expert system: *John is the grandparent of Alex.*

**exploit.** In cybersecurity, a way that a vulnerability can be used to stage a malicious attack.

**extract, transform, load (ETL).** Data processing technique where multiple data sources are pulled from a data source, cleaned or manipulated in some way, and stored in a target system.

# F

**feature.** Predictor variables in a dataset. If observations are arranged in rows, each column (containing one predictor, either categorical or quantitative) represents one feature.

**feature extraction.** Converting data to a more useful format (e.g., extracting phrases from a document, converting unwieldy date/time formats to Julian day, combining features to obtain a new one that is a more powerful predictor).

**feature selection.** Removing unnecessary predictor variables in a dataset.

**field controllers.** Devices that manage information about the state of a system or subsystem (e.g., PLCs, PACs, RTUs, PID controllers, or embedded microcontrollers like Arduino, or Raspberry Pi).

**field devices.** Instruments that drive processes and obtain process data at monitor points (e.g., switches, sensors, valves, meters, actuators, beacons, RFID tags).

**Food Safety Modernization Act (FSMA).** Signed into law by President Obama in 2011, it establishes stronger requirements for food safety, quality, and sustainability. Rollout of FSMA is expected to take several years.

# G

**gemba (現場).** A Japanese term often used in lean management that reflects *where the work is done.* This can be on the shop floor, on a computer screen where a software application is being used, in a conference room, or at a customer site.

**General Data Protection Regulation (GDPR).** Approved by the EU Parliament in 2016 and enacted in May 2018, GDPR provides consistency between data privacy and protection regulations across EU member nations and establishes that information about a person belongs to that person. If your organization collects data that in any way relates to a citizen of the EU, that person should be informed about how you plan to use that information, and kept informed as your organization's data management strategy evolves.

**generative adversarial networks (GAN).** A special kind of neural network that generates new data with the same statistical properties as training data. Can be used to reconstruct 3D models, generate video game worlds, and create fake photos (of people, animals, or any other object) based on collections of real photos.

# H

**hazard.** From *ISO 31000:2018 Risk management—Guidelines*: a source of potential harm. *See also* **risk**.

**hidden layer.** A group of nodes in between the input and output layers of a neural network that exist to transform the inputs into values the output layer can use.

**hidden Markov model (HMM).** A model that describes a sequence of observable events, when one or more of them are not completely observable.

**high dimensional data.** A dataset with so many features (predictors) that calculations become difficult, or where the number of features (predictors) is greater than the number of observations.

**historian.** A machine (or group of machines) that keeps track of historical production status, performance, quality information, tracking and provenance, alarms, and other events. Also used for troubleshooting, regulatory reporting (e.g., for ISO 9001 certification), cost accounting, identifying process improvement opportunities or justifying past process improvements, tracking downtime, and tracking energy consumption.

**horizontal integration.** Connecting systems across functional boundaries (e.g., sales, marketing, development, service), usually with the aid of digital systems.

**human–machine interfaces (HMIs).** The interfaces between human operators and a process (often industrial in nature). HMIs enable the operator to closely monitor production and respond to changing demands in real time. Can be deployed via computer, tablet, smartphone, augmented reality, or wearable.

**hybrid intelligent system.** System that combines or blends techniques from different artificial intelligence domains (e.g., expert systems, computer vision, reinforcement learning) to solve a real-world problem.

**hyperledger.** A blockchain framework developed and managed by IBM and the Linux Foundation, supporting major new proof of concept initiatives like the IBM Food Trust for enhancing the transparency and auditability of the food supply chain.

# I

**industrial control systems (ICSs).** Hardware and software that gather information about a process from its endpoints, interprets that information in the context of production system goals, and facilitates interactions among operators, field controllers, and field devices.

**industrial hygiene.** The study and practice of protecting the health and safety of workers, in particular from chemical, physical, biological, and ergonomic stressors.

**industry 4.0.** Term that emerged from a German government project in 2011, intended to catalyze Germany's industrial economy by implementing smart factories enabled by connected, intelligent, automated technologies. Describes the digital transformation of some industries, including automotive, aviation, chemicals, defense, medical devices, and pharmaceuticals.

**information technology (IT).** Department in many companies that handles internet connectivity, infrastructure, applications for business systems, and software, including finance, HR, and support functions.

**Information Technology Infrastructure Library (ITIL).** A best practices framework for IT service management that originated in the UK.

**infrastructure as a service (IaaS).** The most common and lowest risk cloud service model, in which the cloud service provider supplies computational capabilities, storage, and network management that the customer uses to manage their data and run their applications.

**input layer.** Nodes of a neural network that represent each element of an input array. For example, when an image is used as input to a neural network, the input layer contains one node for each pixel.

**Institute of Electrical and Electronics Engineers (IEEE).** A professional association based in the United States that initially served electrical engineers but has grown to support engineers from all disciplines and scientists that support engineering efforts.

**intelligence.** Ability to think, reason, solve problems, be creative, and apply emotional reasoning; self-awareness.

**intelligent agent.** A computer system situated in a specific environment and capable of autonomous, goal-directed action within this environment. *See also* **agent**.

**intelligent system.** Any system—human, machine, or a combination—that is autonomous, reactive, proactive, social, adaptive, and/or capable of learning.

**internet of things (IoT).** An interconnected collection of cyber-physical systems that can communicate over a network.

**internet of things (IoT) hub.** A cloud-based and typically cloud-managed service that handles bidirectional communication between IoT endpoint devices and the applications that analyze, process, or use that data.

**ISA-95.** A technology-agnostic information model developed in the 1990s by the International Society of Automation (ISA) that describes the relationships between business and production data; intended to ease systems integration (particularly for industrial and manufacturing facilities).

**ISO 8000.** ISO standard for data quality and enterprise master data.

**ISO 9001.** ISO standard for quality management systems.

**ISO 14001.** ISO standard for environmental management systems.

**ISO 31000.** ISO standard for risk management.

**ISO 45001.** ISO standard for occupational health and safety management systems.

**ISO/IEC 15504.** *See* **SPICE**.

**ISO/IEC 27001.** ISO standard for information technology security management systems.

**ISO/IEC 27005.** ISO standard for information technology risk management.

# J

**journey map.** Model for the stages of interaction a company and customer have with one another. Also called *customer journey map*.

**just-in-time (JIT).** A pull-based system in lean production for managing people, materials, and inventory, in which resources arrive or are replenished only when they are needed.

# K

**kaizen (改善).** Change for the good. A Japanese word used to summarize the principles and practices of continuous improvement as embodied by lean management.

**kanban (看板).** A lean method for managing work in process that visually tracks work in terms of what tasks are outstanding, what are currently being worked on, and what has been completed. Means "sign" in Japanese; used to refer to billboards and shop signs. See also **just-in-time (JIT)**.

**key performance indicators (KPIs).** A metric used to assess performance (e.g., product, process, customer satisfaction, customer engagement, workforce capability, workforce capacity, leadership, governance, financial, market, and strategy).

# L

**lagging indicators.** Metrics or KPIs that indicate results that have already occurred. *See also* **leading indicators**.

**latent Dirichlet allocation (LDA).** A statistical technique in natural language processing that models a document as groups of related topics.

**leading indicators.** Predictive factors that can be used to anticipate future changes, scenarios, or events. *See also* **lagging indicators**.

**lean management.** An approach to organizational management and performance improvement that focuses on customer value, defining a value stream, creating flow, establishing pull-based systems, and aiming for perfection.

**linear discriminant analysis (LDA).** A linear transformation technique used to pre-process machine learning data for dimensionality reduction (choosing the most powerful predictors).

**linear regression.** A statistical method used to create predictive models where one dependent variable is predicted by one or more explanatory (independent) variables. Although linear regression is often performed on large datasets and often included in books on machine learning, many practitioners do not believe it qualifies as machine learning, while others argue that it definitely does, especially when there are hundreds of independent variables.

**logistic regression.** A probabilistic model that estimates how likely it is that an observation will fall into one of two categories (binary logistic regression) or one of many categories (ordinal logistic regression). Although logistic regression is often performed on large datasets and often included in books on machine learning, many practitioners do not believe it qualifies as machine learning, while others argue that it definitely does, especially when there are hundreds of independent variables.

# M

**machine learning (ML).** A subset of artificial intelligence (AI) that focuses on the use of algorithms and statistical models that are implemented by machines to generate analytics and insights on big data.

**machine learning algorithm.** Core computational recipes used to carry out machine learning methods that are used to build models.

**machine learning method.** Techniques that leverage machine learning algorithms to create models that can adapt to new data, and revise themselves to increase prediction accuracy. Machine learning methods use one or more machine learning algorithms.

**machine learning model.** A mathematical description of the relationships between inputs and outputs, expressed by implementing a machine learning algorithm.

**Malcolm Baldrige National Quality Award (MBNQA).** The highest level of national recognition for performance excellence that can be obtained by a U.S. organization, administered by the Baldrige Performance Excellence Program (BPEP) at the National Institute of Standards and Technology (NIST). Award applicants are evaluated using the Baldrige Excellence Framework (BEF).

**manual materials handling (MMH).** Tasks carried out in an industrial environment (e.g., lifting, carrying, holding, and placing) that must be designed so the physical requirements do not exceed the physical capabilities of the workers who must perform those tasks.

**Markov chain.** A model that describes a sequence of observable events. *Compare with* **hidden Markov model (HMM).**

**master data.** Data entities that represent parties (people), locations, and things and are critical to business processes.

**master data management (MDM).** Policies, procedures, and guidelines for managing key business data. *See also* **master data**.

**metadata.** Data *about* other data (e.g., timestamp, keywords, storage location).

**model.** A mathematical description of the relationships between inputs and outputs, designed based on beliefs about those relationships.

**model-based software engineering.** The formalized application of information modeling to defining software and system requirements, capturing design, and verifying and validating designs against requirements.

**muda (無駄).** Japanese word for waste. Includes waste associated with transport, inventory, motion, waiting, overproduction, overprocessing, defects, and underutilization of skills.

**multi-armed bandit.** A variation on A/B testing that helps decision makers determine the most beneficial option from a set of choices. *See also* **A/B testing**.

**multi-class.** A classification model that separates results into three or more categories.

**mura (斑).** Japanese word for unevenness; the waste that comes from working too little, and then too much.

**muri (無理).** Japanese word for overwork; the waste that comes from exhaustion or depletion.

**musculoskeletal disorders.** Issues such as lower back injuries, muscle strain, and carpal tunnel syndrome that cost businesses over $50 billion per year in medical expenses, lost time, lost productivity, and workers' compensation claims.

# N

**Naive Bayes.** A supervised or semi-supervised machine learning approach where observations are classified based on previous knowledge about the characteristics of those observations or relationship of the observations to external factors (e.g., an e-mail containing the word "Viagra" is much more likely to be spam than legitimate).

**National Institute of Standards and Technology (NIST).** U.S. government agency that promotes innovation and competitiveness through standards for physical measurements, scientific processes, communications, security, and operations.

**natural language processing (NLP).** A branch of artificial intelligence that focuses on systems for speech recognition, understanding natural language, understanding meaning, and generating language. Chatbots are an example of a technology that heavily leverages NLP.

**neural network.** A computing construct that models systems of inputs, outputs, and their connections to make predictions or classify observations. *See* **artificial neural network (ANN)**.

**NIST Cybersecurity Framework (CSF).** A policy framework consisting of standards, guidelines, and best practices for managing cybersecurity-related risk. *See also* **National Institute of Standards and Technology (NIST)**.

**NoSQL.** A nonrelational database (e.g., CouchDB, MariaDB, MongoDB) designed to store and retrieve information from unstructured data objects like web pages.

# O

**observation.** A row of a dataset where each column corresponds to one (and only one) predictor variable (feature).

**operations technology (OT).** Hardware and software close to a production process, including field devices, field controllers, and human–machine interfaces (HMIs), usually in Level 1 or Level 2 of the ISA-95 automation hierarchy. *See also* **ISA-95**.

**opportunities for improvement.** Possibilities for reducing variation, reducing waste, improving flow, improving effectiveness, or enhancing any other quality or performance outcome. Popularized by the Malcolm Baldrige National Quality Award (MBNQA). Pronounced "oh-fee."

**output layer.** The nodes in a neural network that represent the answer to a prediction or classification problem. For example, the output layer for a neural network that classifies integer digits from 0 through 9 might have ten nodes, one for each potential answer (there are many ways to design neural networks, so this may not always be the case).

**overfitting.** In machine learning, the practice of creating a model from data that so closely describes that data, it is unable to generalize to new incoming observations effectively. As a result, the model overwhelmingly describes the noise in the input data rather than the signal.

# P

**personal protective equipment (PPE).** Special clothing, hats, goggles, boots, or other apparel that is designed to withstand harsh or hazardous work conditions.

**platform as a service (PaaS).** A cloud computing service that provides customers with infrastructure, operating system, and runtime environment, allowing the customer to focus on maintaining only the data and applications.

**precision.** A performance measure for machine learning classifiers that evaluates its ability to correctly classify into one category; the number of true positives divided by all positives. *See also* **recall.**

**predictive analytics.** Metrics that project future states. Can incorporate results from forecasting, model building, descriptive analytics, or diagnostics.

**prescriptive analytics.** Metrics used to identify a recommended course of action, typically created using optimization to identify the best alternatives to achieve an objective.

**principal component analysis (PCA).** Dimensionality reduction technique that extracts "components" (combinations of independent variables) that best describe changes in the dependent variable that is being predicted. See also **linear discriminant analysis (LDA).**

**programmable automation controller (PAC).** A more modern form of a programmable logic controller (PLC) that can have additional memory, data logging, or enhanced I/O capacity and can integrate more easily with databases; used for discrete control.

**programmable logic controller (PLC).** Developed to replace physical relays, each PLC contains electrical inputs and outputs, and the capability to program logic that determines how and when the outputs are triggered based on the inputs; used for discrete control. Invented in 1969, it was the catalyst for the third industrial revolution.

**proportional—integral—derivative (PID) controller.** Measures the gap between an observed value from a field device or remote terminal unit (RTU) and the target value.

# Q

**Q-learning.** A reinforcement learning algorithm that learns what steps an agent must take under what circumstances to maximize rewards. In contrast with other reinforcement learning approaches, Q-learning does not use a model, just information about the relative rewards that can be obtained at each step.

**quality.** The characteristics of an entity (product, service, person, system, project, agent) that bear on its ability to satisfy stated or implied needs (old ISO 8402); fitness for use or purpose (Juran); conformance to requirements (Crosby); the efficient production that the market expects (Deming).

**Quality 4.0.** Enhancing connectedness, intelligence, and/or automation to achieve quality goals and improve performance. Each of these three areas builds on each other to amplify quality and performance benefits. While Industry 4.0 is the digital transformation of certain industries and describes how new and emerging technologies support performance breakthroughs in manufacturing and other industrial processes (e.g., smart factory), Quality 4.0 can serve as an umbrella term for similar practices in *any* industry (e.g., Health 4.0, Lean 4.0, Logistics 4.0, smart agriculture, smart cities, Supply Chain 4.0, Tourism 4.0).

**Quality 4.0 initiative.** Any digital transformation or Industry 4.0 initiative that has been designed to satisfy quality objectives or improve quality or performance, regardless of industry.

**Quality 4.0 strategy.** A digital transformation (or Industry 4.0) strategy driven by quality and performance.

# R

**radio frequency identification (RFID).** Tags that store information electronically and can be affixed to products and assets to track movement.

**random forest.** A supervised machine learning approach that creates an ensemble of decision trees from a set of training observations, and then recommends the classification or regression tree that best represents a consensus view of the decision process.

**recall.** A performance measure for classifiers that describes how many of a particular class are successfully identified from all true members of that class. For example, a medical test that can successfully identify all members of a community infected with a disease so they can be quarantined would have a high recall.

**reinforcement learning.** A machine learning technique that explores a sequence of steps or activities to learn how to maximize reward or quality.

**relational database management system (RDBMS).** A database structure where information is stored in tables that contain fields, and fields are linked to one another to create relationships between the fields.

**remote terminal unit (RTU).** The simplest type of field controller, it collects analog signals from the field and converts them to digital signals. More advanced RTUs have capabilities nearing those of programmable logic controllers (PLCs). Can also be referred to as *remote telemetry unit.*

**risk.** From *ISO 31000:2018 Risk management—Guidelines*: the effect of uncertainty on objectives. *See also* **hazard** *and* **threat**.

**robotic process automation (RPA).** Software programs or scripts that are used to execute repetitive, recurring, or nondeterministic tasks or processes.

# S

**safety instrumented system (SIS).** Robust, hardened, high-reliability industrial control system (ICS) that has one and only one purpose: stopping or shutting down processes if unsafe conditions occur. An SIS protects against random, unintended hardware faults.

**sampling frequency.** Number of observations taken in a particular unit of time. For example, 5G communications will support data transmission at a higher sampling frequency from IoT sensors than 4G or 3G.

**Sarbanes-Oxley (SOX).** A 2002 U.S. law that imposed requirements for transparency and auditability of corporate accounting, intended to protect against fraud and corruption.

**SCADA (Supervisory Control and Data Acquisition).** Often used to refer to the combination and coordination of programmable logic controllers (PLCs) and human–machine interfaces (HMIs); some operators refer to the HMIs alone as "the SCADA system." Performs supervisory gathering and reporting only and does not make decisions. Has distributed intelligence so monitor and control can continue even if communications are temporarily lost. Sometimes called *Master Terminal Unit*.

**sensitivity.** A performance measure for a classifier that describes the proportion of observations for the dominant group (e.g., positives) that are correctly identified (e.g., you correctly identify everyone in a group that has a certain disease).

**simulation.** The act of using a model or collection of models to critically examine or reproduce the behavior of a system.

**simulation-based design.** An Industry 4.0 approach to prototyping and new product development in which simulation is the primary way to evaluate and verify how well design alternatives satisfy specification and design goals. The objective is to eliminate unfit or unsuitable designs as early as possible to prevent waste.

**situation awareness.** A model for examining decision-making scenarios for individual and team readiness. It considers perception of available data (Level 1), comprehension of that data and its context (Level 2), and projection of that knowledge into the awareness of future states (Level 3).

**Six Sigma.** A statistical methodology for reducing defects and errors; to some, a more generalized management philosophy.

**small and medium enterprises (SMEs).** A term used by the World Bank and European Union to describe companies of a certain size (less than 40 million Euro revenue and fewer than 250 employees).

**smart contract.** Business rules that verify the validity of records or transactions before committing them to an immutable blockchain.

**smart product.** Offerings that have connected, intelligent, and sometimes automated capabilities (e.g., personalization, situation awareness, location awareness, proximity management, responsiveness to changing environment or needs).

**software as a service (SaaS).** Delivery model where customers subscribe to use a cloud-based software system rather than purchase and install it on premises.

**spam.** Unsolicited commercial e-mail or unsolicited communications, sometimes initiated with malicious intent.

**sparse.** A matrix or a column in a dataset that contains mostly zeroes.

**specificity.** A performance measure for a classifier that describes the proportion of observations for the nondominant group (e.g., negatives) that are correctly identified (e.g., you correctly identify everyone in a group that has a certain disease).

**SPICE (ISO/IEC 15504).** A model that defines processes and best practices for software development, tailored for the automotive industry, that includes recommendations for acquisition, supplier management, systems engineering, software development, support processes, management processes, process improvement, and software reuse.

**standard work.** The rate at which output must be produced, the steps to produce it, and an outline of the materials required to carry out those steps. *See also* **lean management**.

**statistical process control (SPC).** A quality control technique used to monitor variables or attributes of a process to determine whether special cause variation has occurred, at which point an intervention to restore quality should be taken.

**strong AI.** Artificial intelligence that can apply beliefs, desires, emotions, intentions, or similar higher-level processes attributed to humans to pattern identification, prediction, classification, reasoning, or inference tasks.

**support vector machine (SVM).** A supervised machine learning approach (often used for classification) that separates observations into groups using a plane or hyperplane that is separated by each of the data points in the training set by a maximum margin.

**system of record (SoR).** A software application or data platform that is considered the "gold standard" for a particular data object or entity (e.g., employee, customer, financial transaction). Most organizations will have multiple SoRs but should manage them so that no more than one system is the SoR for a particular data source. If multiple systems need to use a data record, that information should be sourced from the SoR.

**systems thinking.** Consideration of a problem context by examining the components of the system, their interconnections and relationships, and the impact of parts of the system on each other and on the environment within which they are situated.

# T

**threat.** In cybersecurity, a person or event that has the potential to damage or otherwise negatively impact a process, resource, or asset.

**trusted third party (TTP).** From cryptography, an entity that facilitates interactions between two parties. The two parties may not trust each other, but each does trust the third party.

# U

**Unified Modeling Language.** A modeling language used to specify, visualize, and document the design of software systems. Also sometimes used in model-based software engineering as the basis for automated code generation.

# V

**value.** Importance, worth, or the benefit that may be gained from a product or a service; always measured from the point of view of the customer or stakeholder.

**value proposition.** The specific ways you choose to respond to your customer or stakeholder's needs or wants.

**value stream.** The set of all steps in a process that creates value, from ideation through delivery and maintenance of the value proposition.

**value stream map (VSM).** A technique for studying a process that focuses on the value-adding steps and helps teams eliminate steps that do not add value. *See also* **lean management**.

**variable.** One categorical or quantitative measurement that describes an observation (in full or in part). *See also* **feature**.

**variety.** In big data terms, the many different types and formats of data that an organization may need or wish to use in analysis or modeling.

**velocity.** In big data terms, the rate at which new data arrives.

**veracity.** In big data terms, the uncertainty of the incoming data quality.

**vertical integration.** Connecting systems so communication can be supported across logical layers (e.g., ISA-95).

**voice of the customer (VoC).** The continuous process of identifying and interpreting customer needs, measuring or estimating the relative priorities of those needs against the needs of other stakeholders, and applying that knowledge to enhance customer satisfaction.

**volume.** In big data terms, the amount of data an organization needs to manage (which usually keeps growing over time, sometimes very quickly).

**vulnerability.** A weakness in a process, control, or resource. In cybersecurity, an error or weakness that can be leveraged to support a malicious attack.

# W

**weak AI.** Artificial intelligence that carries out a limited pattern identification, prediction, classification, reasoning, or inference task but does not apply beliefs, desires, emotions, intentions or similar higher-level processes attributed to humans.

# X

**XML (eXtensible Markup Language).** A structured, nonrelational mechanism for storing and transmitting data.

# Y

**Yarowsky algorithm.** An unsupervised machine learning approach for disambiguating word meanings based on proximity to other words that provide context.

# Z

**zettabyte.** A billion terabytes of data. International Data Corporation (IDC) anticipates that this amount of data will be generated on an annual basis worldwide by 2025.

# INDEX

Note: Page numbers followed by *f* or *t* refer to figures or tables, respectively.

# ABOUT THE AUTHOR

**Nicole Radziwill** is Senior Vice President of Quality and Strategy at Ultra-nauts, a professional services firm specializing in quality assurance and quality engineering for software, data science, and digital transformation. Formerly VP of the Global Quality and Supply Chain Practice at Intelex Technologies in Toronto, and a tenured Associate Professor of Data Science and Production Systems, she is a fellow of the American Society for Quality (ASQ), editor of *Software Quality Professional*, and a past chair of the ASQ Software Division. She regularly serves as an examiner for the Malcolm Baldrige National Quality Award (MBNQA), and has contributed to the development of ISO 26000 Guidance on Social Responsibility.

Nicole has a PhD in Quality Systems and an MBA; she is an ASQ Certified Manager of Operational Excellence (CMQ/OE) and an ASQ Certified Six Sigma Black Belt (CSSBB). She does research at the intersection of data science, machine learning, and business value. She has 25 years of experience in quality management and data science, from building numerical weather prediction models and new observing systems in meteorology, to enterprise software implementation for telecom and high-tech, to managing big data and analysis pipelines for astronomical research. Her 2019 book, *Statistics (The Easier Way) With R, 3rd Edition* (Tidyversion), has been used for undergraduate and graduate courses in more than 30 schools in 13 countries.